SPRINGER HANDBOOK OF
AUDITORY RESEARCH

Series Editors: Richard R. Fay and Arthur N. Popper

Springer

New York
Berlin
Heidelberg
Barcelona
Hong Kong
London
Milan
Paris
Singapore
Tokyo

SPRINGER HANDBOOK OF AUDITORY RESEARCH

Volume 1: The Mammalian Auditory Pathway: Neuroanatomy
Edited by Douglas B. Webster, Arthur N. Popper, and Richard R. Fay

Volume 2: The Mammalian Auditory Pathway: Neurophysiology
Edited by Arthur N. Popper and Richard R. Fay

Volume 3: Human Psychophysics
Edited by William Yost, Arthur N. Popper, and Richard R. Fay

Volume 4: Comparative Hearing: Mammals
Edited by Richard R. Fay and Arthur N. Popper

Volume 5: Hearing by Bats
Edited by Arthur N. Popper and Richard R. Fay

Volume 6: Auditory Computation
Edited by Harold L. Hawkins, Teresa A. McMullen, Arthur N. Popper, and Richard R. Fay

Volume 7: Clinical Aspects of Hearing
Edited by Thomas R. Van de Water, Arthur N. Popper, and Richard R. Fay

Volume 8: The Cochlea
Edited by Peter Dallos, Arthur N. Popper, and Richard R. Fay

Volume 9: Development of the Auditory System
Edited by Edwin W. Rubel, Arthur N. Popper, and Richard R. Fay

Volume 10: Comparative Hearing: Insects
Edited by Ronald R. Hoy, Arthur N. Popper, and Richard R. Fay

Volume 11: Comparative Hearing: Fish and Amphibians
Edited by Richard R. Fay and Arthur N. Popper

Volume 12: Hearing by Whales and Dolphins
Edited by Whitlow W.L. Au, Arthur N. Popper, and Richard R. Fay

Volume 13: Comparative Hearing: Birds and Reptiles
Edited by Robert J. Dooling, Richard R. Fay, and Arthur N. Popper

---

Forthcoming volume

Speech Processing in the Auditory System
Edited by Steven Greenberg, William Ainsworth, Arthur N. Popper, and Richard R. Fay

Robert J. Dooling
Richard R. Fay
Arthur N. Popper
Editors

Comparative Hearing: Birds and Reptiles

With 114 Illustrations

 Springer

Robert J. Dooling
Department of Psychology
University of Maryland
College Park, MD 20742, USA

Richard R. Fay
Parmly Hearing Institute and
Department of Psychology
Loyola University of Chicago
Chicago, IL 60626, USA

Arthur N. Popper
Department of Biology and
Neuroscience and Cognitive Science Program
University of Maryland
College Park, MD 20742, USA

Series Editors: Richard R. Fay and Arthur N. Popper

Cover illustrations: Turtle ear is from Wever EG: *The Reptile Ear: Its Structure and Function.* Copyright © 1985 by PUP. Reprinted by permission of Princeton University Press. The barn owl is from Knudsen EI: In: *Comparative Studies of Hearing in Vertebrates.* Copyright © 1980 Springer-Verlag New York.

Library of Congress Cataloging-in-Publication Data
Comparative hearing. Birds and reptiles / editors, Robert J. Dooling, Richard R. Fay, Arthur N. Popper.
    p.   cm.–(Springer handbook of auditory research; 13)
    Includes bibliographical references.
    ISBN 0-387-94684-5 (hard cover : alk. paper)
    1. Hearing–Handbooks, manuals, etc. 2. Physiology, Comparative–Handbooks, Manuals, etc. 3. Birds–Physiology–Handbooks, manuals, etc. I. Title: Birds and Reptiles. II. Dooling, Robert J. III. Fay, Richard R. IV. Popper, Arthur N.
    V. Springer handbook of auditory research; v. 13.
    QP461.C633 2000
    573.8′9179–dc21   99-052789

Printed on acid-free paper.

Production managed by Terry Kornak; manufacturing supervised by Erica Bresler.
Typeset by Best-set Typesetter Ltd., Hong Kong.
Printed and bound by Maple-Vail Book Manufacturing Group, York PA.
Printed in the United States of America.

9 8 7 6 5 4 3 2 1

ISBN 0-387-94684-5   Springer-Verlag New York Berlin Heidelberg   SPIN 10527981

Series Preface

The *Springer Handbook of Auditory Research* presents a series of comprehensive and synthetic reviews of the fundamental topics in modern auditory research. The volumes are aimed at all individuals with interests in hearing research, including advanced graduate students, post-doctoral researchers, and clinical investigators. The volumes are intended to introduce new investigators to important aspects of hearing science and to help established investigators to better understand the fundamental theories and data in fields of hearing that they may not normally follow closely.

Each volume is intended to present a particular topic comprehensively, and each chapter will serve as a synthetic overview and guide to the literature. As such, the chapters present neither exhaustive data reviews nor original research that has not yet appeared in peer-reviewed journals. The volumes focus on topics that have developed a solid data and conceptual foundation rather than on those for which a literature is only beginning to develop. New research areas will be covered on a timely basis in the series as they begin to mature.

Each volume in the series consists of five to eight substantial chapters on a particular topic. In some cases, the topics will be ones of traditional interest for which there is a substantial body of data and theory, such as auditory neuroanatomy (Vol. 1) and neurophysiology (Vol. 2). Other volumes in the series will deal with topics which have begun to mature more recently, such as development, plasticity, and computational models of neural processing. In many cases, the series editors will be joined by a co-editor having special expertise in the topic of the volume.

Richard R. Fay, Chicago, IL
Arthur N. Popper, College Park, MD

Preface

Birds and reptiles have long fascinated investigators studying hearing and the auditory system. Both groups of vertebrates have a highly evolved auditory inner ear with many characteristics in common with the ear of mammals. Indeed, it has been suggested that the avian and reptilian basilar papilla, which is often also called the cochlea, is homologous to the mammalian cochlea. Although this relationship has not yet been firmly established, it is clear that many of the functional characteristics found in the mammalian cochlea are also present in birds and reptiles. Thus, both groups have become important models of helping understand the form and function of the vertebrate, and mammalian, auditory systems.

Birds are also fascinating because so much is known about other aspects of their acoustic life style. Birds use complex acoustic signals for communication, mate selection, individual recognition, and vocal learning. Some use sound for finding prey. For this reason, a number of avian species have become important models for studies of complex processing of sound, ontogeny of hearing and vocal communications, and generation and regeneration of hair cells. Reptiles represent somewhat of an enigma. Reptiles have long interested auditory anatomists for the diversity of their inner and middle ear systems. Further, although a good deal of work has been done on the ear of reptiles, very little is known about their use of sound. While the reptilian ear is very well developed, not many species are known to use sound for communication, and even in those species where sounds are produced on a regular basis, the sounds themselves are not nearly as complex as in birds (or even amphibians and fishes).

While birds and reptiles are naturally treated together because of their evolutionary histories and the many similarities in their auditory systems, it becomes apparent in putting together chapters that there is a significant difference in what we know about hearing in the two groups. This difference is heavily related to the way the two groups use sounds, and also in the ability to obtain data on hearing capabilities in the two groups. Birds are highly amenable to behavioral studies, and this has led to a plethora of data on hearing capabilities, much of which is reviewed in this volume. Reptiles, on the other hand, are notoriously difficult to study behaviorally, and consequently, there are few reliable data on their hearing capabilities.

This volume covers the broad range of our knowledge of hearing and acoustic communication in birds and reptiles. In Chapter 1, Dooling and Popper provide an overview of the topic, and point out a number of areas of research that are particularly ripe for future studies. Saunders and his colleagues in Chapter 2 discuss the structure and function of the middle ear in both groups. This is possible because the middle ear is quite similar in birds and reptiles, and there is a large comparative literature. The inner ear of birds and Crocodilia are treated together in Chapter 3 by Gleich and Manley, as their ears are so similar to one another, while Manley, in Chapter 4, discusses the highly derived ears of reptiles.The central auditory systems of the two groups are discussed in detail by Carr and Code in Chapter 5. So little is known about hearing capabilities of reptiles that very little can be said, but what is known is discussed in Chapters 6 and 7. In Chapter 6, Klump discusses the highly sophisticated sound localization capabilities of birds, while Dooling and his colleagues provide a detailed overview of auditory psychophysics in Chapter 7.

As is generally the case with volumes in this series, their is a strong interrelationship with chapters in other volumes. Studies on the middle ear, as described in this volume by Saunders and his colleagues, have been well documented for mammals, as shown in chapters by Rosowski on comparative function in Vol. 4 of this series, *Comparative Hearing: Mammals* and on modeling of the middle ear in Vol. 6, *Auditory Computation*. The structures of the auditory portions of the inner ear in birds and reptiles, as discussed by Gleich and Manley in their chapters, can be compared to the ear of mammals as described in detail by Echteler et al. in Vol. 4 and by Slepecky in Vol. 8, *The Cochlea*, and for bats by Kössl and Vater in Vol. 5, *Hearing by Bats*. The inner ears of fishes has also been dealt with in some detail by Popper and Fay in Vol. 11, *Comparative Hearing: Fish and Amphibians*, and by Lewis and Narins for the amphibians in the same volume. The anatomy of the auditory CNS discussed in this volume by Carr and Code has been well documented for mammals in all of the chapters of Vol. 1 of this series, *The Mammalian Auditory Pathway: Neuroanatomy*, and for bats in a number of chapters in Vol. 5, while it has been discussed for fish and amphibians by McCormick in Vol. 11. Sound source localization, as discussed by Klump in this volume, has been dealt with in a number of volumes, including in a chapter by Wightman and Kistler in Vol. 3, *Human Psychophysics*, and by Brown in Vol. 4. Issues of sound localization by fish and amphibians are discussed by Fay and Megela Simmons in their chapter in Vol. 11. Finally, psychophysics in birds, as described by Dooling and colleagues, should be examined in light of the work in Vol. 3 of this series, while it is also considered from the standpoint of fish and amphibians in the aforementioned chapter by Fay and Megela Simmons and by Moss and Schnitzler for bats in Vol. 5.

Robert J. Dooling, College Park, MD
Richard R. Fay, Chicago, IL
Arthur N. Popper, College Park, MD

Contents

Contributors

Catherine E. Carr
Department of Biology, University of Maryland, College Park, MD 20742, USA

Rebecca A. Code
Department of Biology, Texas Woman's University, Denton, TX 76204-5799, USA

Micheal L. Dent
Department of Psychology, University of Maryland, College Park, MD 20742, USA

Darryl E. Doan
Department of Otorhinolaryngology: Head and Neck Surgery, University of Pennsylvania, Philadelphia, PA 19104, USA

Robert J. Dooling
Department of Psychology, University of Maryland, College Park, MD 20742, USA

R. Keith Duncan
Department of Otorhinolaryngology: Head and Neck Surgery, University of Pennsylvania, Philadelphia, PA 19104, USA

Otto Gleich
HNO-Klinkum, Universitat Regensburg, Franz-Josef-Strauballe 11, 93042 Regensburg, Germany

Christine Köppl
Institut Für Zoologie, Technische Universität München, 85747 Garching, Germany

Georg Klump
Institut Für Zoologie, Technische Universität München, D-85747 Garching,
Germany

Bernard Lohr
Department of Psychology, University of Maryland, College Park, MD
20742, USA

Geoffrey A. Manley
Institut Für Zoologie, Technische Universität München, 85747 Garching,
Germany

Arthur N. Popper
Department of Biology and Neuroscience and Cognitive Science Program,
University of Maryland, College Park, MD 20742, USA

James C. Saunders
Department of Otorhinolaryngology: Head and Neck Surgery, University
of Pennsylvania, Philadelphia, PA 19104, USA

Yehudah L. Werner
Department of Evolution, Systematics and Ecology, The Hebrew University of Jerusalem, 91904 Jerusalem, Israel

1
Hearing in Birds and Reptiles: An Overview

ROBERT J. DOOLING and ARTHUR N. POPPER

1. Introduction

Birds and reptiles have been important subjects for the study of the vertebrate auditory system. These groups provide important insight into the overall structure and function of vertebrate audition, and at the same time, they serve as useful models for delving into comparative issues, particularly those related to mammalian hearing. Indeed, birds and, to a lesser degree, reptiles continue to be fundamentally important models for studies as diverse as regeneration of sensory hair cells, development of hearing, signal processing, and biophysics of sensory hair cells.

The chapters in this volume review much of what is known about the auditory system of birds and reptiles. The emphasis here is on how members of both groups detect and process signals and less on how they serve as models. The two groups are treated in a single volume because there are many similarities in their auditory systems. Yet, there are also differences that will be discussed in a comparative context. But together, these two groups of vertebrates also offer very special opportunities and unique challenges for comparative auditory science. On one hand, the opportunities arise from the extensive anatomic database in reptiles, which reveals an extreme degree of anatomic auditory diversity. On the other hand, the equally extensive database of behavioral thresholds in birds reveals extremely sophisticated acoustic signal processing capabilities paralleling the complex vocal system of birds. In comparing reptiles and birds then, the challenges arise from an almost complete lack of behavioral data in reptiles due to difficulties in doing appropriate studies on these species and the relatively reduced diversity, at least at a gross level, in both structure and function in birds.

Saunders, Duncan, Doan, and Werner (Chapter 2) review the present knowledge of the fundamental properties of the reptilian and avian middle ears. Because the middle ears of reptiles and birds are quite similar in many ways, a fairly clear comparative picture is available of the mechanical principles at work that determine the frequency-dependent characteristics of the middle ear as they transmit power to the inner ear. Because the inner ear of the birds is most similar to the ear in a subset of the reptiles, the crocodiles, Gleich and Manley (Chapter 3) treat the inner ear of these two groups of vertebrates together. Manley (Chapter 4) considers the highly derived ears of the remaining reptile groups and documents the differences in size and structural and physiological variation. Birds exhibit more consistency in their peripheral auditory system both in structure and in function compared to reptiles. In Chapter 5, Carr and Code provide a detailed comparison of the central auditory system of the two groups. Aside from humans, the psychophysics of hearing in birds is more completely known than that in any other mammal and certainly in any other vertebrate group. Dooling, Lohr, and Dent (Chapter 6) summarize what is known about the psychophysics of hearing in birds. In Chapter 7, Klump takes on the special case of sound localization in birds, vertebrates with closely spaced ears and generally poor high-frequency hearing, and compares both behavior and physiology of sound localization in unspecialized birds with that of the sound localization specialists such as the barn owl.

Each chapter examines hearing in both groups where possible. As is most apparent in several chapters, there are significant gaps in our knowledge about the behavioral hearing capabilities of reptiles (Chapters 6 and 7). Thus although these animals have highly derived and diverse ear structures (see Chapter 4) and an auditory central nervous system (CNS) that is relatively similar to that in birds (Chapter 5), even after decades of imaginative experimentation, investigators have yet to find ways to get rigorous behavioral indications of what and how well reptiles hear.

1.1. Comparative Hearing of Reptiles and Birds

Comparative sensory studies often fall into two broad categories: those aimed at developing animal models for normal and pathological processes in humans and those aimed at enriching our understanding of particular biological processes by elucidating the fundamental principles by which anatomical structures determine physiological and behavioral functions over ontogenetic and evolutionary time. For both reasons, the comparative approach is particularly important in the study of hearing (Fay 1994). This is clearly true at a structural and mechanistic level but also in other ways. Our own personal experience tells us that, more than any other sense, the sense of hearing links us to the rest of humanity both at an intimate, emotional level and at the level of acoustic communication through language and speech. The chapters in this volume review what is understood about the similarities and differences in hearing among birds and reptiles both physiologically and behaviorally and provide unique contrasts and similarities with what we know of hearing in humans. This is true for the most peripheral structures that condition and transform sound before it is transduced within the inner ear to the perception of complex sounds and objects by the whole organism.

1.2. Relationship Between the Behavior and Physiology of Hearing

The relationship between the behavior and physiology of hearing becomes more important than usual when comparing reptiles and birds because any comparison of hearing between these two groups requires bridging that gap. Hearing is usually defined as the behavioral response to the form of energy we call sound and is demonstrated and described by analyzing the effects of a sound on behavior (Stebbins 1983; Fay 1994). In nature, hearing also means sound source determination: determining the "who," "what," and "where" of sound. This is obviously an exceedingly complicated operation that is often simplified and constrained in the laboratory.

It is important to remember that nearly all we know about human hearing and auditory perception comes from laboratory psychophysical experiments where human observers make behavioral responses to the occurrence of a sound or a change in sound. Over the past several decades, the equivalent psychoacoustic procedures for animals have become quite sophisticated (see Klump et al. 1995) and, especially in the case of birds, now allow comparison at higher levels of auditory perception between simple detection and discrimination of sounds. Many of these complex perceptual phenomena, such as judgments about timbre, pitch, and category membership have no precise counterpart in physiology.

Studies of the physiology of hearing, on the other hand, provide an understanding of the mechanisms of hearing. Species can be easily compared not only on the behavioral responses to sound but also on the basis of these underlying physiological mechanisms. It is such parallel comparisons that allow us to establish the general principles that apply to all vertebrates, to explore the limits of specializations, and to glimpse the evolution of the acoustic sense of animals and humans. Importantly, for the present purposes, decades of rigorous psychophysical procedures in birds combined with a number of physiological studies involving cochlear microphonic and single-unit studies in the same species, combine to provide reasonable confidence that the only available measures of hearing in the reptiles, physiological measures of bandwidth and sensitivity, provide an acceptable measure of hearing and a reasonable basis for comparison with the birds and other vertebrate groups for which considerably more data exist. Thus although we may justifiably wonder what the auditory world of the reptiles and birds is like and how it relates to our own, there is no longer much doubt about the bandwidth of hearing and the limits of auditory sensitivity in reptiles and birds.

1.3. Auditory Sensitivity and the Bandwidth of Hearing in Reptiles and Birds

In vertebrate evolution, reptiles were the first completely terrestrial vertebrates. Although the fossil record makes clear that reptiles and birds split over 200 million years ago, there are considerable similarities not only in anatomic structure and function but also in the most basic aspects of hearing. Compared with mammals, for instance, reptiles and birds share a relatively restricted range of hearing between a few hundred Hertz to about 10 kHz (only in birds) where sensitivity, especially in the birds, can approach the levels reported for humans and other mammals. We know this from the extensive behavioral studies in birds and the large number of physiological studies in the reptiles many conducted by Wever (1978). Although reptiles and birds share a narrow bandwidth of hearing, the reptile ear is generally tuned to lower frequencies than is the avian ear and is less sensitive. But compared with fishes and amphibians (see Popper and Fay 1999), the evolution of an elongated cochlea in reptiles and birds clearly served to increase the surface area of the basilar papillae, accommodating both an extended range of hearing and better sensitivity.

2. The Mechanisms of Hearing in Reptiles and Birds

2.1. The Middle Ear of Reptiles and Birds

The response of any physical object set in vibration is determined by the physical properties of stiffness, mass and friction. In the case of the middle ears of reptiles and birds, one must consider the vibration of interconnected elements that have complex and dissimilar properties. Saunders, Duncan, Doan, and Werner (Chapter 2) describe the middle ears of reptiles and birds by a "folded seesaw" model that can be used to characterize the mammalian, reptilian, and avian middle ears as well as the essential features of the three types of lizard middle ears. It is now known that a conductive system has emerged on multiple occasions and probably independently in amphibians and reptiles. Saunders, Duncan, Doan, and Werner (Chapter 2) argue that changes in bony structure during the transition from stem reptiles to the ancestors of the various subclasses of reptiles had important consequences for the evolving middle ear. This occurred because the later-arising reptiles had stronger limbs that lifted the head and body from the ground, thereby isolating them from substrate vibrations that were probably transferred directly to the ear in more primitive forms via the skull. This led to a more slender columella, an extracolumella, and the transformation of skin into a tympanic membrane.

The avian middle ear has retained the basic reptilian design but, in contrast to the tremendous diversity in structural organization found in reptiles, the avian interspecific differences among middle ear systems are minimal. However, the middle ear cavity of birds is spacious relative to head size. In addition, in both birds and crocodiles, tubes running through the posterior and ventral portions of the porous, trabeculated skull connect the middle ear cavities on both sides of the head with a common opening to the pharynx. This connection has been referred to as the "interaural pathway." This pathway has been investigated a number of times over the last century, and although its exact role in spatial hearing in birds remains obscure, it certainly could function to aid in sound localization (Larsen et al. 1997; Klump, Chapter 7).

The accessibility of the middle ear structures of reptiles and birds facilitate functional studies such as, for example, velocity transfer functions. Two things are clear from a number of studies in both reptiles and birds. There is considerable functional homology in the band-pass shape of the reptilian and avian band-pass functions. And the band-pass characteristics of the middle ear play a large role in the shape of the audibility curve. The restricted high-frequency hearing in reptiles and birds is in stark contrast to the "high-pass" transfer function in mammals afforded by the ossicular middle ear.

2.2. The Inner Ear of Reptiles and Birds

The great diversity of the group of animals considered under the term reptiles has prompted Manley and Gleich (Chapter 3) to consider the lizards as separate from the crocodiles, alligators, and their relatives that are more closely related to birds than to other reptiles. In addition, most lizards are not vocal, which raises interesting questions about the selective pressures leading to the variety of different ears in lizards. Lizards have evolved two types of hair cell areas containing either a "unidirectional type" or a "bidirectional type" hair cell. These do not appear to be homologous with the two populations of hair cells in the mammalian cochlea (inner and outer hair cells) or with those in the bird basilar papilla (tall and short hair cells), but rather they are thought to have evolved independently of those of other tetrapod groups (Manley 1990; Manley and Köppl 1998; Manley, Chapter 4). There are substantial species differences in the way that localized frequency tuning is achieved in the lizards, with corresponding differences in innervation patterns. As Manley (Chapter 4) points out, this variation in basilar papilla structure in lizards presents a unique opportunity to study structure-function relationships.

Birds and crocodiles are the only group of amniotes other than mammals that have specialized sensory hair cells across the width of their auditory papillae. In crocodiles, there are two anatomic types of hair cells, whereas in birds, there is a more gradual transition from tall to short across the papilla (Fischer 1994a, 1994b; Manley 1996). Tall hair cells found predominantly over the immovable superior cartilaginous plate are the less specialized ones. They also bear similarities to the hair cells of some anamniotes. Short hair cells, by contrast, occupy most of the space over the freely moving basilar papilla. These gradients are species specific as is the pattern of hair cell bundle rotation. In other words, the orientation of the hair cell stereovillar bundles along the width and length of the auditory sensory surface are fixed in mammals but vary systematically in birds, a feature that sets birds apart from all other tetrapods. The innervation of hair cell types also differs from tall hair cells, having large afferent and small efferent connections with the pattern reversing toward the abneural side of the papilla, in some cases leading to hair cells with exclusively efferent innervation (Fischer 1998). In general, more advanced species show a denser afferent innervation, especially on neural hair cells. The papillar morphology of every species is unique, with some cases showing anatomic specializations for low-frequency hearing as in the pigeon (*Columba livia*) and the chicken (*Gallus gallus*) or high-frequency hearing as in the canary (*Serinus canarius*), zebra finch (*Taeniopygia guttata*), and barn owl (*Tyto alba*). Tuning is accomplished by electrical characteristics of the hair cell membrane at low characteristic frequencies and by micro- and macromechanical mechanisms at higher characteristic frequencies (CF), including the number and height of hair cell stereovilli, the mass of the tectorial membrane, and active movements of hair cells as evidenced from the presence of otoacoustic emissions in birds.

Isoresponse curves from single-unit recordings of primary auditory fibers, with a criterion just above the spontaneous rate, generally correspond well to behavioral thresholds, but the tuning curves are typically narrower and more symmetrical than those found in mammals. In some cases, low CF fibers even show a steeper low-frequency slope than a high-frequency slope. Work on the starling (*Sturnus vulgaris*), with a large sample of cells to a single test stimulus, has led to a description of the cochlear excitation pattern that has been confirmed by critical band measurements, leading to perhaps the most complete model excitation patterns in any vertebrate (Gleich 1989; Buus et al. 1995). But gradients in sensitivity and frequency selectivity of afferent fibers innervating hair cells at different positions across the basilar papilla in birds are species specific so the exact nature of the transduction process remains obscure.

2.3. The Central Auditory System of Reptiles and Birds

The central auditory system of both reptiles (including crocodiles) and birds are organized along a common plan. The cochlear nuclei of crocodiles are tonotopically organized in a fashion similar to that of birds, with similar primary-like single-unit responses. There is segregation in the encoding of time versus level information between the nucleus magnocellularis and nucleus angularis (equivalent to the mammalian cochlear nucleus) which is prominent in the barn owl but exists to a lesser extent in the chicken. Carr and Code (Chapter 5) hypothesize that because interaural level differences become more prominent at higher frequencies, the pairing of time and level may have provided the selective pressure for the evolution of phase locking to high frequencies in the barn owl. Compared with mammals, there are a number of other specializations associated with phase locking in birds, including large principal cells of the nucleus magnocellularis, fewer dendrites, short time constants, and AMPA-subtype glutamate receptors, with rapid desensitization throughout the auditory hindbrain.

The cochlear efferent system is better characterized in birds than in reptiles (Code 1997). Neurons that project to the basilar papilla are located binaurally in the hindbrain and can be divided into two spatially distinct cell groups: the ventrolateral and the dorsomedial. Because some hair cells in the avian basilar papilla have only efferent innervation, it is critical to determine the function of efferent innervation. The most comprehensive study to date is by Kaiser and Manley (1994) showing that cochlear efferents are more sensitive to contralateral than to ipsilateral stimulation and have poor frequency selectivity and low spontaneous activity. Data from reptiles are sparse but, in general, the anatomic distribution of avian cochlear efferent neurons has more features in common with crocodiles than with turtles.

Controversy exists as to whether the primary ascending thalamic targets in mammals and sauropsids are homologous to those of birds and reptiles, but the ovoidalis complex has been homologized to the mammalian medial geniculate nucleus. In the barn owl, there are neurons that respond to interaural time differences, levels, differences, and stimulus frequency, but there is no systematic representation of sound localization cues. In the starling, few units were observed with complex tuning properties, no neurons preferred song, and responses to complex stimuli can be predicted on the basis of responses to simple sounds.

Field L is the principal target of ascending input from the auditory dorsal thalamus and it is now known to consist of a number of layers and sublayers. Although field L provides auditory inputs to the songbird vocal control system, it is not the main source of auditory information to the vocal system in budgerigars (*Melopsittacus undulatus*). Instead, auditory inputs to the vocal control system appear to derive from the nucleus basalis and the frontal neostriatum (Durand et al. 1997; Striedter et al. 1997). Despite independent evolution, the song control circuits of oscines and psittacines share a number of similar design features (Durand et al. 1997). Vocal control in these two groups of birds differs from other sauropods because the pre-existing projection from the archistriatum to the intercollicular region appears to have been modified so that the robust nucleus of the archistiatum (RA) in birds projects directly to the motor nucleus of the hypoglossal nerve in the medulla in addition to a region of the nucleus intercollicularis.

3. Localization, Discrimination, and Perception of Sound in Birds

3.1. Sound Localization and the Formation of Auditory Objects

Hearing is a primary means for vertebrates to acquire information about their world. On this point, there is a wealth of behavioral information from birds and virtually none from reptiles. In the case of birds, the ear and the auditory system are part of a complex acoustic communication system designed to receive and decode species-specific messages. An even more primary role of hearing is to first gain insight into the overall environment (Popper and Fay 1997), followed by sound source determination (Pumphrey 1950; Yost and Sheft 1993), now referred to as object perception or auditory image or scene analysis. It involves at least the process of identifying the location of a sound source so that an auditory object can be formed. Klump (Chapter 7) reviews the importance of sound localization for birds, the special problems resulting from small interaural distances and poor high-frequency hearing, and the special adaptations in nocturnal predators such as the barn owl. To date, sound localization has been investigated with a number of behavioral methodologies in a wide variety of species from small songbirds to relatively large nocturnal predators such as the owl.

Compared with other terrestrial vertebrates, birds have developed some new solutions to solve some of the acoustic problems that accrue from living in a physical world. The small interaural distances and the existence of interaural pathways in birds have suggested that the ears of birds may function as sound-pressure difference receivers (Klump and Larsen 1992). Klump (Chapter 7) reviews the evidence for and against this case in birds as well as several other strategies that some birds might use to enhance interaural differences. These include modifying the shape and size of the opening of the auditory meatus, contracting the stapedius muscle, and controlling the air pressure in the middle ear space through the tracheopharyngeal tube (Larsen et al. 1997). Several things emerge from this review. Large interspecies differences exist in the resolving capability and mechanisms of sound localization in birds. Birds such as the barn owl and other birds with asymmetric external ears that appear specialized for sound localization are different from unspecialized birds. The barn owl, at present, still provides the best foundation for understanding the physiological mechanisms underlying avian sound localization abilities. Because the barn owl is specialized for sound localization, it is not clear that it represents a very good model for sound localization in unspecialized birds. We know from field studies, however, that small songbirds have a remarkable ability not only to detect but also to judge both the direction and the distance of sound sources in their habitat (Klump 1996; Nelson and Stoddard 1998).

3.2. Discrimination and Auditory Perception in Birds

As a group, birds hear best between about 1 and 5 kHz, with absolute sensitivity approaching 0- to 10-dB SPL (sound pressure level) and a frequency of best hearing typically around 2–3 kHz. Within this narrow frequency region of best hearing, the abilities of birds to discriminate among both simple and complex sounds has been investigated extensively over the years (Dooling, Lohr, and Dent, Chapter 6).

3.2.1. Masking and Frequency Analysis

All vertebrates analyze simple and complex sounds into their frequency components. There are a number of different measures of frequency resolution and the most common are those involving masking. From a wide variety of masking experiments including critical ratios, directly measured critical bands, psychophysical tuning curves, and masking by notched noises, a picture is emerging about how the avian ear resolves frequency and extracts signals from noise. In general birds do almost as well, and sometimes better than, many mammals (Dooling 1982; Dooling, Lohr, and Dent, Chapter 6). Most birds show increasing filter bandwidths with frequency, but some such as the budgerigar, great tit (*Parus major*), and barn owl do not. In some but not all cases, there is a close correspondence among the different estimates of frequency resolution obtained from different masking measurements, suggesting a common underlying process. In aggregate, however, sufficient species differences exist so that one must conclude that the exact mechanisms of frequency resolution in birds remains unknown.

3.2.2. Frequency and Intensity Discrimination

Studies of frequency discrimination in a number of birds support the notion that these species are quite sensitive to changes in the frequency of acoustic signals with Weber fractions falling in the range of 0.008–0.020. In other words, birds can discriminate better than about a 1% change in frequency in the spectral region of 1–4 kHz. They approach the sensitivity of humans in this regard, which places them better than most other vertebrates that have been tested. By contrast, data from a number of species show quite clearly that birds are not particularly sensitive to changes in intensity and are certainly not more sensitive than humans. In general, birds show intensity discrimination thresholds that are similar to, if not worse than, those measured in other nonhuman vertebrates, falling in the range of 1–4 dB.

On many tests of the discrimination of complex sounds including species-specific vocalizations, human speech sounds, and other complex sounds such as noises with rippled spectra and cosine noise, birds do very well, thereby suggesting that the auditory world of birds approaches the richness and complexity of our own. It is not too surprising, then, that a number of higher-order perceptual phenomena including perceptual constancy, categorical perception, comodulation masking release, and auditory scene analysis have recently been demonstrated in birds. Some of these phenomena are quite evident in the perception of speech sounds by birds, lending even more support to the notion that birds, to a great extent, may hear the world as we do.

3.2.3. Temporal Integration and Resolution

For many measures of temporal resolution involving envelope cues such as minimum and maximum temporal integration, duration discrimination, and gap detection, birds are also similar to mammals (Fay 1988). This in itself is surprising given the structural differences between the avian and mammalian peripheral auditory systems. However, on other temporal measures involving changes in temporal fine structure of complex waveforms such as timbre discrimination, detection of a mistuned harmonic, and discrimination of changes in the phase structure of harmonic complexes, birds may exceed the capabilities of humans. If such enhanced temporal resolution is engaged when birds perceive complex species-specific communication sounds, it is interesting to ponder what that means for our traditional ways of defining, characterizing, and quantifying the vocal repertoire of many avian virtuosos. Perhaps some birds hear fine temporal detail in such sounds and have a much larger repertoire than previously thought.

4. Hearing in Reptiles and Birds: Solved and Unsolved Problems

As pointed out at the beginning of this chapter, the study of birds and reptiles is of interest not only for itself but also because these groups have provided important insight into broad areas of sound detection and processing that have implications for the understanding of hearing in all vertebrate groups. Although the importance of these groups for understanding the mammalian auditory system is not emphasized in this volume, these include issues related to hair cell biophysics, the development of hearing, the ontogeny of auditory mechanisms, and the relationship between hearing and vocal production. Because of the evidence for widespread vocal learning and hair cell regeneration in birds, this group plays a uniquely special role in providing models for both normal and pathological hearing-related processes in humans related to hair cell degeneration and regeneration, auditory system plasticity, and the recovery of vocal precision with the return of auditory function (Salvi et al. 1996; Dooling et al. 1997). There are many remaining fundamental questions raised in these chapters that are unresolved in these groups of vertebrates in both the auditory periphery and in the CNS, including the relationship between electrical and mechanical tuning of hair cells, the function of hair cell orientation patterns, the role of the tectorial membrane in transduction in birds, the selective pressures that led to hearing organ diversity in lizards, the evolution of the dorsal octaval columns, the effects of hair cell regeneration on the CNS, and the evolution of different auditory-vocal circuits in different groups of birds. One of the biggest challenges, however, remains the lack of data and insight into the auditory world of reptiles.

References

Buus S, Klump GM, Gleich O, Langemann U (1995) An excitation-pattern model for the starling (*Sturnus vulgaris*). J Acoust Soc Am 98:112–124.

Code RA (1997) The avian cochlear efferent system. Poultry Avian Biol Rev 8:1–8.

Dooling RJ (1982) Auditory perception in birds. In: Kroodsma DE, Miller EH (eds) Acoustic Communication in Birds, Vol 1. New York: Academic Press, pp. 95–130.

Dooling RJ, Ryals BM, Manabe K (1997) Recovery of hearing and vocal behavior after hair cell regeneration. Proc Natl Acad Sci USA 94:14206–14210.

Durand SE, Heaton T, Amatea SK, Brauth SE (1997) Vocal control pathways through the anterior forebrain of a parrot (*Melopsittacus undulatus*). J Comp Neurol 877:179–206.

Fay RR (1988) Hearing in Vertebrates: A Psychophysics Databook. Winnetka, IL: Hill-Fay Associates.

Fay RR (1994) Comparative auditory research. In: Fay RR, Popper AN (eds) Comparative Hearing: Mammals. New York: Springer-Verlag, pp. 1–17.

Fischer FP (1994a) Quantitative TEM analysis of the barn owl basilar papilla. Hear Res 73:1–15.

Fischer FP (1994b) General pattern and morphological specializations of the avian cochlea. Scanning Microsc 8:351–364.

Fischer FP (1998) Hair cell morphology and innervation in the basilar papilla of the emu (*Dromaius novaehollandiae*). Hear Res 121:112–124.

Gleich O (1989) Auditory primary afferents in the starling: correlation of function and morphology. Hear Res 37:255–268.

Kaiser A, Manley GA (1994) Physiology of putative single cochlear efferents in the chicken. J Neurophysiol 72:2966–2979.

Klump GM (1996) Bird communication in the noisy world. In: Kroodsma DE, Miller EH (eds) Ecology and Evolution of Acoustic Communication in Birds. Ithaca, NY: Cornell University Press, pp. 321–338.

Klump GM, Larsen ON (1992) Azimuthal sound localization in the European starling (*Sturnus vulgaris*). I. Physical binaural cues. J Comp Physiol A 170: 243–251.

Klump GM, Dooling RJ, Fay RR, Stebbins WC (eds) (1995) Methods in Comparative Psychoacoustics. Basel: Birkhauser-Verlag.

Larsen ON, Dooling RJ, Ryals BM (1997) Roles of intracranial air pressure on hearing in birds. In: Diversity in Auditory Mechanics. Singapore: World Scientific Publishers, pp. 253–259.

Manley GA (1990) Peripheral Hearing Mechanisms in Reptiles and Birds. New York: Springer-Verlag.

Manley GA (1996) Ontogeny of frequency mapping in the peripheral auditory system of birds and mammals: a critical review. Aud Neurosci 2:199–214.

Manley GA, Köppl C (1998) Phylogenetic development of the cochlea and its innervation. Curr Opin Neurobiol 8:468–474.

Nelson BS, Stoddard PK (1998) Accuracy of auditory perception of distance and azimuth by a passerine bird. Anim Behav 56:467–477.

Popper AN, Fay RR (1997) Evolution of the ear and hearing: issues and questions. Brain Behav Evol 50:213–221.

Popper AN, Fay RR (1999) The auditory periphery in fishes. In: Fay RR, Popper AN (eds) Comparative Hearing: Fish and Amphibians. New York: Springer-Verlag, pp. 43–100.

Pumphrey RJ (1950) Hearing. Symp Soc Exp Biol 4:1–18.

Salvi RJ, Henderson D, Fiorino F, Colletti V (eds) (1996) Auditory System Plasticity and Regeneration. New York: Thieme Medical Publishers.

Stebbins WC (1983) The Acoustic Sense of Animals. Cambridge, MA: Harvard University Press.

Striedter GF, Marchant TA, Beydler S (1998) The "neostriatum" develops as part of the lateral pallium in birds. J Neurosci 18:5839–5849.

Wever EG (1978) The Reptile Ear: Its Structure and Function. Princeton, NJ: Princeton University Press.

Yost WA, Sheft S (1993) Auditory perception. In: Yost WA, Popper AN, Fay RR (eds) Human Psychophysics. New York: Springer-Verlag, pp. 193–236.

2
The Middle Ear of Reptiles and Birds

James C. Saunders, R. Keith Duncan, Daryl E. Doan, and Yehudah L. Werner

1. Introduction

The middle-ear system of all vertebrates improves the efficiency of sound transmission from the surrounding medium, be it air, water, or ground, to the inner ear. The process by which this is achieved is similar across both mammalian and nonmammalian forms. The specific structures and mechanisms that have evolved to accomplish this task, however, vary considerably from species to species. In this chapter we hope to develop an appreciation of how the middle-ear system is organized, how it operates, and how it contributes to hearing in reptiles and birds. The chapter begins by examining how the middle ear is studied and how it functions. A brief exposition of middle-ear evolution is followed by a consideration of structure and function in the reptilian and avian middle ears. The contribution of middle-ear muscle contraction as well as middle-ear development is then presented. Finally, the chapter concludes with a discussion of the contribution of the middle ear to the overall process of hearing in these species.

2. Middle-Ear Function

The response of any object set in vibration is determined by the physical properties of stiffness, mass, and friction. If the object consists of a uniform material whose density and bulk modulus (stiffness) are known, then the equations that describe the interaction of these properties predict accurately many parameters of the response. This includes the natural or resonant frequency, the absorption and transmission of sound, and the amplitude and phase of vibration at different frequencies. When dealing with a middle-ear system, the vibration of simple objects gives way to the vibration of interconnected elements, which have complex and dissimilar physical properties. For example, the stiffness and mass of the tympanic membrane (TM), the ossicles, and suspensory ligaments are all different from each other. Investigators of the middle ear are frequently interested in those processes that act as barriers to the flow of vibratory energy, and these barriers determine the impedance properties of the middle ear.

Middle-ear impedance at low frequencies is determined largely by the *stiffness reactance*, which describes the elastic properties of the system. The magnitude of this reactance is negative and decreases toward zero as frequency rises. Stiffness reactance becomes smaller (more negative), for example, as middle-ear cavity volume or TM surface area increases, and either of these conditions would result in improved low-frequency sound conduction.

The *mass reactance*, as part of the inertial response of the middle ear, controls the high-frequency response of the system. As the amount of mass increases, this reactance increases. Similarly, mass reactance grows as frequency rises. Thus, the reduction in middle-ear sound conduction at high frequencies is attributable, in part, to the mass of the middle-ear apparatus. A reduction in the effects of mass will result in improved high-frequency sound conduction, and this can be achieved by smaller or lighter ossicles, or by compaction of ossicular mass about the axis of rotation to reduce the moment of inertia.

At some frequency, the mass and stiffness reactances will have the same magnitude but one will be positive (mass) and the other negative (stiffness). As a consequence they will cancel each other out. There is minimum impedance to energy flow through the middle ear at this frequency, and only the frictional forces of resistance dominate the response. Resonance occurs at this frequency and the conduction of energy through the system is most efficient.

It should be apparent, even in this simple description, that determining the middle-ear response is difficult based on the physical properties of the system. Further aspects of middle-ear physics can be found elsewhere (Møller 1974a, 1974b; Relkin 1988; Rosowski 1994, 1996).

2.1. Measuring the Middle-Ear Responses

There are several ways of studying the middle ear, and one of these considers the overall response of the system by estimating the flow of energy through the conductive apparatus. Energy flow is related to the impedance of the system (or its reciprocal admittance). Acoustic impedance (Z) can be measured at the input (the tympanic membrane) and at the output (the oval window) of the middle ear, and may be defined as the ratio of acoustic pressure (P) to volume velocity (U) of the moving object. Knowledge of the input and output impedance allows many properties of the intervening system to be described. Measures of middle-ear impedance have been used with great success to define and model many aspects of the conductive apparatus (Rosowski 1994, 1996).

A second approach measures directly the motion of middle-ear structures at various locations along the conductive pathway. Several technologies are available for precisely determining the velocity or displacement of a vibrating object, and these are capable of measuring motion at levels on the order of nanometers. One of these employs Mössbauer spectroscopy to measure Doppler shifts in the energy level of gamma-ray particles emitted by a radioactive source secured to the vibrating object. In addition, miniature capacitive probes also have been developed to measure the changes in charge between the probe tip and a vibrating object. Finally, laser interferometry measures the changes in light level when an incident beam interacts with a reflected beam from a vibrating object.

There are advantages and problems associated with these techniques and their considerations are beyond the scope of this presentation. Nevertheless, all these methods have been applied successfully to study the middle-ear response of amphibians, reptiles, birds, and mammals. These responses, obtained from the TM surface or the columella footplate, are measured as either velocity or displacement, and plotted as a transfer function. This function traces the magnitude and phase of the response, across frequency, when the input stimulus to the system (at the TM) is held constant. This function defines the efficiency of energy transfer between air particle motion and motion of the TM. Such a function treats the various tissue compartments between the point of stimulation and the point of measurement as a "black box," meaning that the contribution of elastic, inertial, or frictional forces do not have to be specified. This approach is appropriate as long as the parameters within the "box" respond linearly. When the sound pressure level (SPL) at the TM is constant, the transfer function measured at the stapes footplate characterizes the behavior of the entire conductive system independent of the mechanical properties of any individual component.

As previously noted, these methods return a measure of either velocity or displacement, and both these measures are mathematically related to one another. The time-averaged velocity of sinusoidal motion is related to displacement by Equation 1, where v is velocity, d is displacement, and f is frequency in Hertz.

$$v = 2 fd \qquad (1)$$

Air particles stimulated at 100 dB SPL across frequency show different motion characteristics when described in displacement or velocity terms. Particle displacement declines at a rate of 6 dB/octave, which means that for each doubling of frequency, there is a halving of displacement amplitude. Equation 1 shows that when frequency rises and displacement diminishes at the same rate, air-particle velocity remains constant. Transfer functions most clearly define the system response when the input to the system is constant. Thus the relationship between stimulus input and system response is best described by plotting the response in velocity terms. All of the transfer functions in this presentation are described in terms of the peak-to-peak velocity response for a constant input stimulus of 100 dB SPL. This required a recalculation of the original data in many instances. This normalization was important because it allowed the relative sensitivity and shape of the middle-ear response to be compared among different species, preparations, and procedures.

2.2. The Middle-Ear Transformer

The middle-ear system serves as a transformer designed to match the low impedance of air to the higher impedance of inner-ear fluids. The classic view of the middle ear, prevalent to the mid-1960s, was that improvements in the efficiency of sound transmission from air to the cochlear fluids were achieved by the transformer action of several pressure amplifiers inherent in the conductive apparatus. The "folded seesaw" model has been used to describe these amplifiers. The seesaw is folded in mammals because the fulcrum is represented by the axis of rotation of the conductive apparatus (see Fig. 2.1A), and describes two mechanisms that act as amplifiers to increase the pressure applied to the fluids of the cochlea at the stapes footplate. The description that follows is modified from Relkin (1988).

The area of the eardrum (A_d) and that of the stapes footplate (A_s) can be viewed as piston heads at two ends of a hydraulic system. The pistons are connected to each other through the middle-ear ossicles. In mammals, the long arm of the malleus (the manubrium) is attached to the TM, while the long arm of the incus (the crus) is attached to the head of the stapes and hence the stapes footplate. When sound pressure is applied to the TM, a force is generated at the larger piston head. This force is transmitted to the footplate through the ossicular chain. Amplification in the pressure delivered at the stapes footplate results from the differences in the areas of the two pistons and the lengths of the two arms of the ossicles. The force applied to the drum membrane equals the sound pressure level (P_d) times the effective area of the TM (A_d). The product of the force at the piston head of the TM ($P_d \cdot A_d$) and the lever arm of the malleus (l_m) must be the same as the product of the pressure at the footplate (P_s) and the footplate area (A_s) and the incus lever arm (l_i). This relationship is seen in Equation 2.

$$P_d \cdot A_d \cdot l_m = P_s \cdot A_s \cdot l_i \qquad (2)$$

P_d and P_s are the pressures at the drum membrane and stapes footplate, respectively. The consequence of this equality is that a given pressure distributed over the larger surface of the TM with its longer lever arm results in a substantial increase in the pressure applied at the smaller footplate surface. The equalities of Equation 2 can be rearranged with regard to the ratio of foot plate and tympanic membrane pressures (Eq. 3).

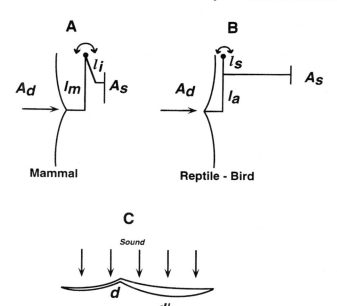

FIGURE 2.1. The illustrations in **A** and **B** show a schematic organization of the "folded seesaw" model of the mammalian and reptilian/avian middle ears. The large dot with the circular arrows above it indicates the fulcrum and axis of rotation. A_d = drum area; A_s = stapes area; l_m = length of the manubrium; l_i = length of the incus arm; l_a = anchorage length; l_s = length of pars superior. Panel **C** shows the action of the drum lever. The posterior surface of the tympanum undergoes a large displacement (d') that results in a smaller displacement at the conical tip of the reptilian/avian TM (d). Thus a lever advantage is realized at d and this is communicated through the conductive apparatus.

$$P_s/P_d = (A_d/A_s) \cdot (l_m/l_i) \tag{3}$$

Thus the pressure transformation of the middle ear equals the *area ratio* (A_d/A_s) times the *lever ratio* (l_m/l_i). In all tetrapod forms the area of the TM and the length of the lever arm associated with the TM (l_m) is always greater than the area of the footplate and the length of the lever arm driving the footplate (l_i). Consequently, the pressure ratio is greater than one, indicating a larger pressure at the stapes footplate than at the TM. The elements of the conductive system that define the area and lever ratios in mammals are sketched in Figure 2.1A. The conductive apparatus is suspended by ligaments that attach the drum membrane and footplate to the surrounding bony framework. Additional ligaments that stabilize the entire system support the ossicles. Finally, middle-ear muscles attach to various components of the conductive apparatus (e.g., the ossicles or the TM), and contraction of these muscles can significantly alter middle-ear mechanics (see Section 6).

The organization of the lizard and avian middle ear is diagramed in Figure 2.1B, and the exact anatomical structures involved will be described in subsequent sections. For now, it is only important to recognize that the hydraulic and mechanical levers serve virtually the same function as that in the mammalian middle ear. The illustrations indicate that both middle-ear systems operate as second-class levers with the fulcrum located at the axis of rotation. The hydraulic gain in pressure is derived from the ratio of areas between the TM and footplate. Additional mechanical gain is achieved through the differing length of the lever arms.

Another amplifier is found in the mechanical lever of the TM (Fig. 2.1C). The conical lever of the TM was first proposed by Helmholtz (1873) and was considered further by others (Wever and Lawrence 1954; Tonndorf and Khanna 1970; Manley 1990). The basis of this lever is the "catenary" principle. The TM is essentially secured at its perimeter, and large motions in the body of the membrane (d' in Fig. 2.1C) will result in smaller movements at the ossicle attached to the membrane (d), and this yields a mechanical advantage. Analytic treatments of the middle ear like that in Equation 2 do not take the membrane lever into account because its anatomical basis is not easily defined.

There is a flaw in the analytic treatment of the middle ear described in equations 2 and 3. The considerations of pressure gain were based on the static behavior of the conductive apparatus, and motion has yet to be taken into account. Work, or the transfer of energy, can occur only when pressure leads to motion. The idea of motion in the middle ear was introduced through the concept of impedance (Zwislocki 1962; Møller 1963; Mundie 1963).

As noted above, impedance (Z) is the ratio of the pressure (P) to the volume velocity (U) of a vibrating surface. The concept of volume velocity has been considered in detail by Relkin (1988), and it is easily understood as the product of a piston area (A) times the velocity (v) of its motion (Eq. 4).

$$U = A \cdot v \qquad (4)$$

An examination of Figure 2.1A reveals that the velocity of motion at the tip of the manubrium and at the long arm of the incus (the crus) must necessarily be different. Both points are moving at the same time, but length differences in the lever arms mean that the tip of manubrium (l_m) is traveling a longer distance than the end of the crus (l_i). Since velocity is related to distance and time, the tip of the longer lever arm (l_m) has a greater velocity of motion than that of the shorter lever arm (l_i). In fact, as the manubrium and crus rotate about their axis of rotation, the ratio of the velocities at their tips will equal the ratio of their lever arms. This means that the velocity ratio (v_m/v_i) and the lever ratio (l_m/l_i) are interchangeable, and one could be substituted for the other in Equation 4. Moreover, pressure (P) will then equal the product of impedance (Z), area (A), and velocity (v). This relationship can be substituted for the pressure transformation in Equation 2 to obtain the impedance transformation in Equation 5:

$$Z_d/Z_s = (A_s/A_d)^2 \cdot (l_i/l_m)^2 = (P_d/P_s)^2 \tag{5}$$

Thus the square of the pressure ratio (Eq. 3) equals the acoustic impedance transformer ratio of the middle ear. The impedance transformer ratio in Equation 5 relates the output impedance at the stapes to the input impedance at the drum membrane. The value is less than one because the stapes footplate is smaller than the TM and the length of the crus is shorter than that of the manubrium. This means that high impedance at the footplate is transformed by the middle-ear system to lower impedance at the drum. Conversely, we could compare the input impedance of the drum to the output impedance of the footplate. In this situation, the low impedance of air is raised at the footplate to more closely match the high input impedance of the cochlear fluids. The acoustic impedance is measured in MKS ohms (Kinsler and Frey 1962). The specific input impedance of the human cochlea has been estimated to be around 112,000 acoustic ohms based on measures of cochlear cross-sectional area, basilar membrane compliance, and the impedance of the cochlear spaces (Zwislocki 1975; Lynch et al. 1982; Puria and Allen 1991). Since these variables vary with each species, there is no single cochlear input impedance that applies to all tetrapods. The impedance transformation ratio in Equation 5 is calculated to be about 0.016 in humans. Multiplying the cochlear input impedance by the impedance transformation ratio reveals that the impedance of the TM is about 190 ohms, which is close to the acoustic impedance of air (415 ohms). The important point to remember is that the impedance value itself is less important than the match between impedances at the interface between two media. The 190-ohm impedance of the human TM is considered closely matched to that of air. If the output impedance of the middle ear is substantially higher or lower than the input impedance of the cochlea, then the efficiency of energy transfer across the interface will deteriorate.

Even the impedance transformation model, however, is an incomplete description of the middle-ear response. It does not take into account gain produced by the "catenary" action of the TM as discussed above. In addition, the TM does not work as a simple piston moving back and forth as a rigid plate. At frequencies above 2.0 kHz, different locations on the membrane vibrate with different displacement amplitude, and this is true in the mammalian, avian, reptilian, and amphibian TM. Moreover, it was noted in the previous section that elastic and inertial force controls vibrating systems, and that opposition to motion, from the reactance properties of these forces, is frequency dependent. Impedance is also a frequency-dependent property. However, these simple models do not take frequency into account. Multicompartment models have been utilized to predict the behavior of the sound-conducting system. One such approach uses the equations for mass and stiffness reactance as well as frictional resistance to derive the middle-ear response as a function of frequency (Relkin 1988; Doan et al. 1994).

2.2.1. Measuring the Transformer

The middle-ear transformer results presented here come from published data and were derived from either anatomic or cochlear microphonic measures. The morphologically measured area and lever ratios are largely self-explanatory. Many of the published area ratios are limited to planar surfaces of the TM, and do not take into account the added area due to the curved shape of the membrane. Modern computer methods make it possible to measure areas with great accuracy, and the added area, due to the "tent-like" appearance of the membrane, produces a 3–7% increase over the surface measured as a flat plate. The anatomy of the middle-ear lever in reptilian and avian ears is complicated by the geometry of the tympanic processes of the extracolumella. These are the "legs" of the extracolumella that connect the columella (the long single shaft in Fig. 2.1B) to the inner surface of the TM.

Cochlear microphonic responses can be used to estimate the overall advantage of the middle-ear system. Among other things, the microphonic potential is thought to result from a voltage drop across the hair cells as a consequence of sound-driven basilar membrane movements (Durrant and Lovrinic 1995). Over a considerable range of intensities, this potential is linear with respect to sound pressure at the TM surface. The significance and role of this potential has been a source of argument since its discovery. Regardless, it can serve as a powerful tool for assessing the middle ear due to the fact that the microphonic response is both stable and reproducible. Changes in the sensitivity of this potential, before and after manipulations of the middle ear, can be used to study various properties of the conductive apparatus. In practice, stimulus intensity at the TM is varied across frequency to produce an iso-response contour (e.g., a 0.1-µV response) of the cochlear microphonic (measured at the round window membrane). This contour has been referred to as a cochlear microphonic "sensitivity curve" or "audiogram." In a procedure pioneered by Wever and Werner (1970) the microphonic sensitivity curve is measured first with closed-field stimulation of the TM. The TM is then removed and the columella shaft clipped near the footplate. A closed-tube acoustic system is then sealed over the oval window and the stump of the footplate stimulated. The level of acoustic stimulation needed to achieve the criterion microphonic response was greater when sound was delivered to the footplate than when it was presented directly to the TM. The difference in decibel between the two microphonic sensitivity curves (when plotted as a ratio) provided an indication of the total contribution of the middle-ear transformer (P_s/P_d in Eq. 3). This measure includes the middle-ear pressure amplifiers due to the drum lever, the ossicular lever, and the area ratio. The square of this ratio is the middle-ear impedance transformer (Eq. 5).

3. Evolution of the Reptilian and Avian Middle Ear

A number of presentations consider in detail the evolution of hearing and the vertebrate ear (Henson 1974; Manley 1974, 1990; Wever 1974; Webster, Fay, and Popper 1992). Other articles have focused on the evolution of the middle-ear system (Henson 1974; Fleischer 1978; Kühne and Lewis 1985; Lombard and Bolt 1988; Manley 1990; Bolt and Lombard 1992; Clack 1992; Fritzsch 1992; Hetherington 1992). These relatively recent presentations, as well as a literature that spans more than a century of study, have yet to resolve endless controversy about the emergence and progression of the apparatus for hearing.

3.1. Methods of Study

Living animals may be used to construct groupings of species in which there are common structural relationships that grade themselves in an orderly fashion. This phylogenetic organization of the animal kingdom has been systematically refined over the last 200 years. Paleontology, or the study of the fossil record, is then used to verify the phylogenetic relationships, and extend the history of living animals backward into ancient times.

There are difficulties with both these approaches. For one, the phylogeny of species may be far from clear, and debate has emerged concerning the grading of structures that differentiate one species from another. Moreover, the fossil record, consisting of bony replicates or impressions in stone, provides only a blurred image of the original structural organization. The advent of modern statistical treatments (cladistic analysis) has helped grade the relationships among extant and fossil species, and has served to clarify somewhat the sequence of progressive changes. Cladistic analysis is now in widespread use, but also requires a degree of consensus over the structural parameters examined, and this is often difficult to achieve given the gaps in the fossil record.

Perhaps the most interesting aspect of any evolutionary description is the assumed relationship between selective pressures and resulting structural changes; for example, the process of natural selection. Considerations of the selective pressures leading to the emergence of adaptive characteristics retained in the genetic history of the animal is the most speculative and challenging aspect of any evolutionary scheme. Reconstruction of these scenarios are abundant in the literature, and they may be clever, brilliant, and perhaps even close to the truth (e.g., Gould 1989)! Nevertheless, they need to be regarded with a healthy level of skepticism since the emergence and evolution of any biologic system can never be known with complete certainty.

3.2. The Reptilian Path

The theoretical basis for the evolution of species was formulated during the middle decades of the nineteenth century, and by the late 1800s, it was widely held that all organ systems evolved along a continuum. The ear was seen as just another biologic example to which this principle applied. Descriptions of middle-ear evolution up to the middle of this century attempted to place the development of the conductive apparatus along an unbroken path that extended from the earliest fish to the mammals. The notion that middle-ear progression from primitive to modern forms occurred in a smooth series of steps proved to be an impossible hypothesis to confirm, largely because of gaps in the paleontologic record. The recent literature, nevertheless, makes it abundantly clear that the middle ear emerged on multiple occasions. It is currently argued that the conductive system as we know it today most likely evolved independently in fish, amphibians, and reptiles (Wever 1974; Lombard and Bolt 1979, 1988; Bolt and Lombard 1985, 1992; Manley 1990; Schellart and Popper 1992). In spite of these concerns, the story of the amniote middle ear remains one of the most clearly documented instances of a functional transition in vertebrate evolution (Lombard and Bolt 1979, 1988; Carroll 1987).

Figure 2.2 presents an evolutionary tree of amniote life beginning with the rhipidistian fish at the water–land transition of vertebrate life (Starck 1978). A newcomer reading the literature on the evolution of the ear encounters a head-spinning array of unfamiliar terms, many of which appear in Figure 2.2. As we progress through this section, we will try to use more generic terminology wherever possible, and define in a clear manner those terms that would otherwise be unfamiliar to most students of hearing.

The earliest adaptations for substrate and aerial hearing may have emerged in those fish that achieved the capacity to live a portion of the time on land (rhipidistians). The labyrinthodonts showed additional specializations of the middle ear that led to the amphibian (temnospondyls) and reptilian (Suropsida) lines of evolution. The reptiles branched into subclasses differing in the number of temporal openings in the skull. The Anapsida (no openings) became the turtles (Testudinia or Chelonia) of today. The Synapsida (one temporal opening in the skull) passed through a series of steps to become the mammals. The Diapsida (two temporal openings) separated into two additional subclasses; the scaly reptiles (Lepidosauria) and the "grand" reptiles (Archosauria). The Lepidosauria comprise two orders, the lizards and snakes (Squamata) and the snout-headed reptiles (Rhynchocephalia) represented by the tuatara. The Archosauria evolved into many orders including the pre-dinosaurs (Theocondontia) and dinosaurs

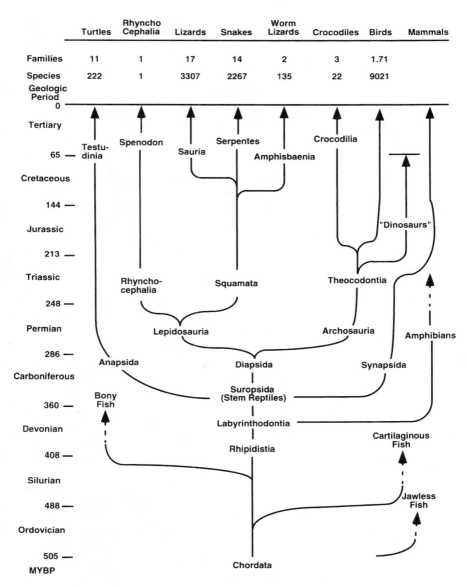

FIGURE 2.2. The evolutionary tree of amniotes is depicted in relation to other Cordata. The number of orders, families, and species are also indicated (Parker 1982). The left-hand axis shows both geologic time and periods. MYBP = million years before present.

like tyrannosaurus (Saurischia), triceratops (Ornithishia), and the peri-
dactyles (Petrosauria). Another suborder was the Crocodilia, which
included the modern alligators, caimans, and crocodiles. There are various
schools of thought, one of which suggests that birds emerged from the crocodilian line in the mid-Triassic period. Birds, however, are so unique in body
structure from their reptilian predecessors that they are traditionally considered a class unto themselves (aves). There are several other subclasses
of the reptilian line that are not included in Figure 2.2 (e.g., the large aquatic
reptiles) because they died out during the great extinction at the
Cretaceous–Tertiary boundary.

Figure 2.2 provides us with a conceptual framework for understanding
where living orders come from. This offers a rational basis for organizing
the structural and functional aspects of the reptilian and avian middle ears
that follow. Within living reptilian and avian classes there are 34 orders and,
as Figure 2.2 shows, 219 families and about 15,000 species.

3.3. The Rhipidistian Fish

There is a progression in the early fish from cartilaginous rods, which were
part of the gill-support arches, to the appearance of a quadrate bone and
hyomandibular arch. These rods and the cartilaginous and bony components of the arch contributed to the emergence of the single-ossicle middle
ear. The ancient lobe-finned fish (rhipidistia) are believed by some investigators to have had descendants that had the capacity to live out of water
part of the time. Such a water-to-land transition must have involved profound biologic changes in the mechanisms of locomotion and respiration as
well as in the nervous system to coordinate these functions. These animals
had a small air-filled chamber that was isolated from the outside world, and
was associated with a diverticulum of the first gill or spiracle pouch (Hotton
1959; van Bergeijk 1966, 1967; Schmalhausen 1968). The hyomandibular
bone was located in this chamber and was a component of the hyoid arch.
The arch was a series of bones and ligaments that provided structural
support between the junction of the upper and lower jaws and the cranium.
The hyomandibular bone supported the paleoquadrate of the upper jaw
and articulated with the operculum (the bony covering that partially pro-
tected the gill flaps and slits), as well as the bony structures that articulated
with the lower jaw (the quadrate bone).

There is evidence that the proximal or central connections of the
hyomandibular bone communicated with the brain case at two places,
and the more ventral of these was adjacent to the saccular chamber of
the labyrinth. At this stage of evolution an oval window was absent in the
cranium. However, the relative thinness of the cranial bone over the saccular chamber suggests that bone flex could have resulted in the transmission of a pressure wave to the fluids of the ancient labyrinth. The distal or
peripheral portion communicated with both the operculum (via the gill

slits) and the quadrate. The former may have served as a detector of aerial sound while the latter may have been sensitive to ground vibrations.

The insertion of the hyomandibular bone adjacent to the labyrinth at one end, and the surface of the skull at the other, is one of the critical events in middle-ear evolution. The hyomandibular bone is widely held to be the precursor of the stapes or columella found in more advanced middle ears, while the air-filled chamber of the first-gill pouch most likely became the middle-ear cavity. The rhipidistian middle ear was retained in many of the early tetrapods and the hyoid arch and the spiracle pouches were to change many times as they evolved from structures with a primary suspensory function to one with a primary auditory function. While the "efficiency" of this middle ear can be discussed only fancifully, it may not yet have been particularly effective for detecting aerial sound.

3.4. The Labyrinthodonts

The labyrinthodonts were primitive tetrapods that evolved from the rhipidistian fish. These early Paleozoic amphibians were perhaps the first animals to possess a structure that we would readily recognize today as a middle ear. It is thought that they exhibited three adaptations that improved impedance matching (Henson 1974). These were: (1) a membrane (the tympanum) located toward the back of the skull, that was sensitive to sound pressure at its surface; (2) a conductive system capable of conveying, with efficiency, the motion of the tympanum to the fluids of the labyrinth; and (3) an air-filled chamber that permitted the sound-conducting apparatus to operate with a minimum of frictional resistance. Labyrinthodonts also possessed an oval window located above the sacculus of the inner ear. The evolution of the hyomandibular bone into a stapes may be related to the forward migration of the quadrate bone and the loss of the operculum. Hotton (1959) speculated that these changes in the quadrate and operculum arose from adaptations in the skull and body skeleton brought about by changes in food-gathering behaviors, which favored a more mobile head.

The proximal end of this primitive stapes contained several otic processes, one of which connected the stapes to the inner ear via the oval window. Other processes coupled the stapes to the bones of the cranium and upper jaw. As the labyrinthodonts evolved, it appears that all but the connection to the oval window were lost. The stapes was oriented upward in a dorsolateral direction and, communicated with an epithelial tissue (the tympanum) at its distal end that may have been more sensitive to sound pressure vibrations at its surface than the surrounding skin.

The tympanum in Labyrinthodontia was located in the otic notch. This notch was frequently found posteriorly, at the extreme dorsal-lateral extent of the skull, but was highly variable among the fossil specimens of early amphibians. The tympanum occupied the anterior portion of the notch and was bordered rostrally by the squamosol bone, dorsally by the supra temporal bone, and ventrally by the tabular bone. The membrane was bordered posteriorly by muscle and connective tissue. The presence of an otic notch and its importance to the middle ear has been the subject of some debate. At present, there is no definitive proof (other than the posterior orientation of the amphibian stapes) that the notch actually contained a functional tympanum (Allin and Hopson 1992; Bolt and Lombard 1992).

3.5. The Stem Reptiles

Labyrinthodontia gave rise to the stem reptiles, and the most striking feature in these fossil specimens is the absence of the otic notch or any indication of a tympanum at the side of the head. Some authors have argued that the tympanum never existed in the otic notch of earlier animals (Watson 1953; Tumarkin 1968) or that it became much smaller and migrated to a more ventral position on the skull where it might function better for directional hearing. Nevertheless, this is an example where the fossil record shows a discontinuity. A tympanum was evident in the labyrinthodonts but is now absent in the stem reptiles.

The stapes at this stage of development appeared to slant obliquely downward and posterior from the oval window to the articular region of the quadrate. As the quadrate assumed a vertical orientation it served to more efficiently connect the upper and lower jaws. In this position, the quadratic also contacted the outer skin covering at the side of the head, and it is possible that aerial vibrations detected by the skin were communicated to the quadrate and hence the stapes. We will see that such a middle-ear design exists in the snakes of today (Section 4.2.1). It has been suggested that loss of the notch and the tympanum may again be related to feeding mechanisms associated with more powerful and rapid lower jaw closures arising from rearrangements of the bony suspension (Hotton 1959; Henson 1974; Manley 1990).

The stapes in many stem reptiles was massive and resembled that seen in the labyrinthodonts. This suggests that the middle ear was still a relatively poor detector of aerial sound. It may be that these middle ears, with their massive stapes and tiny or absent tympanums, functioned more as inertial detectors. In this situation, low-frequency vibrations would cause the head to move as a whole while the stapes remained at relative rest. Such motion would lead to stimulation of the inner ear (Manley 1990).

3.6. Origin of the Modern Middle Ear

Changes in bony structure during the transition from stem reptiles to the various sub classes of reptiles had important consequences for the evolving middle ear. It is possible that stronger limbs resulted in the head and body being lifted from the ground. Isolation of the head from substrate vibrations represented a source of selective pressure that could have led to improvements in aerial sound detection. These improvements are expressed in a more slender columella with greatly reduced mass, an extracolumella that couples the bony columella to the tympanum, and the transformation of the skin surface into an eardrum. The stapes rotated to a more horizontal orientation and the tympanum came to lie posterior to the quadrate bone and at the distal end of the stapes. The consequences of these changes would be a considerable improvement in sound conduction and hearing (Hotton 1959; Henson 1974; Manley 1990).

All orders of living reptiles evolved from the stem reptiles, and each shows unique middle-ear designs. Some show specialization for sound transmission from substrate vibrations (the Amphisbaenia), for sound transmitted in an aquatic environment (the turtles), for both water- and airborne sound transmission (the alligators), and for sound transmitted through the air (the lizards). The birds, arising as they do from the "grand" reptiles (Archosauria), exhibit middle-ear organization that is remarkable uniform across all avian species.

4. Structure and Function of the Reptilian Middle Ear

The evolutionary pathways illustrated in Figure 2.2 show that there are four orders of living reptiles, represented by the Testudinata, Crocodilia, Squamata, and the Rhynchocephalia. These orders comprise over 5,900 reptilian species and the middle ears of only a fraction of these animals have been examined in any detail. In this section we consider the structure and function of representative animals in three of these orders. The Rhynchocephalia have been described elsewhere, and the reader is referred to descriptions by Henson (1974) and Wever (1978) for further details.

4.1. The Saurian Middle Ear

The Squamata (scaly reptiles) comprise three suborders; the lizards (Sauria), the snakes (serpents), and the worm lizards (Amphisbaenia), and we begin our discussion with the middle ear of lizards. In his monumental work on the reptilian ear, Wever (1978) distinguished four types of lizard middle ears: iguanid, gecko, skink, and divergent. This section considers only the first three of these, and the reader is referred elsewhere for information on the divergent type (Wever 1973, 1978).

4.1.1. General Features

Versluys (1898) originally described the lizard middle ear in detail, and the accounts that follow were additionally extracted from Henson (1974), Wever (1978), Manley (1990), and various other original publications. The

TM in some species is slightly depressed at the side of the head (the Iguanidae), while in others, like the geckos and skinks (the Gekkonidae and Scincidae), there is a meatal cavity that terminates at the surface of the membrane. In some lizard families (e.g., the Chameleontidae), the ear opening is absent altogether and there is no tympanic membrane.

In those species with a well-defined external meatus there may be a muscle present capable of constricting the meatus. This "meatal closure muscle" is quite prevalent in the Gekkonidae where it takes two forms. The L-type, found in the tokay gecko (*Gekko gecko*), has the muscle following the posterior edge of the meatus as a compacted circular bundle. On the ventral side of the meatus the muscle bends rostrally (thus forming the "L") and fans out as it runs forward. This muscle varies greatly in size among different species. In the subfamily Diplodactylinae the second or "loop"-type closure muscle is found. As the name implies, this muscle encircles the outer perimeter of the external meatus.

Contraction of the "L"-type muscle pulls the posterior and dorsal wall of the meatus forward and occludes the opening of the ear canal. Contraction of the "loop" muscle constricts the ear canal opening. Either type of muscle action is clearly protective in that it prevents foreign objects from entering the meatus. Reflex closure of the muscle could also reduce acoustic input to the TM and protect the ear from overstimulation by self-made vocalizations. The relation between vocalization and the presence of a closure muscle, however, is inconsistent with the presence of well-developed closure muscles in several nonvocal species.

The "C"-shaped quadrate bone supports the anterior margin of the TM in most lizards. The retroarticular process of the mandible and a portion of the squamosol bone also communicate with the membrane. All of these elements are associated with movements of the jaws, and indicate that a frame capable of some movement supports the tympanic membrane. This is very different from the immobile bony framework surrounding the TM of birds and mammals.

Medial to the membrane is the middle-ear (tympanic) cavity whose posterolateral wall is composed of the depressor and abductor mandibular muscles. The paraoccipital process forms the roof of the cavity, while the cranial bones peripheral to the labyrinth form the dorsomedial wall. An opening on the ventromedial wall communicates with the pharynx (the eustachian tube). The round window is found on the medial wall in the area of the metotic fissure dorsal to the sphenoid bone.

The TM communicates with the oval window of the cochlea through two structures: the extracolumella and the columella. The columella is seen as a more or less slender bony shaft that terminates on its proximal end with the columella footplate. The footplate inserts into the oval window and is supported at its periphery by the annular ligament. The columella communicates at the distal end with the cartilaginous extracolumella. The distinction between columella and extracolumella may be abrupt or gradual. In some forms the joint appears to have a degree of mobility, while in others there is a fusion between the two elements supported by a sheath of connective tissue. In these latter forms a "joint action" appears unlikely.

The extracolumella communicates with the inner surface of the tympanic membrane through the tympanic process, which consists of a number of spreading struts (as many as four). These struts are firmly attached to the fibrous middle layer of the TM. The anchorage of the extracolumella is the largest component of the tympanic process and extends from the upper (dorsal) edge of the TM ventrally toward the center of the membrane for a third to a half of its diameter. The shaft of the extracolumella articulates with the anchorage and defines two segments referred to as the pars superior (upper segment) and pars inferior (lower segment). The distal end of the pars superior attaches via the extrastapedial ligament to the bony frame supporting the dorsal edge of the TM (the cephalic condyle of the quadrate bone). This ligament provides a degree of stability to the tympanic process by anchoring the pars superior to bone. In some lizard middle ears (the gekkonoid type) there is a muscle (the extracolumella muscle) attached to the tip of pars superior at one end and to the ceratohyal process at the other end. Contraction of this muscle may produce tension on the tympanic process of the extracolumella (see Section 6.1).

The shaft of the extracolumella presses on the anchorage causing it to rotate about the distal end of pars superior. As a consequence, the tip of pars inferior pushes into the TM and creates the convex appearance of the drum membrane when viewed externally. An anterior and posterior tympanic process may also be found extending along the inner surface of the TM. These processes secure the extracolumella to the TM, stabilize it on the membrane surface, and help tense the membrane.

4.1.2. The Iguanid Middle Ear

The iguanid type of middle ear is illustrated in the top panels of Figure 2.3 and the collared lizard, *Crotaphytus collaris*, serves as an example. Many features described in the preceding paragraphs can be seen. The quadrate bone provides support for the TM in the anterior and dorsal regions of the membrane. The retroarticular process of the mandible supports the ventral side while fascia and skin support the posterior border. The three legs of the extracolumella are in contact with the fibrous middle layer of the TM. The extracolumellar ligament stabilizes further the tympanic process. At the junction of the extracolumella and bony columella, an additional cartilage extension, the internal process, communicates with the quadrate bone. This is the distinguishing feature of the iguanid middle ear.

The role of the internal process was studied by measuring cochlear microphonic sensitivity curves (see Section 2.2.1) before and after severing this structure. Microphonic sensitivity across frequency was the same in both

30   J.C. Saunders et al.

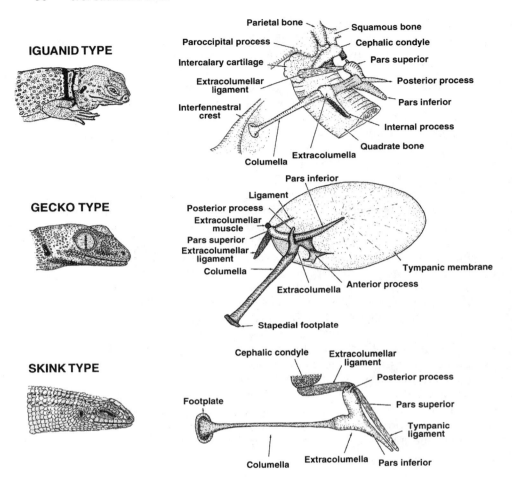

FIGURE 2.3. The essential features of three types of lizard middle ears. The iguanid type has the ossicular chain supported by the internal process. The gekkonoid type lacks the internal process, but has the extracolumellar muscle. The skink type has neither the internal process nor the extracolumellar muscle. (Modified from Wever 1978.)

conditions, suggesting that this structure contributed little to sound conduction (Wever and Werner 1970). It may be that the internal process protects the ossicles against injury during vigorous head activity such as chewing or eating (Wever 1978).

The total transformer action of the middle ear was examined by first stimulating the tympanic membrane with a closed-field sound system and measuring the cochlear microphonic sensitivity curve (from the round window). The columella shaft was then severed near the footplate and removed along with the TM. The stump of the footplate was then directly stimulated and another sensitivity curve measureed. As described in Section 2.2.1, the differences between the footplate and TM sensitivity curves represent the total transformer action of the middle-ear system. The results in Figure 2.4A indicate that the footplate required greater acoustic stimulation than the TM to achieve the criterion microphonic response. This makes sense because the pressure amplifiers of the middle ear were bypassed completely when the footplate was stimulated directly. The difference in decibels between the two curves, or the transformer function, is presented in panel B of Figure 2.4. As frequency increases from 0.2 to 2.0kHz the transformer effect reaches a maximum of approximately 38dB. The average difference in decibels across frequency represents a transformer ratio of 25:1. Recall from Equation 5 that the impedance ratio is the square of the transformer ratio, and in this example that is 625. Thus, the impedance of air, which is 415

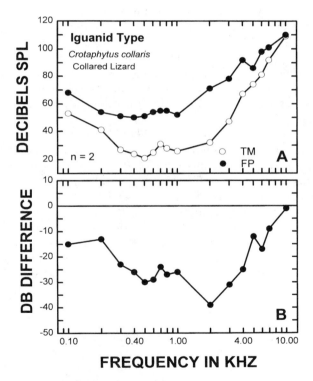

FIGURE 2.4. (A) Cochlear microphonic sensitivity curves in the collared lizard during closed-field stimulation of the tympanic membrane (TM) or footplate (FP). (B) The difference between the sensitivity curves. A negative value indicates a reduction in sensitivity with oval window stimulation. The results in B represent the total transformer action of the middle ear. (Data from Wever and Werner 1970; Wever 1978).

acoustic ohms (per square meter), is made to look like 259,380 acoustic ohms at the cochlea of the collared lizard.

Wever and Werner (1970) estimated the mechanical lever ratio of the collared lizard to be about 2:1. This estimate was based on cochlear microphonic measures of direct stimulation at the tip of pars inferior (on the TM surface) and at the stump of the columella footplate (using a vibrating needle). The value of this mechanical lever was similar to the anatomically measured lever ratio (the ratio between the length of the anchorage divided by the length of the pars superior). The area of the TM was estimated to be 9.7 mm^2 while that of the footplate was 0.48 mm^2 (Wever 1978). Using 66% of the TM as its *effective area* (Békésy 1960), an area ratio of 13.3 was calculated. The transformer ratio arising from the area and lever ratios (13.3 × 2) was then 26.6, which translates into an impedance ratio of 708. This impedance ratio times the mechanical impedance of air (415 ohms) yielded a value of 293,820 acoustic ohms, which is the output impedance of the middle-ear system calculated from the morphologic data.

The output impedance values derived from anatomic measures of the area and mechanical lever ratios were nearly the same as those from the comparison of microphonic curves following TM or footplate acoustic stimulation (Fig. 2.4B). The latter estimate of output impedance included the contribution of the drum membrane lever as well as the area and lever ratios. It would appear that the contribution of the drum lever in this species might be negligible.

Mössbauer measures of motion at the TM (tip of the inferior process) and footplate have been obtained for an iguanid middle ear, and are presented as velocity transfer functions for the dragon lizard *Amphibolurus (Ctenophorus) reticulata* in Figure 2.5A. The response of the footplate increased steadily in the low frequencies at a rate of 5.1 dB/octave reaching a peak around 2.0 kHz. In the high frequencies the footplate response declined at a rate of 11.5 dB/octave. The difference between the velocity functions at these two locations describes the mechanical lever ratio, and this is plotted in Figure 2.5B. The lever ratio remains fairly constant at 4.6:1 between 0.1 and 2.0 kHz, and then rises to 8.6:1 at 10.0 kHz.

4.1.3. The Gekko-Type Middle Ear

The features of the gekkonoid middle ear are also seen in Figure 2.3. The four legs of the tympanic process are present, however, the posterior process appears shortened while the anterior process is elongated. The anchorage begins with the pars superior at the dorsal edge of the TM and tapers to the pars inferior at the approximate center of the membrane. The extracolumella ligament angles from the pars superior in an anterosuperior direction. Opposite this ligament is a smaller ligament that passes over the inner surface of the drum membrane. The distinguishing feature of the gecko middle ear is the extracolumellar muscle, which is attached to the

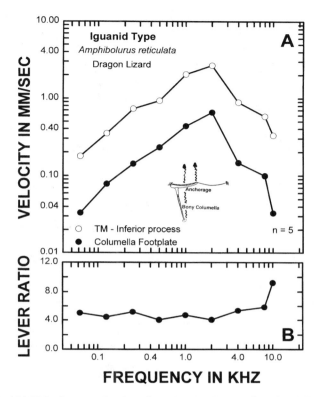

FIGURE 2.5. (**A**) Velocity transfer functions for the dragon lizard obtained from the TM and columella footplate at 100 dB SPL. The insert indicates the locations of the Mössbauer sources on the TM and footplate. (**B**) The ratio between these two locations represents the mechanical lever action of the middle ear. (Data from Saunders and Johnstone 1972.)

distal tip of the pars superior and terminates on the ceratohyal process. The size of this muscle varies widely with species.

Transfer functions at the TM, measured with laser interferometry, have been reported for two gekko species (Werner et al. 1998). Figures 2.6A and C illustrate these functions for the leopard gecko (*Eublepharis macularius*) and the marbled velvet gecko (*Oedura marmorata*). The results in each panel show the velocity responses at three different locations on the membrane. These locations are immediately opposite the insertion of the extracolumella (Col), at the tip of the conical TM over the inferior process (Pars Inf), and in the center of the TM between the pars inferior and the outer edge of the membrane (TM). The rate of low- and high-frequency change in each of these species was about the same for the three locations and on average was 10.9 and 12.7 dB/octave. The resonant peaks differed

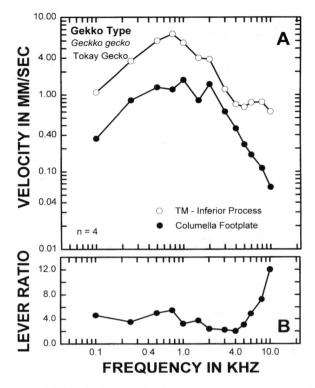

FIGURE 2.6. (**A**) and (**C**) Velocity transfer functions for two gecko species measured at three locations on the tympanic membrane (see inset in **A**). (**B**) and (**D**) The difference between the pars inferior and columella velocity representing the lever ratio. (Data from Werner et al. 1998.)

among these species with *Eublepharis* and *Oedura* being 2.3 and 2.5 kHz, respectively.

The TM velocity was always greater than that of the pars inferior, which, in turn, was larger than that at the columella location. These locations can be used to estimate the middle-ear lever ratio (the difference between the columella and pars inferior locations) as well as the lever action of the drum membrane (the difference between the TM and pars inferior locations). Figures 2.6B and D show the velocity difference between the pars inferior and columella responses expressed as a lever ratio. From 0.15 and 4.0 kHz, this ratio averaged 2.1:1 and 2.0:1 in these panels. The ratio appeared to increase above 4.0 kHz, and above 10 kHz was very erratic. The velocity difference from 0.15 to 4.0 kHz between the drum membrane (TM) and the tip of the pars inferior (Pars Inf) provided a lever advantage that averaged 1.6:1 in both species.

The velocity results above 6.0 to 8.0 kHz, for the TM locations in Figure 2.6, show that the drum membrane was responding in a complex pattern. This is not so surprising since it has been reported in mammalian species that the eardrum does not move as a simple plate at high frequencies, but rather shows an increasing number of nodes on the membrane, each moving at different amplitudes relative to each other (Khanna and Tonndorf 1972). The spatial responses of the TM have been mapped at various frequencies in the tokay gecko using Mössbauer methods, and the membrane above 4.0–6.0 kHz showed regional movements. This regional motion of the TM became increasingly complex as frequency rose (Manley 1972b), and more than likely accounts for the high-frequency behavior of the membrane in Figures 2.6A and C.

Mössbauer velocity transfer functions at the TM and columella footplate have also been reported for the tokay gecko (Manley 1972a). Between 0.1 and 1.0 kHz, the footplate function rose at a rate of 4.1 dB/octave, reaching a maximum of 1.2 mm/sec at 1.0 kHz. On the high-frequency side the response declined at 11.9 dB/octave. The lever ratio representing the velocity difference between the TM and footplate responses averaged about 3.8:1 between 0.1 and 4.0 kHz. In the higher frequencies it rose to around 12:1 at 10.0 kHz. Additional results showed that the middle ear responded linearly over the intensity range from 72 to 96 dB SPL at all frequencies tested (Manley 1972a).

4.1.4. The Skink-Type Middle Ear

Figure 2.3 also illustrates the skink middle ear. This is the simplest of the three designs because it lacks both the extracolumellar muscle and the internal process. The pars superior and pars inferior of the extracolumella, as in the other middle-ear types, secures the anchorage to the fibrous layers of the TM. The posterior and anterior processes are short and arise as appendages from the pars superior. The extracolumellar ligament forms a separate structure from the tympanic ligament, and is attached to the head of the pars superior, extending dorsally to terminate on the cephalic condyle. The shaft of the extracolumella is also short and attached to the bony columella with a distinct demarcation between cartilage and bone. A "thin and delicate" bony columella shaft is found, which may be flexible in its mid portion.

Velocity transfer functions measured at the TM or footplate for two skink species, the shingle-back lizard (Scincidae: *Tilgua (Trachydosaurus) rugosa*) and the alligator lizard (Anguidae: *Gerrhonotous (Elgaria) multicarinatas*), appear in Figures 2.7A and B. The results from the shingle-back lizard are expressed in relative velocity units because of the method used to obtain these data. The peak frequency in this function was 1.6 kHz, and the low- and high-frequency rates of change were 5.0 and 9.3 dB/octave. The peak frequency of the alligator lizard was 1.9 kHz, and the low- and high-

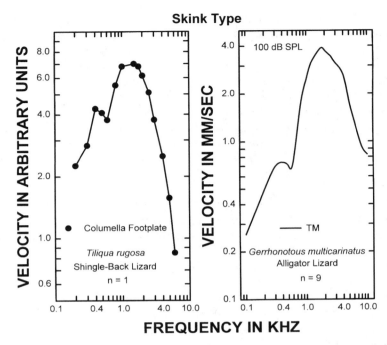

FIGURE 2.7. The left and right panels show transfer functions for the shingle-back and alligator lizards. The data from the shingle-back was obtained from the columella footplate and plotted as velocity in arbitrary units. The results from the alligator lizard come from the TM. Both functions were obtained for a constant stimulus of 100 dB SPL at the TM. (Data from Rosowski et al. 1984; Rosowski et al. 1985; Manley et al. 1988.)

frequency roll-off was 5.9 and 6.6 dB/octave. The mechanical lever ratio of the alligator lizard, based on anatomical estimates of the anchorage and pars superior lengths, was about 3.0:1 (Rosowski et al. 1985).

4.2. The Serpentine and Amphisbaenian Middle Ears

There are estimated to be over 2,200 and 135 species respectively in the serpentine (snakes) and amphisbaenian (worm lizards) suborders of squamata (Fig. 2.2). Relatively little is known about the auditory system and middle ear in these animals, and what information is available comes largely from Wever and his colleagues.

4.2.1. The Middle Ear of Snakes

Figure 2.8 illustrates the middle-ear system of the northern water snake *Natrix (Neroda) s. sipedon*. All species of snake lack an outer ear and tym-

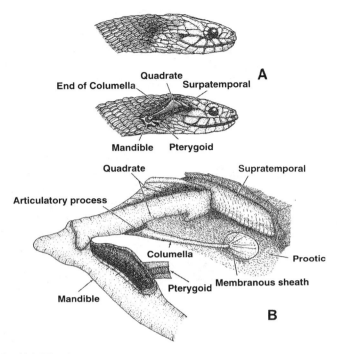

FIGURE 2.8. (**A**) The head of the snake and the location of the quadrate bone. (**B**) The middle-ear apparatus of the northern water snake. (Modified from Wever 1978.)

panic membrane. The middle-ear mechanism lies beneath the skin surface at the side of the head. One end of the quadrate bone communicates with the mandible of the lower jaw and is held in position by ligaments. At the other (anterodorsal) end, the quadrate communicates with the edge of the supratemporal ligament and hence to the prootic bone of the skull. The middle-ear cavity is a narrow air-filled space revealed only by the most delicate of dissections. Within this cavity lies the long, tapered shaft of the columella, oriented in an anterior-posterior direction. The anterior (proximal) end expands to form a rather large footplate, while the posterior (distal) end communicates with the articulatory process of the quadrate. The cartilaginous connection between the quadrate and the columella shows considerable diversity among species. The footplate enters a deep cleft in the otic capsule and is covered by connective tissue that forms a sheath about the base of the columella shaft.

Acoustic stimulation at the side of the head is transmitted to the quadrate bone and hence to the shaft of the columella and the inner ear. Removal of the skin overlying the quadrate bone, however, does not appreciably alter the sensitivity of cochlear microphonic functions. This observation suggested that the overlying tissue was unimportant, and that the response to sound was dominated by the surface of the quadrate bone. Sectioning the ligaments between the quadrate and both the mandible and the supratemporal, revealed little change in the microphonic response, suggesting further that the columella was not set in motion by vibrations of the skull or lower jaw. However, a substantial loss in microphonic sensitivity occurred when the quadrate was removed altogether with the columella shaft left in position. The quadrate thus acts like a "floating plate" vibrating as a unitary structure in response to sound.

The articulations of the quadrate with both the skull and the mandible argues that snakes should have good hearing for substrate vibrations. However, experiments applying a vibratory stimulus directly to various portions of the skull and jaw do not necessarily show microphonic responses better than aerial stimulation of the quadrate. The loose connections between the skull and mandible may dampen vibrations from reaching the quadrate. While it is possible for snakes to receive substrate vibrations, the greater efficiency of aerial stimulation of the quadrate makes this the more likely mode of hearing in these animals (Wever 1978).

4.2.2. The Amphisbaenian Middle Ear

This suborder consists of a small group of worm lizards. These animals are characterized by their burrowing behavior and, in some species, by the capacity to manipulate body scales in such a way that they can move forward or backward freely. Some species rarely, if ever, leave their tunnels, while others do so only in dim light or at night. As a consequence, these animals are not well understood and even lack common names in English (Wever 1978).

Figure 2.9A shows a drawing of the South American species *Amphisbaena alba*. The skull of *Amphisbaena manni* is presented in Figure 2.9B, and the oval window is seen to occupy a posterior position in the skull. The footplate is rather large and occupies most of the medial surface of the otic capsule. The shaft of the bony columella projects only slightly beyond the rim of the oval window. Figure 2.9C is a frontal section and the most striking feature is the long extracolumella. The extracolumella communicates with the stapedial shaft and extends anteriorly for a considerable distance to articulate with a thickening of skin lying over the lower jaw. The end of the extracolumella makes a firm connection with the inner layer of this skin. In its course the extracolumella penetrates connective tissue and muscle.

In a variety of species including *A. manni*, severing the extracolumella from the bony columella produced a reduction in cochlear microphonic sensitivity (during acoustic stimulation of the lower jaw region) on the order of 30–40 dB between 0.2 and 0.5 kHz (Gans and Wever 1972; Wever 1978). These observations indicated that the extracolumella was capable of detect-

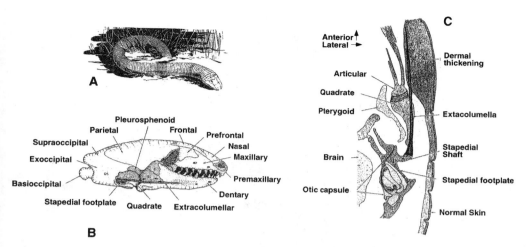

FIGURE 2.9. (**A**) The worm lizard *Amphisbaena alba*. (**B**) Lateral view of the *Amphisbaena manni* skull. (**C**) Horizontal section cut through the ear region. Note the extended length of the extracolumella. (Modified from Wever and Gans 1973; Wever 1978.)

ing sound vibrations presented to the facial skin and of conducting these vibrations to the otic capsule. Nevertheless, the nature of this system does not appear to be particularly well suited to the conduction of high frequencies.

4.3. The Middle Ear in Testudines and Crocodilia

4.3.1. The Turtle Middle Ear

The conductive apparatus of a turtle appears in Figure 2.10. The TM is not particularly evident at the side of the head because the outer layer of skin covers the membrane with its surface pattern unbroken. Nevertheless, the skin over the drum is thinned, perhaps slightly recessed, and when depressed reveals a "soft and yielding" area beneath.

The tympanic process of the extracolumella is a disk-shaped plate attached to the TM through the posterior ligament. The lateral portion of the columella and the inside of the TM is housed in the middle-ear cavity. An eustachian tube connecting the cavity with the pharynx is found in the floor of this space. The extracolumella, as depicted in Figure 2.10B protrudes only slightly into the middle-ear cavity where it articulates with the bony columella. The shaft of the columella is long, thin, and curved in its lateral segment. The medial portion penetrates the quadrate bone to enter the fluid-filled pericapsular recess. The columella is sheathed in connective tissue as it traverses the canal between the middle-ear cavity and the peri-

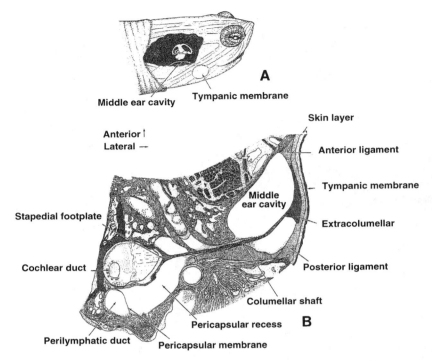

Middle ear cavity   Tympanic membrane   **A**

Skin layer

Anterior ↑
Lateral —

Anterior ligament

Tympanic membrane

Middle
ear cavity

Stapedial footplate

Extracolumellar

Cochlear duct

Posterior ligament

Columellar shaft

Pericapsular recess   **B**

Perilymphatic duct   Pericapsular membrane

FIGURE 2.10. (**A**) The location of the labyrinth in the turtle head. (**B**) A horizontal section of the middle ear. Note the curvature of the columella shaft. (Modified from Wever 1978. Copyright © 1985 by Princeton University Press. Reprinted by permission of Princeton University Press.)

capsular recess. This canal permits back and forth movement of the shaft but restricts lateral motion. A large "funnel-shaped" footplate characterizes the proximal end of the columella. Figure 2.10B shows that there is no contact between the oval window area and the middle-ear cavity, and this is a defining characteristic of the turtle middle ear.

Cochlear microphonic sensitivity in the red-eared slider (*Chrysemys (Trachemys) scripta*), using acoustic stimulation of the TM was obtained before and after a small section of the columella was cut in the middle-ear cavity. With the columella clipped there was on average a 42-dB loss in the microphonic response between 0.1 and 1.0 kHz. Above 1.0 kHz the efficiency of the middle ear deteriorated even further (Wever and Vernon 1956). Nearly similar changes were observed in behavioral thresholds from *C. scripta* before and after the columella was severed from the extracolumella (Patterson 1966).

The TM velocity transfer function in the red-eared slider appears in Figure 2.11. Peak sensitivity lies between 0.5 and 0.6 kHz with the low-frequency response exhibiting a roll-off of about 10 dB/octave (between 2.0

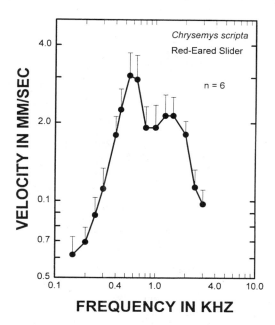

FIGURE 2.11. The tympanic membrane velocity transfer function for 100 dB SPL stimulation, averaged over six ears, is presented for the red-eared slider. The error bars show one standard error. (Data from Moffat and Capranica 1978.)

and 4.0 kHz). The high-frequency side of the curve shows a double peak with the roll-off between 1.75 and 3.0 kHz on the order of 9 dB/octave. Compared to the other reptilian velocity transfer functions presented previously, the slider data exhibited a similar magnitude of sensitivity at the peak of the response (3.0 mm/sec at 0.5 kHz).

4.3.2. The Crocodile Middle Ear

The middle-ear apparatus in the crocodilian *Caiman crocodilus* is illustrated in Figure 2.12A. The TM in this ear is located on the floor of a shallow depression just behind the eye. A pair of muscular flaps that extend in a rostral-caudal orientation with one above and the other below the opening to the external meatus (Fig. 2.12A) cover the depression. These muscles are referred to as "ear lids" because of their appearance. Above water, the animal is seen with these lids relaxed and the TM exposed to the external environment. The lids, however, reflexively close whenever the head goes under water. There also appears to be a close relation between eye- and ear-lid "blinking." The cochlear microphonic response was used to measure the attenuation of sound transmission when the ear lids were closed, and between 0.1 and 3.0 kHz there was on average a 15-dB reduction in microphonic sensitivity relative to the open condition (Wever and Vernon 1957).

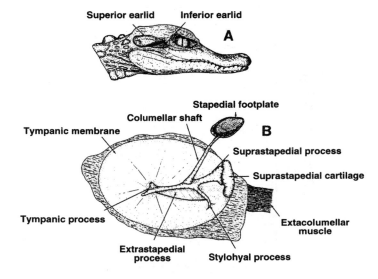

FIGURE 2.12. (**A**) The head of the caiman with the position of the earlids noted. (**B**) The organization of the middle ear. (Modified from Wever 1978.)

The organization of the conductive apparatus was originally described by Retzius (1884) and is illustrated in Figure 2.12B. The anchorage separates into a superior (suprastapedial) and inferior (tympanic) process, with a third process (the stylohyal) stabilizing the tympanic process on the TM. The extracolumellar muscle is attached to the head of the stapedial process and the underlying TM. The bony columella is slender and articulates with the extracolumella at the anchorage. At its proximal end the columella expands to a relatively large footplate.

Figure 2.13 shows the columella footplate velocity response in the caiman. This function has a resonant frequency of about 0.6 kHz, with a decline in sensitivity of 5.6 dB/octave between 0.05 and 0.4 kHz and 6.7 dB/octave between 1.0 and 4.0 kHz. The filter properties of the conductive apparatus in this example do not seem very sharp compared to the transfer functions in the lizard species presented previously.

4.4. Summary

This section has dealt with only a tiny fraction of reptilian life, and the range of structural diversity in the middle ear of these animals has yet to be appreciated fully (Wever 1978). Variations and subtleties of middle-ear design were seen in the examples provided above. Nevertheless, the basic design of all reptilian middle ears centers on a conductive apparatus that contains a single ossicle, and transfer functions that appear to exhibit a bandpass

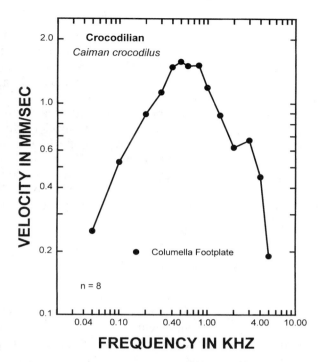

FIGURE 2.13. The columella footplate velocity (for a constant stimulation of 100 dB SPL) is plotted for the caiman. (Data from Wilson et al. 1985.)

function. The peak frequency and rates of high- and low-frequency roll-off depend on the specific mechanical properties of the middle-ear apparatus in each species, however, they are more similar than dissimilar in the examples presented here. The interplay of elastic, inertial, and frictional forces, and specific structural design, needs to be considered further in future investigations of the reptilian conductive apparatus. Another factor that we have not considered, but one that can effect middle-ear and auditory sensitivity, is the preferred temperature of each species. Further information on this topic can be found elsewhere (Werner 1983).

5. Structure and Function of the Avian Middle Ear

There are 28 orders, 171 families, and over 9,000 individual species of birds. The diversity of avian life is wondrous, and outside of the tallest peaks or the polar regions, birds occupy almost every physical niche on the planet.

The avian middle ear retained the basic reptilian design, but, in contrast, the variations in structural organization are at best incremental among the different families. The uniformity of design may be attributed, in part, to the evolutionary history of birds in that the adaptive pressures brought about by flight focused on the need to detect aerial sounds. This latter point is reinforced by the fact that birds are bipedal creatures, and with the head raised well above the ground, there is no need for detecting substrate vibrations. The structure of the avian middle ear has been considered in a number of presentations (Henson 1974; Saito 1980; Kühne and Lewis 1985; Saunders 1985; Saunders and Henry 1988; Manley 1990; Manley and Gleich 1992) and the material presented here is condensed from these and other articles. Moreover, various articles in the earlier literature offer illustrative material that is particularly instructive (Krause 1901; Smith 1904; Breuer 1908; Pohlman 1921; Stellbogen 1930; Müller 1960).

5.1. Structure of the Bird Middle Ear

5.1.1. Organization of the Ear Canal

The walls of the external auditory meatus are defined caudally, ventrally, and dorsally by muscles associated with the jaws and rostrally by the quadrate bone and the end of the lower jawbone. The external opening of the canal is typically smaller than the area of the TM, which is oriented obliquely at the base of the canal. As in reptiles, the epidermal layer of the canal wall is continuous with the outer surface of the TM. The ear canal varies in shape considerably. It is straight in the chicken (*Gallus domesticus*) and great blue heron (*Ardea herodias*). A relatively large meatus may be found in the parrots and songbirds, while in others (e.g., the silver-gray fulmar, *Fulmerus glacialoides*) the meatus may be divided by a ridge into an internal and external segment.

The external meatus clearly serves to protect the TM, and this may have allowed the membrane to become thinner with a somewhat larger surface area than that seen in the reptilian counterpart. A larger and less massive TM would be expected to improve high-frequency sensitivity.

The ear canal of birds is typically short (2–7 mm), and the wavelength of sound that would resonate in canals with these lengths is between 43 and 12 kHz, respectively. Given that birds in general do not hear above 10 kHz, it is safe to assume that the ear canal provides no "meaningful" pressure amplification at the TM through resonance. Nevertheless, sound transmission in the ear canal was measured in the quail (the Phasianidae) and found to exhibit lowpass characteristics (Hill et al. 1980) with a roll-off, above 0.8 kHz, of 8 dB/octave. In other birds such as the loons (the Gaviidae), there appeared to be no ear canal gain, while in birds of prey (the Falconiformes), the crows and jays (the Corvidae), and owls (Strigiformes), a slight 1 to 4 dB gain was found from 0.5 to 6.0 kHz (Iljitschew and Izwekowa

1961). The gain observed in these short canals is an unusual observation and needs further examination.

5.1.2. Organization of the Avian Middle Ear

The middle-ear or tympanic cavity is spacious relative to head size, and has a very irregular shape. It is bordered by bony structures, with the exception of the ventrolateral wall where the TM is found. The medial wall of the cavity is formed from the squamosol, prootic, opisthotic, and exoccipital bones. Ventrally and rostrally, the basisphenoid and orbitosphenoid bones form the walls. This is quite unlike that seen in the lizards where the borders were largely constructed of soft tissue (see Section 4.1.1).

As previously noted, the middle ear of most reptilian species communicates directly with the pharynx through a relatively large eustachian tube. In the crocodiles, however, there are tubes running through the posterior and ventral portion of the skull. These tubes connect the middle-ear cavities on both sides of the head with each other and with a common opening to the pharynx. The same is true of birds. The pharyngotympanic tube, whose distal opening is found in the rostroventral part of the cavity, courses slightly rostrally around the base of the cranium and joins with the tube from the opposite cavity at the midline. A rostral extension enters the roof of the pharynx where muscle structure controls venting into the oral cavity. This tube in the skull has been referred to as the "interaural pathway" because it acoustically connects the left and right tympanic cavities and has functional significance for sound localization (Coles et al. 1980; Hill et al. 1980; Rosowski and Saunders 1980). A number of other air-filled spaces contained in the posterior and ventral skull region communicate with the middle-ear chamber, and Kühne and Lewis (1985) have reviewed these.

The TM is oval with the longer axis oriented nearly vertically. The perimeter of the drum membrane is supported by the squamosol, exoccipital, and basisphenoid bones, and rostrally by a ligament that extends from the otic process of the quadrate. These connections with the skull render the TM immobile with respect to movements of the lower jaw. A series of drumtubal ligaments attach the drum to the bony perimeter (Fig. 2.14) and serve to keep the membrane taut.

The tympanic process of the extracolumella consists of three prominent legs. The names of these processes are somewhat confusing in the literature. Older presentations refer to the extra-, superior-, and inferior-stapedial processes (Pohlman 1921). Gaudin (1968) referred to the same structures as the extra-, supra-, and inferior-columella. Kühne and Lewis (1985) call them the rostral-, caudal-, and ventral-cartilaginous processes of the extracolumella respectively. We will adopt the nomenclature of Pohlman (1921) here.

The three tympanic processes of the extracolumella form a "Y"-like support on the surface of the TM (Fig. 2.14). This design stabilizes the col-

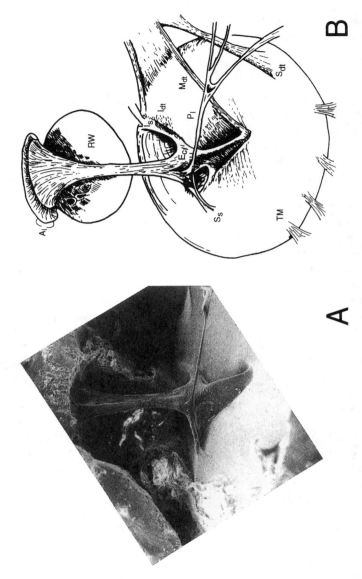

FIGURE 2.14. (**A**) A drawing of the parakeet middle ear. (**B**) Scanning electron micrograph of the same scene (modified from Saunders 1985). A_l = annular ligament; RW = round window; E_c = extracolumella; S_s = suprastapedius; I_s = infrastapedius; TM = tympanic membrane; P_l = Platner's ligament; S_{dt} = superior drum-tubal ligament; I_{dt} = inferior drum-tubal ligament; M_{dt} = medial drum-tubal ligament. The intrinsic drum ligaments are seen in the lower margin of the TM.

umella shaft and tenses the TM. The lower leg of the "Y" is the extrastape-dial, and the central tip of this process presses into the TM giving it a convex appearance (as seen from the external meatus). The inferior and superior processes arise from the shaft of the extrastapedial process, and may extend to the edge of the TM. The angle formed by these two arms of the "Y" varies greatly among species, and it can be as large as 90°. The extrastapedial is not in direct contact with the TM, but communicates with its surface through the ascending ligament. This ligament extends from the caudal edge of the TM, passes between the inferior and superior processes, and projects to the center of the membrane under the extrastapedial process (see Fig. 2.14). The ligament acts to protect and stabilize the TM in the region of the extracolumella, and might also serve to transfer motion from the membrane to the extrastapedial process.

Platner's ligament arises at the junction between the superior and extrastapedial process and extends across the middle-ear cavity. It termi-nates on a bony ridge formed by the junction of the prootic and quadrate bones (Fig. 2.14) and additionally stabilizes the columella system. Platner's ligament serves much the same function as the internal process did in the iguanid middle ear (see Section 4.1.2).

Considerable variability among species is found in the tympanic processes of the extracolumella (Freye-Zumpf 1953). For example, in the boreal owl (*Aegolius funereas*) the inferior and superior processes are attached to the bony margin of the tympanum and act as hinge points for motion at the tip of the extrastapedial process. The same arrangement is found in the falcons. In song birds (the passerines), the tympanic processes are delicate and terminate at the marginal edge of the TM.

The appearance of the columella and its footplate varies considerably and has been described for 70 species by Krause (1901). In the canary (*Serinus canarius*), parakeet (*Melopsittacus undulatus*), and cowbird (*Molothrus ater*), for example, the columella shaft communicates with the footplate through multiple crura. A more "stumplike" footplate has been reported in the chicken (*G. domesticus*) and pigeon (*Columba livia*) (see Fig. 2.15). The columella consists of a bony shaft or scapus whose proximal expansion is called the clipeolus or footplate. The footplate has an oval appearance, and is supported in the oval window by the annular ligament. The shaft has a hollow interior containing little marrow, thus reducing mass, while the ridges, holes, and grooves along the surface of the shaft allow it to retain strength. The shaft of the columella is typically offset on the surface of the footplate toward the anterior edge. The posterior margin of the angular ligament is widest, and it has been suggested that this acts like a hinge allowing the footplate to rock back and forth (Gaudin 1968). The undersurface of the footplate may be a flat plate—as in the eurasian jay (*Garrulus glandarius*) and the common murre (*Uria aalge*)—while in other species it has a rounded "knoblike" base—as in the tawney owl (*Strix aluco*) and parakeet (*M. undulatus*).

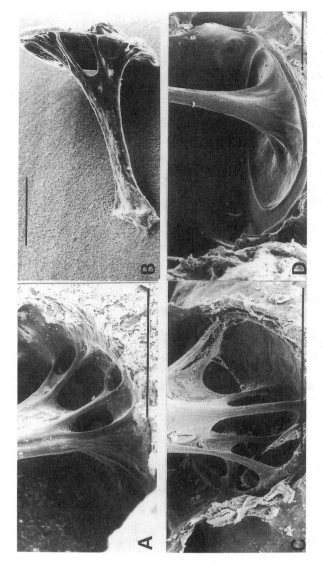

FIGURE 2.15. Three examples of the columella footplate are seen. (**A**) and (**B**) parakeet, (**C**) canary, (**D**) pigeon. The scale bars show 0.5 mm. (Modified from Saunders 1985.)

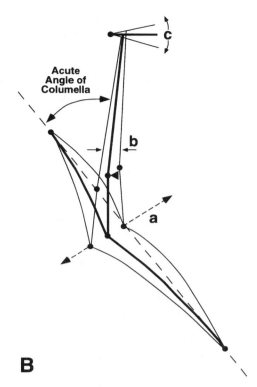

A   **B**

FIGURE 2.16. (**A**) A scanning micrograph of the acute angle formed in the parakeet between the plane of the tympanic membrane (TM) surface and that of the extra stapedius and columella. (**B**) Movements of the TM (a) result in transverse motion of the columella (b), and rocking motion of the footplate (c). (After Gaudin 1968.)

The avian columella lies at an acute angle (30° or less) to the plane of the tympanic ring (Fig. 2.16) and in the direction of the posterior edge of the membrane. The bony columella is bound to the cartilaginous extracolumella by a synchondrosis.

5.2. Middle-Ear Function in Birds

The ratio between tympanic membrane and columella footplate area has been measured and shown to vary between 11:1 and 40:1 in 47 species (Schwartzkopff 1952, 1957). The smallest ratios were found in young chicks (11:1) (*G. domesticus*), and in the grebes (15:1) (*Podiceps cristatus* and *Podiceps ruficollis*). Many fowl-like birds (Galliformes), gulls (Laridae), birds of prey (Falconiformes), and the songbirds have ratios between 20:1 and 30:1. The owls (Strigiformes) have the highest ratios, being as large as 40:1.

It is thought that there may be a relation between the area ratio and hearing sensitivity, with greater sensitivity associated with higher ratios. This is certainly true to a point, but the purpose of the middle ear is to match the impedance of air to that of the cochlear fluids. Too much pressure amplification creates a mismatch that can be as ineffective as too little amplification. Area ratios alone provide an incomplete picture of middle-ear function.

The mechanical motion of the conductive system in birds is different from that in reptiles largely because of the acute angle formed between the plane of the tympanic ring and the shaft of the columella. In reptiles, the columella approximated a 90°-angle relative to the plane of the TM (see Fig. 2.3). Thus a lever advantage is clearly formed between the length of the extracolumella anchorage and the length of the pars superior. The tympanic process of the avian extracolumella has a "Y" shape, and the length of the two legs including the angle formed between them varies greatly among species. The portion of the extrastapedius that articulates with the bony columella is an inward extension at the confluence of the three legs (Fig. 2.16A). The axis of rotation for this ossicular system is a line drawn between the tips of the inferior and superior stapedius. Thus, motion at the tip of the extrastapedius causes the tympanic process to move up and down with respect to the hingepoints of the inferior and superior stapedius. As with the lizards, the lever advantage ought to be determined by the length of the lever arm (the anchorage of the tympanic process) and the position of the bony columella along this arm. The lever advantage in birds, however, results from a complex trigonometric function between the motion at the tip of the extrastapedius and the angle of the bony columella relative to the TM (Fig. 2.16A). Moreover, the acute angle of the extrastapedius probably produces both rotational and translational motion components at the conical tip of the TM.

Two proposals concerning the lever advantage of the bird middle ear have been made, and both use the anatomical relations noted above. The earlier proposal is based on high-pressure, low-frequency stimulation of the TM and direct visual observation of TM, columella, and footplate motion (Gaudin 1968). The movement of the columella observed under these conditions was not longitudinal or piston-like but rather side to side (Fig. 2.16B). That is, the shaft of the columella appeared to rotate about an axis defined by the posterior edge of the footplate. Because of the eccentric insertion of the columella shaft on the footplate, there was a rocking motion of the anterior part of the footplate. This motion, according to Gaudin (1968) was like that of a "musician's foot tapping as it rests on the heel." The lever arms in this model were the tympanic membrane and the tympanic process of the extracolumella complex on the distal end, and the action of the footplate on the proximal end. Gaudin (1968) measured motion in the conductive system of several avian species and predicted a high mechanical lever ratio as a consequence of this arrangement.

Another model used middle-ear geometry to estimate movements of the extracolumella and the TM in the boreal owl (*Ageolius funereus*) (Norberg 1978). The rotational motion of the extracolumella due to TM displacements at the tip of the extrastapedius was seen as nearly parallel to the shaft of the bony columella and yielded a piston-like motion at the footplate. The purely back-and-forth motion at the footplate was aided by a bending action at the joint between the extrastapedius and the columella shaft. The lever advantage in this model was 1.6:1.

Motion of the TM (at the tip of the extrastapedius) and columella footplate has been measured in the ringed turtledove (*Streptopelia risora*) and pigeon (*Columba livia*). Figures 2.17A and C plot velocity transfer functions for these species, and in both the footplate response is smaller than

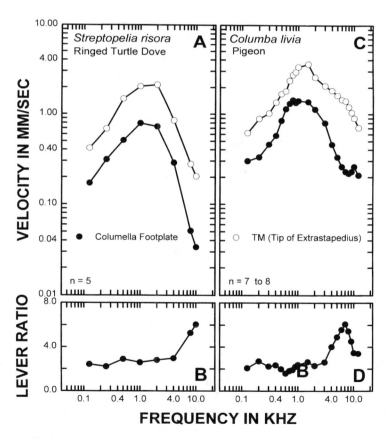

FIGURE 2.17. (**A**) Velocity transfer functions measured at 100 dB SPL in the dove at the tympanic membrane (TM) and footplate. (**B**) The lever ratio is determined from the difference between the functions in panel **A** (Saunders and Johnstone 1972). (**C**) and (**D**) Similar data for the pigeon (Gummer et al. 1989a, 1989b).

that of the TM, indicating a lever action. Panels B and D illustrate the respective difference between the TM and footplate responses, and the lever ratios were between 2.2:1 and 2.5:1 from 0.125 to 4.0 kHz in both species.

Velocity transfer functions obtained from the TM at the tip of the extrastapedius in the chick (*G. domesticus*) and parakeet (*M. undulatus*) have also been reported (Saunders 1985). At the most sensitive frequencies (1.5 to 3.0 kHz), the parakeet velocity appeared to be about 6 dB larger than that of the chick. Figure 2.15 illustrated the columella in these two animals, and it is clear that the parakeet ossicle is much more delicate than that of the chick, which may contribute to the larger TM response. In addition, the bandpass function of the chick is broader than that of the parakeet and has less sharp skirts on the high- and low-frequency side. By comparison the low-frequency roll-off in the chick was 6.6 dB/octave and 10.1 dB/octave for the parakeet. On the high-frequency side it was more similar in both species being 16.2 and 17.0 dB/octave, respectively. As in the lizard middle ear, the relation between TM velocity and stimulus intensity was linear in both these species to levels as high as 120 dB SPL.

5.3. Summary

The middle ear of birds exhibits the same general structural organization as that described for the reptiles in the previous section. As a consequence, the shape of the middle-ear transfer function has the same bandpass characteristic seen in reptiles. The peak frequency of the bandpass function, however, appears to be higher than that seen in the reptiles, and this may be due to the more delicate construction of the columella and TM. A mechanical lever ratio exists in birds and it is similar in magnitude to that seen in the reptiles. Finally, there is uniformity to the avian middle-ear design that may reflect adaptations for detection of aerial sounds.

6. The Middle-Ear Muscle in Reptiles and Birds

6.1. The Middle-Ear Muscle in Reptiles

The presence of an extracolumellar muscle in the lizard and crocodile middle ear was described in Section 4.3.2 and its contraction would serve to tense the tympanic process of the extracolumella and the TM. Either of these actions would alter the mechanical properties of the conductive apparatus, and this, in turn, would change the effectiveness of sound conduction. The actual contribution of this muscle to signal processing in the reptilian auditory system, however, remains unknown at the moment. Nevertheless, it tends to be found in those species capable of vocal communication among members (e.g., the geckos), thus suggesting that it could be activated reflexively during vocalizations.

6.2. The Middle-Ear Muscle in Birds

Birds possess only a single homologue of the mammalian tensor tympani and stapedius muscles. The bird middle-ear muscle has been called the tympanic muscle, stapedius muscle, tensor tympani, and laxor tympani. However, current convention, which we will follow, identifies it as the columellar muscle, *m. columellae* (Evans 1979; Kühne and Lewis 1985).

6.2.1. Structural Organization of the Columellar Muscle

The contractile portion of the columellar muscle lies in a groove in the lateral temporal and occipital bones, near the occipital condyle. This location allows access to the muscle body without interfering with either middle-ear sound transmission or blood supply. Like the mammalian stapedius muscle, the extracolumella is innervated by the facial (VII cranial) nerve.

The muscle tendon passes laterally through a caudoventral opening in the middle-ear cavity and splits into several tendons that progress to the margin of the TM. One tendon attaches to the posterior edge of the TM between the superior and inferior extracolumellar processes, while the other extends across the inner surface of the TM and attaches to the tympanic process of the extracolumella (Fig. 2.14).

Histochemical analysis reveals that the muscle is composed entirely of type II fast fibers, which exhibit three levels of myofibrillar ATPase activity (Counter et al. 1987). The presence of high ATPase activity fibers as well as the occurrence of "en plaque" neuromuscular junctions, characterize it as a fast twitch muscle whose fibers contract rapidly, are fatigue resistant, and are capable of sustained contractions (Borg et al. 1979).

The 65 motoneurons that innervate the columellar muscle of *Gallus* originate on the dorsolateral border of the ventral facial nuclear complex (Wong et al. 1992). This number of neurons, if similar to that in other bird species, is less than the number of fibers in various mammalian species, which range from 100 to over 1,000 neurons (Joseph et al. 1985).

6.2.2. Contraction Properties

Electrical stimulation of the chicken columellar muscle increases the volume of the middle-ear cavity due to a net outward displacement of the tympanic membrane. Muscle contraction results in increased TM tension to the upper and rostral quadrants, while the ventral portions become more compliant. Simultaneously, the joint between the columella and extra-columella moves toward the tympanic membrane, thus pulling the footplate away from the oval window. Increased tension on the columella complex more than likely increases the input impedance of the middle-ear system.

Measures of muscle contraction and relaxation time (during electrical pulse stimulation) revealed responses as fast as 15 to 22 ms that were independent of stimulus strength, and are among the fastest known in the animal kingdom. The tension exerted by a single electrical pulse of stimulation produced a maximum muscle tension in chickens that averaged about 46 mN. Repetitive pulses at a frequency of 160 Hz fused the muscle contraction with peak tensions ranging from 150 to 200 mN (Counter et al. 1981; Borg et al. 1982).

6.2.3. The Physiological Consequences of Muscle Contraction

Cochlear microphonic potentials have been measured as a function of columellar muscle tension in chickens. Increasing tension attenuated the microphonic response across all frequencies with a slightly larger attenuation in the higher frequencies (Counter and Borg 1982). Increasing muscle tension in the starling from 50 mN to 400 mN caused a 20-dB reduction in sound reaching the cochlea that was greatest between 2.0 and 3.2 kHz. These frequencies were in the most sensitive hearing range and encompassed the communication (song) frequencies in this species.

6.2.4. Functional Significance of the Avian Middle-Ear Muscle

Many possible roles have been suggested for the middle-ear muscle including activation as part of an acoustic reflex, protection from self-vocalization, attenuation of masking frequencies for communication, protection from low-frequency wind noise during flight, and providing necessary feedback for vocal development. Some of these possibilities have been demonstrated, while others are purely speculative.

A reflex contraction of the middle-ear muscles in response to loud sound has been demonstrated in numerous mammalian species (Møller 1974b) and in owls (Golubeva 1972). However, an acoustic reflex appears to be absent in the pigeon, chicken, and starling (Wada 1924; Counter and Borg 1979; Oeckinghaus and Schwartzkopff 1983). This latter observation is surprising since the middle-ear muscle contracts in response to self-vocalization in both chickens and starlings. The distress calls and crowing in chickens range from 70 to 103 dB SPL when measured a few centimeters from the beak, and can last for more than one second. This reflex to self-vocalization may serve as a protective mechanism that prevents hearing disruption during self-vocalizations.

Counter and Borg (1979) examined columellar activity during spontaneous calls in the chicken. The latencies of muscle activity lagged behind vocalizations at low vocal intensities, but at higher vocal levels the muscle response actually preceded sound emission. Simultaneous columellar activity with self-vocalization has also been demonstrated during electrical stimulation of the mesencephalic (midbrain) calling area. This suggested that middle-ear muscle contractions during vocalization arise from a reflexive pathway originating either from vocal centers in the deep central nervous systems or from a more local pathway triggered by sensory receptors in the syrinx (voice box). Various experimental manipulations support the notion that middle-ear muscle contraction during vocalization more than likely arises from central mechanisms that coactivated motor nuclei controlling the syrinx (the XIIth cranial nerve) and the extracolumellar muscle (the VIIth cranial nerve) (Grassi et al. 1988).

The presence of a single middle-ear muscle seems to be conserved across all species of birds, and while the ideas concerning its function are interesting, a fuller understanding about its contribution to hearing remains to be achieved.

7. Middle-Ear Development in Lizards and Birds

7.1. Development in the Lizard Middle Ear

Many aspects of middle-ear development in mammals and birds have been reviewed elsewhere, and for more information the reader is referred to several sources (Saunders et al. 1983; Saunders et al. 1993). All reptiles exhibit an advanced state of sensory motor development at the time of hatching, and, like other lower vertebrates, show continued body growth throughout life. This growth includes the skull, and so the possibility exists that components of the reptilian conductive apparatus also grow throughout life. The question, of course, is whether such an expansion in middle-ear structure results in improved auditory function. Growth of the middle-ear apparatus has been positively correlated with increased body size in the alligator lizard (*Gerrhonotous multicarinatus*) (Rosowski et al. 1988). These changes in the conductive apparatus were accompanied by improvements in middle-ear sound admittance measured at the TM. It remains to be seen, however, if improvements in admittance result in better hearing. A recent presentation (Werner et al. 1998) has examined middle-ear function and anatomy in juvenile and adult specimens of *Eublepharis macularius* and *Oedura marmorata*. From juvenile to adult, the body weight in the sampled animals grew by 379% and 149% in the two species, respectively. The middle-ear structures of both species also grew significantly, and the area ratio increased by 76% in *Eublepharis* and 20% in *Oedura*. The mechanical lever ratio was also examined in these species, but unlike the area ratio, showed no change with age.

Tympanic membrane velocity transfer functions in adults and juveniles showed similar patterns of change in both species. Figures 2.18A and C present these data for *Eublepharis*, and it is apparent that the resonant peak in the juvenile is broader than in the adult. These transfer functions were measured at three locations on the TM, and as noted in Section 4, the difference between the pars inferior and columella locations provides an indi-

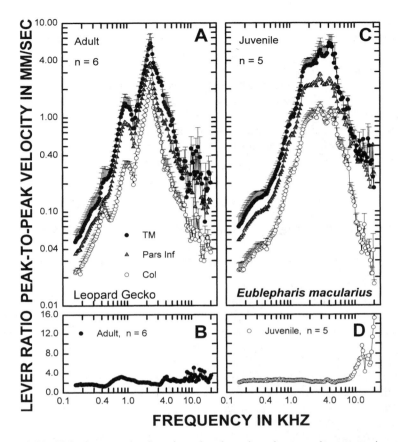

FIGURE 2.18. Velocity transfer functions for three locations on the tympanic membrane (TM) are presented for juvenile (**A**) and adult (**C**) leopard gecko (*Eublepharis macularius*) (Werner et al. 1998). (**B**) and (**D**) The difference between the velocity responses at the pars inferior and the columella positions, which approximate the lever ratio.

cation of the mechanical lever ratio. Panels B and D compare the lever ratio in the juvenile and the adult. The functions are very similar, having values around 2.2:1 from 0.15 to 4.0–6.0 kHz. Above 8.0 kHz the lever ratio in the juvenile ear rises sharply.

The efficiency of middle-ear sound transmission was examined further by first applying acoustic stimulation to the TM and then directly to the columella footplate. Footplate simulation occurred after the bony columella shaft was cut and removed from the middle ear along with the TM. Threshold levels of the compound action potential (CAP) were recorded from the round-window membrane at various frequencies with both modes of stimulation. The CAP is a neural potential evoked by sound that originates in the auditory nerve. The results showed that sound transmission was significantly more efficient through adult middle ears than through the juvenile ears, being on the order of 6 dB across frequency for *Eublepharis* and 15 dB for *Oedura* (Werner et al. 1998).

7.2. Development in the Chick Middle Ear

Viewed from the external auditory meatus, the TM of the chick has a convex appearance, and from embryonic day 10 (E10) to 70 days post-hatch (P70), the height of the cone nearly triples. During the same period the columella and extrastapedius increase in length by a factor of four. The surface area of the tympanic membrane was 1.5 mm^2 at E10, and by P70 it had grown to 33.5 mm^2. The area ratio, 10 days prior to hatching, was 5:1, and increased to 25:1 at P70. This represented a 400% change in the area ratio over this interval (Cohen et al. 1992).

Developmental changes in the fine structure of the TM have also been reported (Jaskoll and Maderson 1978; Chin et al. 1997). The membrane consists of three layers: the outer epidermal layer, the middle lamina propria, and the inner respiratory payer. The bulk of the TM is composed of the lamina propria within which there is a network of collagen fibers arranged in two distinct populations. The network of radial fibers extended from he center of the membrane (at the tip of the extrastapedius) to the periphery, and these fibers occupied the outer layer of the lamina propria, near the epidermal layer. Circular collagen fibers are located on the inner side of lamina propria, near the respiratory layer. From E11 to P52, lamina propria doubled in thickness near the center of the tympanic membrane, increasing from about 40 to 80 μm during this period. The peripheral regions of lamina propria were thinner than the central regions at E11 and actually thinned from 30 to 25 μm by P52 (Fig. 2.19A). In addition, the collagen fibers were thick and widely separated in the embryos. With aging, the radial fibers became thinner, more numerous, and more densely packed. As Figure 2.19B illustrates, these changes in the collagen network were most pronounced near the periphery of the TM.

These structural changes are associated with improved function during middle-ear development. Measures of middle-ear admittance (the reciprocal of impedance) at the chick TM have shown a 20-fold increase from P1 to P70 (Saunders et al. 1986). Tympanic membrane velocity, measured at the tip of the extrastapedius, increased by a factor of 3.6 during the same period (Cohen et al. 1992). The changes in middle-ear admittance were most likely due to improvements in the pressure amplifier as the growth of the tympanic membrane outstripped that of the footplate. Indeed, up to 45% of the admittance increase in the chick may be the result of an increasing area ratio.

Changes in the collagen structure within the tympanic membrane were also suggested as possible factors influencing middle-ear admittance.

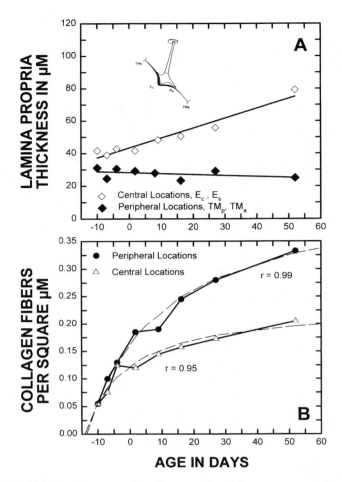

FIGURE 2.19. (**A**) Development of lamina propria thickness averaged for the anterior and posterior edge (TM$_a$, TM$_p$), and for two positions at the center (E$_c$, E$_s$) of the chick TM from E10 to P54. The inset in the panel indicates the locations from which the results were obtained. (**B**) The changes in radial collagen fiber density over the same interval. The dashed lines are fitted exponetial functions to the data and the time to achieve 90% of the maximum response in these functions is indicated. (Data from Chin et al. 1997.)

Another important factor is the drainage of middle-ear fluid. In the chick, a sudden improvement in evoked-response sensitivity was reported between the E18 and E19, and was attributed to the onset of respiration and the draining of fluid from the middle-ear cavity (Saunders et al. 1973).

8. Conclusions

The reptilian and avian middle-ear story reveals several interesting aspects that deserve further comment. This last section relates middle-ear function to the process of hearing, to differences between columella and ossicular (mammalian) middle ears, and to the consistent deterioration of the columella middle-ear response above 4.0 kHz. Finally, issues not addressed in this chapter need to be mentioned in passing in order to complete the presentation.

Figure 2.20 presents data from the tokay gecko and the parakeet that relate velocity transfer functions (at the footplate in Panel A and TM in Panel B) to audibility curves in both species. The threshold curve for the gecko was constructed from the characteristic frequency of single-unit tuning curves recorded from the cochlear nucleus (Manley 1972c) and auditory nerve (Eatock et al. 1981). The parakeet thresholds are from behavioral studies with this species (Saunders et al. 1979). The threshold data in both species were normalized with 0 dB being the most sensitive frequency.

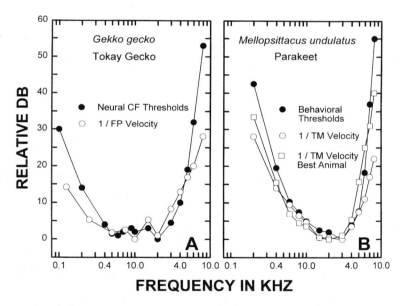

FIGURE 2.20. (**A**) The relation between tokay gecko neural thresholds (derived from the characteristic frequencies of tuning curves obtained from the auditory nerve and cochlear nucleus) and the inverse of the footplate transfer function in Figure 2.6. The data are normalized by plotting the results relative to 0 dB at the most sensitive frequency (Data from Manley 1972b, 1972c; Eatock et al. 1981). (**B**) The same procedure was used to compare parakeet behavioral thresholds and the TM transfer function (Data from Saunders et al. 1979; Saunders 1985).

TABLE 2.1. The peak TM velocity response at 100 dB SPL

Species	Frequency (kHz)	Peak velocity (mm/sec)
Reptiles		
Amphibolurus reticulata	1.0–2.0	2.6
Caiman crocodilus	0.5–0.7	8.4
Chrysemys scripta	0.55–0.65	3.0
Eublepharis macularius		
Juvenile (Pars Inf)	1.6–5.0	2.4
Adult (Pars Inf)	2.2	3.6
Gekko gecko	0.5–1.0	6.5
Gerrhonotous		
multicarinatus	1.9	3.9
Oedura marmorata		
Juvenile (Pars Inf)	2.1–5.5	1.9
Adult (Pars Inf)	2.5	4.1
Birds		
Columbo livia	0.8–1.5	3.6
Gallus domesticus	1.4–5.0	5.0
Melopsittacus undulatus	1.5–3.0	7.1
Streptophelia risor	0.8–2.0	2.2

The TM velocity transfer functions for the gecko and parakeet were also normalized with 0 dB being the maximum velocity response. The middle-ear functions were then inverted so that they could be plotted against the respective threshold curve. As Figure 2.20 indicates, the inverse of the velocity transfer function in both cases approximated the shape of the audibility curve over much of its frequency range. Justification for this comparison is based on the linearity of the tetrapod middle-ear response (Saunders 1985). This linearity permits a direct extrapolation from the iso-stimulus transfer functions to the iso-response contours at the threshold level of signal detection. The important conclusion gained from the resolts in Figure 2.20 is that the bandpass characteristic of the middle ear determines, to a large extent, the shape of the audibility curve.

The reptilian and avian transfer funcions measured at either the TM or columella footplate exhibited functional consistency between all of the examples presented here. The universal nature of this function lies in its bandpass characteristic. The limits of high-frequency hearing in reptiles rarely exceeded 5.0 to 8.0 kHz, while that in birds is restricted to 8.0 to 10.0 kHz (the owls and oilbird being exceptions). The center frequency of the transfer functions varied among species, and these differences arise from the species-specific middle-ear morphology that influences the conductive mechanics.

The peak response of all TM transfer functions exhibited velocities within 5 dB of each other (Table 2.1). This observation is remarkable when the methodologic and procedural differences used to obtain the original data are taken into account. Figure 2.21 presents the relation between the peak

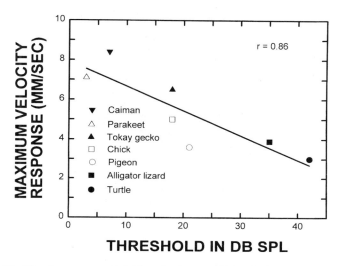

FIGURE 2.21. The frequency exhibiting the largest TM velocity response (at 100 dB SPL) was determined in seven species for which there was complementary neural or behavioral threshold data at the same frequency. The fitted line has a significant slope indicating that higher TM velocity was associated with lower thresholds.

velocity response and threshold sensitivity (neural or behavioral) at the same frequency in seven species. The correlation between these variables is significant and indicates that as maximum velocity increases the hearing threshold at the same frequency becomes more sensitive. This observation suggests that the magnitude of velocity may contribute to the sensitivity of the hearing threshold curve.

The observations presented in the paragraphs above suggest that there is considerable functional homology in the bandpass shape of the reptilian and avian transfer functions. This may not be so surprising given the overall uniformity of structural design seen across orders and species. There is, nevertheless, a startling difference between the performance of the columella and ossicular middle ear. Three representative TM transfer functions, from species with a columella middle ear, are plotted in Figure 2.22 along with two TM functions from species with an ossicular middle ear. It is interesting that the slopes on the low-frequency side in the both types are much the same, suggesting that system compliance may be similar. The high-frequency responses of these two middle-ear designs, however, were very different from each other. The mammalian middle ear had a much more efficient high-frequency response. One implication of the "highpass" transfer function in mammals is that it enables "ultrasonic" hearing in these species. Similar sound conduction is absent in reptiles and birds, and these species rarely hear above 10.0 kHz. This conclusion is totally independent

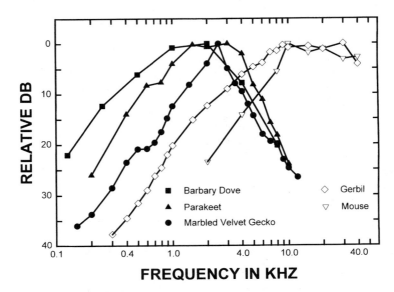

FIGURE 2.22. Examples of transfer functions in columella and ossicle middle ears. The data are plotted in dB relative to the largest velocity response. The highpass function of the mammals is obvious as is the bandpass function of the nonmammals. Data replotted from Werner et al. 1998 (gecko); Saunders and Johnstone 1972 (dove); Saunders 1985 (parakeet); Saunders and Summers 1982 (mouse); and Cohen et al. 1993 (gerbil).

of the signal processing capacity of the inner ear or hair cells because hair cells cannot process signals that they do not receive.

The important question, of course, is what accounts for these differences on the high-frequency side of the transfer function. Section 2.0 indicated that mass reactance controlled the high-frequency response of a vibrating system. The axis of rotation in the mammalian middle ear passes through the malleus and incus. In mammalian systems, the mass of these two bones is more or less compacted about the axis of rotation and this greatly reduces the moment of inertia. As a result, the high-frequency response of the system is significantly improved (Fleischer 1978). The axis of rotation in the reptile and bird, as discussed above, lies at the edge of the TM either at the end of the pars superior (reptiles) or between the tips of the inferior and superior stapedial process (birds). As a consequence, the TM, extra-columella, bony columella, and footplate are considerably removed from the axis of rotation. Thus, the moment of inertia is great and the high-frequency response severely curtailed.

Another curious observation, revealed in both the mechanical lever ratios and transformer functions, was the apparent deterioration in middle-ear responses above 3.0–4.0 kHz. The transformer functions always showed a loss of efficiency, while the lever ratios exhibited a sudden increase above these frequencies. This occurred in every case examined in this chapter (See Figs. 2.4, 2.5, 2.6, and 2.18 as examples). We have repeatedly emphasized that an increase in the lever ratio does not mean greater energy transfer to the cochlea. A mismatch, either greater or smaller, between the output impedance of the middle ear and the input impedance of the cochlea will reduce energy transfer across the interface.

There are several possibilities that might explain the deterioration of middle ear sound conduction above 3.0–4.0 kHz. One might be the response of the TM itself. It was previously noted that TM behavior changes in the high frequencies. Indeed, the evidence from geckos (Manley 1972b) suggests that the TM response above 3.0–4.0 kHz shifts from a simple piston-like motion (in the low frequencies) to one that exhibits regional nodal and antinodal motion (at high frequencies). As a consequence, there may be significant changes in the high-frequency motion at the conical tip of the TM. Another possibility could arise from a flex in the lizard pars inferior (and perhaps even the extrastapedial process in birds) at frequencies above 4.0 kHz (Manley 1972b, 1990). This flex is due to the cartilaginous nature of the tympanic process and results in a considerable absorption of the driving force between the TM and the shaft of the bony columella. Finally, unusual behaviors have been found in the response of the columella footplate and extrastapedius of the alligator lizard and pigeon conductive systems. It has been suggested that nonlinear behavior in the inner ear may produce reflected vibrations that interact with the acoustic response of the middle ear, particularly at low intensity. The nature of this interaction, however, remains to be determined (Rosowski et al. 1984; Manley 1990; Gummer et al. 1989a, 1989b).

This chapter has not considered the role of the reptilian and avian middle ear in the process of sound localization. The problem confronted by these animals is that they have small heads and high-frequency hearing that rarely exceeds 10.0 kHz. The wavelengths of sound are too large, even at the upper frequencies, to permit an effective sound shadow, and this limits the use of interaural intensity differences. With a small head, the interaural time differences are also greatly reduced. Birds and reptiles have overcome this problem by using a differential pressure detector at the TM. Differential pressure at the TM arises through two sound-conducting pathways, one acting directly on the outer surface of the TM, and the other on the inner surface via the interconnecting pathway between the two middle ears (e.g., the interaural pathway in birds or the large eustachian tubes in lizards). The mechanisms of this differential pressure detector have been discussed in detail elsewhere (Coles et al. 1980; Hill et al. 1980; Rosowski and Saunders 1980; Saunders and Henry 1988).

Finally, the interested reader might wish to explore further the unique middle-ear organization of the chameleon (Wever 1978) and woodpecker (Kohllöffel 1984). The latter system appears to protect the conductive apparatus from impulsive shocks to the skull. As a last point, the human middle-ear ossicles can be replaced with prosthetic reconstructions called tympanoplasties. Some of these prosthetics rebuild the conductive apparatus in the form of a columella. Thus, there is clinical interest in the mechanics of a single-ossicle conductive system, and understanding function in ears with a natural columella may provide insight for improving reconstructive surgery in the human middle ear (Mills 1994).

Acknowledgments. Research by the first author on the middle ear has been generously supported over the past 20 years by awards from the NIDCD, the National Organization for Hearing Research, the Deafness Research Foundation, and the Pennsylvania Lions Hearing Research Foundation. The assistance of Ms. Rachel Kurian and Ms. Amy Lieberman is gratefully acknowledged.

References

Allin ER, Hopson JA (1992) Evolution of the auditory system in synapsida ("mammal-like reptiles" and primitive mammals) as seen in the fossil record. In: Webster DB, Fay RR, Popper AN (eds) The Evolutionary Biology of Hearing. New York: Springer-Verlag, pp. 587–614.

Bolt JR, Lombard RE (1985) Evolution of the amphibian tympanic ear and the origin of frogs. Biol J Linnean Soc 24:83–99.

Bolt JR, Lombard RE (1992) Nature and quality of the fossil evidence for otic evolution in early tetrapods. In: Webster DB, Fay RR, Popper AN (eds) The Evolutionary Biology of Hearing. New York: Springer-Verlag, pp. 377–403.

Borg E, Counter SA, Rydqvist B (1979) Contraction properties and functional morphology of the avian stapedius muscle. Acta Otolaryngol 88:20–26.

Borg E, Counter SA, Lännergren J (1982) Analysis of the avian middle ear muscle contraction by strain gauge and volume and impedance change measures. Comp Biochem Physiol 71A:619–621.

Breuer J (1908) Über das Gehrögan der Vögel. Berlin: Stzber Wiener Akad 116:249–297.

Carroll RL (1987) Vertebrate Paleontology and Evolution. New York: Freeman and Co.

Chin K, Kurian R, Saunders JC (1997) The maturation of tympanic membrane layers and collagen in the embryonic and post-hatch chick. J Morphol 233:257–266.

Clack JA (1992) The stapes of *Acanthostega gunnari* and the role of the stapes in early tetrapods. In: Webster DB, Fay RR, Popper AN (eds) The Evolutionary Biology of Hearing. New York: Springer-Verlag, pp. 405–420.

Cohen YE, Hernandez HN, Saunders JC (1992) Middle-ear development. II: Structural development of the chick middle ear. J Morphol 212:257–267.

Cohen YE, Rubin DM, Saunders JC (1992) Middle-ear development. I: Extra-stapedius response in the neonatal chick. Hear Res 58:1–8.

Cohen YE, Doan DE, Rubin DM, Saunders JC (1993) Middle-ear development. V: Development of umbo sensitivity in the gerbil. Am J Otolaryngol 14:191–198.

Coles RB, Lewis DB, Hill KG, Hutchings ME, Grower DM (1980) Directional hearing in the Japanese quail (*Coturnix coturnix japonica*). II. Cochlear physiology. J Exp Biol 86:153–170.

Counter SA, Borg E (1979) Physiological activation of the stapedius muscle in *Gallus gallus*. Acta Otolaryngol 88:13–19.

Counter SA, Borg E (1982) The avian stapedius muscle. Influence on auditory sensitivity and sound transmission. Acta Otolaryngol 94:267–274.

Counter SA, Borg E, Lännergren J (1981) Basic contraction properties of the avian stapedius muscle. Acta Physiol Scand 111:105–108.

Counter SA, Hellstrand E, Borg E (1987) A histochemical characterization of muscle fiber types in the avian *m. stapedius*. Comp Biochem Physiol 86A(1): 185–187.

Doan DE, Cohen YE, Saunders JC (1994) Middle-ear development: IV. Umbo motion in neonatal mice. J Comp Physiol A 174:103–110.

Durrant JD, Lovrinic JH (1995) Basis of Hearing Science. Williams and Wilkins: Philadelphia, p. 333.

Eatock RA, Manley GA, Pawson L (1981) Auditory nerve fiber activity in the tokay gecko: I. Implications for cochlear processing. J Comp Physiol A 142:203–218.

Evans HE (1979) Organa sensoria. In: Baumel JJ (ed) Nomina Anatomica Avium. New York: Academic Press, pp. 505–526.

Fleischer G (1978) Evolutionary principles of the mammalian middle ear. Adv Anat Embryo Cell Biol 55:1–70.

Frey-Zumpfe H (1953) Befunde im Mittelohr der Vögel. Wiss Martin-Luther-Univ Halle Wittenb 2:445–461.

Fritzsch B (1992) The water-to-land transition: evolution of the tetrapod basilar papilla, middle ear and auditory nuclei. In: Webster DB, Fay RR, Popper AN (eds) The Evolutionary Biology of Hearing. New York: Springer-Verlag, pp. 351–375.

Gans C, Wever EG (1972) The ear and hearing in Amphisbaenia (Reptilia). J Exp Zool 179:17–34.

Gaudin EP (1968) On the middle ear of birds. Acta Otolaryngol 65:316–326.

Gould SJ (1989) Wonderful Life: The Burgess Shale and the Nature of History. New York: Norton.

Golubeva TB (1972) The reflex activity of the tympanal muscle in the owl *Asio otus*. Zhurn Evol Biol Fisiol 8:173–181.

Gummer AW, Smolders JWT, Klinke R (1989a) Mechanics of a single-ossicle ear: I. The extra-stapedius of the pigeon. Hear Res 39:1–14.

Gummer AW, Smolders JWT, Klinke R (1989b) Mechanics of a single-ossicle ear. II. The columella footplate of the pigeon. Hear Res 39:15–26.

Grassi S, Magni F, Ottaviani F (1988) Mechanisms controlling vocalization-evoked stapedius muscle activity in chickens (*Gallus gallus*). J Comp Physiol A 162:525–532.

Helmholtz HLF (1873) The mechanism of the ossicles of the ear and membrana tympani. Buck AH, Smith N (trans) New York: William Wood and Co.

Henson GW Jr. (1974) Comparative anatomy of the middle ear. In: Keidel WD, Neff WD (eds) Handbook of Sensory Physiology. V/I: Auditory System. Berlin: Springer-Verlag, pp. 40–110.

Hetherington TE (1992) The effects of body size on the evolution of the amphibian ear. In: Webster DB, Fay RR, Popper AN (eds) The Evolutionary Biology of Hearing. New York: Springer-Verlag, pp. 421–437.

Hill KG, Lewis DB, Hutchings ME, Coles RB (1980) Directional hearing in the Japanese quail (*Coturnix coturnix japonica*). I. Acoustic properties of the auditory system. J Exp Biol 86:135–151.

Hotton N (1959) The pelycosaur tympanum and early evolution of the middle ear. Evolution 13:99–121.

Iljitschew WD, Izwekowa LM (1961) Some peculiarities of the function of the auditory analyzer in birds. Zool Zh 40:1704–1714.

Jaskoll TF, Maderson PFA (1978) A histological study of the development of the avian middle-ear and tympanum. Anat Rec 190:177–200.

Joseph MP, Guinan JJ Jr, Fullerton BC, Norris BE, Kiang NYS (1985) Number and distribution of stapedius motorneurons in cats. J Comp Neurol 232:43–54.

Khanna SM, Tonndorf J (1972) Tympanic membrane vibrations in cats studied by time-averaged holography. J Acoust Soc Am 51:1904–1920.

Kinsler LE, Frey AR (1962) Fundamentals of Acoustics. 2nd ed. New York: John Wiley and Sons.

Kohllöffel LUE (1984) Notes on the comparative mechanics of hearing. I. A shock-proof ear. Hear Res 13:73–76.

Krause G (1901) Die Columella der Vögel (*Columella Auris Avium*). Ihr Bau und dessen Einfluss auf die Finhrigkeit. Berlin, Friedlnder.

Kühne R, Lewis B (1985) External and middle ears. In: King AS, McCelland J (eds) Form and Function in Birds. New York: Academic Press, pp. 227–271.

Lombard RE, Bolt JR (1979) Evolution of the tetrapod ear: an analysis and reinterpretation. Biol J Linn Soc 11:19–76.

Lombard RE, Bolt JR (1988) Evolution of the stapes in Paleozoic tetrapods. In: Fritzsch B, Ryan M, Wikzynski W, Hetherington T, Walkowiak W (eds) The Evolution of the Amphibian Auditory System. New York: Wiley and Sons, pp. 37–67.

Lynch TJ III, Nedzelnitsky V, Peake WT (1982) Input impedance of the cochlea in cat. J Acoust Soc Am 72:108–130.

Manley GA (1972a) Frequency response of the ear of the tokay gecko. J Exp Zool 181:159–168.

Manley GA (1972b) The middle ear of the tokay gecko. J Comp Physiol 81:239–250.

Manley GA (1972c) Frequency response of the ear of the tokay gecko. J Exp Zool 181:159–168.

Manley GA (1974) A review of some current concepts of the functional evolution of the ear in terrestrial vertebrates. Evolution 26:608–621.

Manley GA (1990) Peripheral Hearing Mechanisms in Reptiles and Birds. New York: Springer-Verlag.

Manley GA, Gleich O (1992) Evolution and specialization of function in the avian auditory periphery. In: Webster DB, Fay RR, Popper AN (eds) The Evolutionary Biology of Hearing. New York: Springer-Verlag, pp. 405–420.

Manley GA, Yates GK, Köppl C (1988) Auditory peripheral tuning: Evidence for a simple resonance phenomenon in the lizard *Tiliqua*. Hear Res 33:181–190.

Mills R (1994) Applied comparative anatomy of the avian middle ear. J Roy Soc Med 87:222–223.

Moffat AJM, Capranica RR (1978) Middle ear sensitivity in anurans and reptiles measures by light scattering spectroscopy. J Comp Physiol 127:97–107.

Møller AR (1963) Transfer function of the middle ear. J Acoust Soc Am 35:1526–1534.

Møller AR (1974a) Function of the middle ear. In: Keidel WD, Neff WD (eds) Handbook of Sensory Physiology, Vol V/1: Auditory System. Berlin: Springer-Verlag, pp. 491–517.

Møller AR (1974b) The acoustic middle ear muscle reflex. In: Keidel WD, Neff WD (eds) Handbook of Sensory Physiology, Vol V/1: Auditory System. Berlin, Springer-Verlag, pp. 519–548.

Müller HJ (1960) Die Morphologie und Entwicklung des Craniums von *Rea americana* Linné. Z Wiss Zool 165:221–319.

Mundie JR (1963) The impedance of the ear—a variable quantity. In: Fletcher JL (ed) Middle Ear Function Seminar. US Army Medical Research Laboratory Report, Dept. 567, Wright-Patterson AFB, Ohio, pp. 63–85.

Norberg RA (1978) Skull assymetry, ear structure and function, and auditory localization in Tengmalm's owl, *Aegolius funereus* (Linn). Phil Trans R Soc 282B:325–410.

Oeckinghaus H, Schwartzkopff J (1983) Electrical and acoustical activation of the middle ear muscle in a songbird. J Comp Physiol 150:61–67.

Parker SP (1982) Synopsis and Classification of Living Organisms. Vol. 2. McGraw Hill: New York.

Patterson WC (1966) Hearing in the turtle. J Aud Res 6:453–464.

Pohlman AG (1921) The position and functional interpretation of the elastic ligaments in the middle-ear of *Gallus*. J Morphol 35:229–262.

Puria S, Allen JB (1991) A parametric study of cochlear input impedance. J Acoust Soc Am 89:287–309.

Relkin EM (1988) Introduction to the analysis of middle-ear function. In: Jahn AF, Santos-Sacchi J (eds) Physiology of the Ear. New York: Raven Press, pp. 103–123.

Retzius G (1884) Das Gehörorgan der Wirbethiere. Vol II. Das Gehörogan der Reptilien, der Vögel und der Säugethiere. Stockholm: Samson and Wallin.

Rosowski JJ (1994) Outer and middle ears. In: Fay RR, Popper AN (eds) Comparative Hearing: Mammals. New York: Springer-Verlag, pp. 172–247.

Rosowski JJ (1996) Models of external- and middle-ear function. In: Hawkins HL, McMullen TA, Popper AN, Fay RR (1996) Auditory Computation. New York: Springer-Verlag, pp. 15–61.

Rosowski JJ, Saunders JC (1980) Sound transmission through the avian interaural pathway. J Comp Physiol 136:183–190.

Rosowski JJ, Ketten DR, Peake WT (1988) Allometric correlations of middle-ear structure and function in one species—the alligator lizard. Abs Assoc Res Otolaryngol 12:55.

Rosowski JJ, Peake WT, Lynch TJ III (1984) Acoustic input-admittance of the alligator-lizard ear. Nonlinear features. Hear Res 16:205–223.

Rosowski JJ, Peake WT, Lynch TJ III, Weiss TF (1985) A model for signal transmission in an ear having hair cells with free-standing stereocilia. II. Micromechanical stage. Hear Res 20:139–155.

Saito N (1980) Structure and function of the avian ear. In: Popper AN, Fay RR (eds) Comparative Studies of Hearing in Vertebrates. New York: Springer-Verlag, pp. 241–260.

Saunders JC (1985) Auditory structure and function in the bird middle ear: an evaluation by SEM and capacitive probe. Hear Res 18:253–268.

Saunders JC, Henry WJ (1988) The peripheral auditory system in birds: structural and functional contributions to auditory perception. In: Dooling RJ, Hulse WM (eds) Contributions to Auditory Perception in Animals. Collingwood, NJ: Erlbaum, pp. 31–62.

Saunders JC, Johnstone BM (1972) A comparative analysis of middle-ear function in non-mammalian vertebrates. Acta Otolaryngol 73:353–361.

Saunders JC, Summers RM (1982) Auditory structure and function in the mouse middle ear: an evaluation by SEM and capacitive probe. J Comp Physiol 146:517–525.

Saunders JC, Coles RB, Gates GR (1973) The development of auditory evoked responses in the cochlea and cochlear nuclei of the chick. Brain Res 63:59–74.

Saunders JC, Rintelman WF, Bock GR (1979) Frequency selectivity in bird and man: a comparison among critical ratios, critical bands, and psychophysical tuning curves. Hear Res 1:303–323.

Saunders JC, Kaltenbach JA, Relkin EM (1983) The structural and functional development of the outer and middle ear. In: Romand R (ed) Development of Auditory and Vestibular Systems. New York: Academic Press, pp. 3–25.

Saunders JC, Relkin EM, Rosowski JJ, Bahl C (1986) Changes in middle-ear input admittance during postnatal auditory development in chicks. Hear Res 24:277–235.

Saunders JC, Doan DE, Cohen YE (1993) The contribution of middle-ear sound conduction to auditory development. Comp Biochem Physiol 106A:7–13.

Schellart NAM, Popper AN (1992) Functional aspects of the evolution of the auditory system in Actinopterygian fish. In: Webster DB, Fay RR, Popper AN (eds) The Evolutionary Biology of Hearing. New York: Springer-Verlag, pp. 295–322.

Schmalhausen II (1968) The Origin of Terrestrial Vertebrates. New York: Academic.

Schwartzkopff J (1952) Untersuchungen über die Arbeitsweise des Mittelohres und das Richtungshren der Singvögel unter Verwendung von Cochlea-Potentialen. Z vergl Physiol 34:46–68.

Schwartzkopff J (1957) Die Görssenverhaltnisse von Trommelfell, Columella-Fussplatte und Schnecke bei Vögeln verschiedenen Gewichts. Z Morph Ökologie Tiere 45:365–378.

Smith G (1904) The middle ear and columella of birds. Quart J Micros Sci 48:11–22.

Starck D (1978) Vergleichende Anatomie der Wirbeltiere, Band 1. Berlin: Springer-Verlag.

Stellbogen E (1930) Über das äussere und mittlere Ohr des Waldkauzes (Syrnium aluco L.) Z Morph kol Tiere 19:686–731.

Tonndorf J, Khanna SM (1970) The role of the tympanic membrane in middle-ear transmission. Ann Otol Rhinol Laryngol 79:743–753.

Tumarkin A (1968) Evolution of the auditory conducting apparatus in terrestrial vertebrates. In: De Reuck AVS, Knight J (eds) Hearing Mechanisms in Vertebrates. Ciba Foundation Symposium. Boston: Little Brown, pp. 18–36.

van Bergeijk (1966) Evolution of the sense of hearing in vertebrates. Am Zool 6:371–377.

van Bergeijk (1967) The evolution of vertebrate hearing. In: Neff WD (ed) Contributions to Sensory Physiology. Berlin: Academic Press, pp. 1–49.

Versluys J Jr (1898) Die mittlere und äussere Ohrsphäre der Lacertilia und Rhynchocephalia. Zoo Jharb Abt Anat 12:161–406.

von Békésy G (1960) Experiments in Hearing. New York: McGraw-Hill.

Wada Y (1924) Beiträge zur vergleichenden Physiologie des Gehörorganes. Pflugers Arch ges Physiol 202:46–69.

Watson DMS (1953) The evolution of the mammalian ear. Evolution 7:159–177.

Webster DB, Fay RR, Popper AN (1992) The Evolutionary Biology of Hearing. New York: Springer-Verlag.

Werner YH (1983) Temperature effects on cochlear function in reptiles: a personal review incorporating data. In: Fay RF, Gourevitch G (eds) Hearing and Other Senses: Presentations in Honor of E. G. Wever. Groton, CT: Amphora Press, pp. 149–174.

Werner YL, Montgomery LG, Safford SD, Igic P, Saunders JC (1998) How changes in body size relate to middle ear structure and function and auditory sensitivity in Gekkonoid lizards. J Exp Biol 201:487–502.

Wever EG (1973) The function of the middle ear in lizards: divergent types. J Exp Zool 184:97–126.

Wever EG (1974) Evolution of vertebrate hearing. In: Keidel WD, Neff WD (eds) Handbook of Sensory Physiology. V/I: Auditory System. Berlin: Springer-Verlag, pp. 423–454.

Wever EG (1978) The Reptile Ear: Its Structure and Function. Princeton, NJ: Princeton University Press.

Wever EG, Gans C (1973) The ear in Amphisbaenia (Reptilia): further anatomical considerations. J Zool Lond 171:189–206.

Wever EG, Lawrence M (1954) Physiological Acoustics. Princeton, NJ: Princeton University Press.

Wever EG, Vernon JA (1956) Sound transmission in the turtle's ear. Proc Natl Acad Sci USA 42:292–299.

Wever EG, Vernon JA (1957) Auditory responses in the spectacled caiman. J Cell Comp Physiol 50:333–339.

Wever EG, Werner YL (1970) The function of the middle ear in lizards: *Crotaphytus collaris*. J Exp Zool 175:327–342.

Wilson JP, Smolders JWT, Klinke R (1985) Mechanics of the basilar membrane in *Caiman crocodilus*. Hear Res 18:1–14.

Wong CJH, To EC, Schwarz DWF (1992) Location of motoneurons innervating the middle ear muscle of the chicken (*Gallus domesticus*). Hear Res 61:31–34.

Zwislocki J (1962) Analysis of middle-ear function. Part I: Input impedance. J Acoust Soc Am 34:1514–1523.

Zwislocki J (1975) The role of the external and middle ear in sound transmission. In: Tower DB (ed) The Nervous System, Vol. 3: Human Communication and Its Disorders. New York: Raven press, pp. 445–455.

3
The Hearing Organ of Birds and Crocodilia

OTTO GLEICH and GEOFFREY A. MANLEY

1. The Hearing Organ of Birds and Crocodilia

Among the vertebrates, birds are one of the most vocal groups. Many birds (especially the passerines, or song birds) rely strongly on their sense of hearing for communication in territorial, social, and sexual behavior and for alarm signals. Their relatives, the Crocodilia (crocodiles, alligators, and gavials) are also vocal—a rare trait in reptiles. They are known to use several kinds of vocalization as communication signals in different behavioral contexts both as adults and as young, even within the egg (e.g., Garrick et al. 1978). In addition, some birds use their hearing for passive sound localization of prey (e.g., owls, Konishi 1973) or for active echolocation in their cave habitats (e.g., oil birds and cave swiftlets, Konishi and Knudsen 1979). Thus the sense of hearing is critically important in the life of many birds and Crocodilia, and selection pressures have produced an excellent sensitivity to sound (in birds as good as in mammals) in the frequency range covered (few birds hear higher frequencies than about 10–12 kHz).

The hearing abilities of birds and Crocodilia are, however, not only interesting because they are among the most vocal of animals. In addition, they are extremely important in regard to comparative studies of the evolution and function of complex auditory organs. As mammals, they have specialized sensory hair cells across the auditory papillae. Thus the study of their auditory papillae can provide information on the principles underlying particular patterns of structural organization and thus be very useful in understanding mammalian audition.

This short review is not intended to be exhaustive. Instead, we will try to give the reader a general impression of the anatomical substrate of hearing at the peripheral level—in the hearing organ, the basilar papilla itself. We will also briefly discuss the response characteristics of the sensory hair cells and their afferent nerve fibers that pass information to the brain on the frequency, phase, intensity, and time structure of the sounds analyzed.

1.1. Introduction to the Systematics and Ancestry of Birds and Crocodilians

Birds and crocodilians are the only living representatives of a group known as Archosauromorpha, which also includes the dinosaurs of the Mesozoic period. This group of reptiles had its origin in the diapsid Thecodonts, which are known from early Triassic times (about 230 million years ago; Hennig 1983). True crocodiles are known from the late Triassic, and these coexisted with the ancestors of the true birds that are first known from the Jurassic (140 million years; Carroll 1988). Since birds do not fossilize well, evolution within the class Aves is still poorly understood (Carroll 1988; Padian and Chiappe 1998). The main source of data on the evolution of the soft structures of the hearing organ are comparative anatomical and neurophysiological studies on living species, especially those closely related to putative ancestral groups. These studies suggest that at the time of the divergence of the evolutionary line leading to both birds and Crocodilia, the auditory papilla was still a relatively unspecialized structure showing weak, if any, hair cell specializations (Manley and Köppl 1998). The kind of hair cell specializations we see in modern archosaurs must have developed early, however, soon after their common ancestor diverged from the ancestors of the diapsid lepidosaurs such as lizards and snakes (which show a different kind of hair cell specialization), but before the divergence of the avian ancestors from those of the Crocodilia, since the common pattern of hair cell specialization we see in these groups is unique (Manley 1990). In the continuing course of evolution, birds and the Crocodilia further refined their auditory papillae, specializing two groups of sensory hair cells organized across the papilla and their innervation pattern in a similar manner to the (independent) evolution of the hearing organ of mammals. The structural similarity is even greater to the noncoiled cochleae of monotreme mammals (Ladhams and Pickles 1996). We have previously discussed evidence for important functional parallels between these auditory specializations in birds and mammals (Manley et al. 1989; Manley 1990). Study of the hearing organs of modern archosaurs may thus be instructional in our attempt to understand the mechanisms underlying hair cell functional specializations in general.

Due to the relatively poor fossil record of birds, there are uncertainties with regard to the details of relationships between different avian groups. Apart from the flightless Ratitae (whose status as a monophyletic group is undecided [Hennig 1983; Carroll 1988; Padian and Chiappe 1998] we will refer here mainly to work on the emu, *Dromaius novaehollandiae* [a list of the scientific and common names of species mentioned in this chapter can be found in Table 3.1]), the other birds can be divided into a (more primitive) water bird assemblage and a (more advanced) land bird assemblage (Feduccia 1980; Carroll 1988). Of the birds we will refer to, the ducks and seagulls belong in the water bird assemblage. The pigeon (*Columba livia*)

TABLE 3.1. Scientific and common names of the species mentioned in Chapter 3

Scientific Name	Common Name
Agelaius phoeniceus	Redwing blackbird
Anas platyrhynchos	Domestic duck
Aythya fuligula	Tufted duck
Baleana mysticetus	Bowhead whale
Caiman crocodilus	Spectacled caiman
Casuarius casuarius	Cassowary
Cavia procellus	Guinea pig
Columba livia	Pigeon
Dromaius novaehollandiae	Emu
Gallus gallus domesticus	Chicken
Gekko gecko	Gecko
Homo sapiens	Human
Larus argentatus	Herring gull
Larus marinus	Great black-backed gull
Melopsittacus undulatus	Budgerigar
Mus musculus	Mouse
Rhea americana	Nandu
Serinus canarius	Canary
Struthio camelus	Ostrich
Sturnus vulgaris	European starling
Taeniopygia guttata	Zebra finch
Tiliqua rugosa	Bobtail lizard
Tyto alba	Barn owl

and chicken (*Gallus gallus domesticus*) belong to a primitive subdivision of the land bird assemblage. The barn owl *Tyto alba* and the passerines or song birds (the largest bird family, here represented by the starling, canary, and zebra finch) belong to more derived group of land birds.

2. Morphology of the Archosaur Auditory Papilla

2.1. General Patterns in Papillar Shape and Size

There is a strong similarity in the structural arrangement of the cochlear ducts of birds and Crocodilia, each containing receptor areas both of the lagenar macula and the basilar papilla (Fig. 3.1). The lagenar macula is an otolithic organ present in all nonmammalian vertebrates and in monotreme mammals. In birds and crocodiles it is situated at the apical end of the cochlear duct, separated from, but close to the apex of the auditory basilar papilla. Morphologically, the lagenar macula appears to be a vestibular organ and there is no evidence for any auditory function (Manley et al. 1991). The relatively short duct containing the hearing organ is not coiled as in therian mammals, but bent and somewhat twisted. In the long cochlear

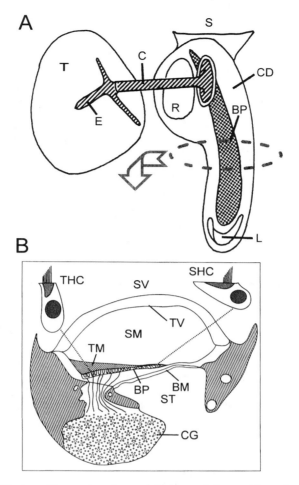

FIGURE 3.1. Diagram illustrating the morphology of the cochlear duct and basilar papilla of archosaurs. (**A**) Schematic diagram of the auditory apparatus. The tympanic membrane (T) is connected to an extracolumella (E) and through it to the columella (C) that ends with its footplate inserted into the basal region of the cochlear duct (CD) next to the round window (R). The elongated cochlear duct has the lagenar macula (L) at its apical end and a connection to the sacculus (S) basally. The broken ring indicates the approximate level of section shown in the framed drawing below. (**B**) Diagrammatic transverse section of the cochlear duct of a bird, approximately in its mid-region (see arrow in A). The shaded areas are sectioned limbic material, the hatched area (TM) is a cross section through the tectorial membrane that lies over the basilar papilla (BP). The tegmentum vasculosum (TV) separates the endolymph-containing scala media (SM) from the perilymph-containing scala vestibuli (SV). The perilymphatic scala tympani (ST) is located below the basilar membrane (BM), which supports the basilar papilla (BP). The cochlear ganglion (CG, labeled by open dots) contains the cell bodies of the afferent neurons that send dendrites toward the hair cells. The side of the papilla where the nerve fibers enter is termed neural and the other side that lies over the free basilar membrane is termed abneural. Enlarged examples of a tall hair cell (THC) from the neural side and a short hair cell (SHC) from the abneural side of the papilla are shown at the top of the framed drawing; their nuclei and cuticular plates are shown stippled.

ducts of owls, this twisting is quite complex (Schwartzkopff and Winter 1960; Fischer et al. 1988). The overall structure of the hearing organ conforms to the general amniote pattern (Manley 1990), having an endolymphatic tube (the cochlear duct or scala media) with the auditory organ forming its lower border. This tube is limited above and below by perilymphatic spaces known as scala vestibuli and scala tympani, respectively. The cochlear ganglion, which contains the somata of the primary sensory neurons of the cochlear and lagenar portion of the eighth nerve, lies on the medial side of scala tympani in close proximity to the brain. The afferent nerve fibers from the lagenar macula (whose numbers vary between 850 and 2,050 across species; Köppl 1997b; Gleich et al. 1998; Köppl and Wegscheider 1998) have their cell bodies more in the apical region of the cochlear ganglion (Manley et al. 1985; Manley et al. 1991). The cell bodies of the auditory afferents (whose numbers vary greatly across species) lie more basally (Fischer et al. 1994).

In general, the average avian auditory sensory epithelium is shorter (mostly less than 4–5 mm in length, maximally 11 mm; Fig. 3.2) and wider than that of a typical mammal. In mammals, basilar membrane length varies from approximately 6 mm in the mouse (*Mus musculus*) and 30 mm in humans (*Homo sapiens*) to 60 mm in the elephant (no generic name of the described specimen available) and the bowhead whale (*Baleana mysticetus*; reviews in Fay 1992 and Ketten 1992). The shortest avian papillae investigated are those of the canary *Serinus canarius* and the zebra finch *Taeniopygia guttata*, that are about 2 mm long in the living state (Gleich et al. 1994). The longest avian papilla described in detail is that of *T. alba*, with a length of 11 mm (Smith et al. 1985; Fischer et al. 1988). The length of the papilla in ratites is also similar to the avian average. In *D. novaehollandiae*, the length in young animals is about 5.5 mm in the living state (Köppl and Manley 1997). Werner (1938) reports that in the cassowary *Casuarius casuarius* and in the ostrich *Struthio camelus* the basilar membrane is 5.1 mm and 4.3 mm long, respectively. The basilar papilla of the spectacled caiman (*Caiman crocodilus*) is approximately 5 mm in length in the unfixed state (von Düring et al. 1974; Leake 1977).

A thick tectorial membrane covers the entire papilla in all species, and all hair cell stereovillar bundles are firmly attached to it (Fig. 3.1). The tectorial membrane is much thicker at its origin on the neural side of the papilla, and tapers toward the abneural edge. Although in the fixed state, the tectorial membrane shrinks and shows holes, in the unfixed state its upper surface is quite smooth (Runhaar 1989). The hair cells are not arranged in neat rows along or across the papilla, as they are in the mammalian organ of Corti, but instead they form a complex mosaic (Fig. 3.3). They are surrounded on all sides and below by supporting cells. The basilar papilla of modern archosaurs contains no cells with prominent stiffening fibers, such as the pillar cells of mammals that lie between the inner and

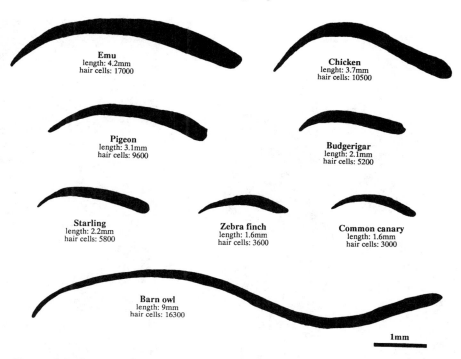

FIGURE 3.2. Gross-morphological view of the papillar dimensions for eight species of birds. The black object represents the hair cell field of the epithelium as seen from above and as if it were flat. For each species, the total length of the fixed, dried papilla is given, together with the average number of hair cells counted (Fischer et al. 1988; Gleich and Manley 1988; Manley et al. 1993; Gleich et al. 1994; Manley et al. 1996; and unpublished data).

outer hair cells and form the tunnel of Corti. Neither are there stiffened cells to support the hair cells, such as mammalian Deiter's cells. Thus the entire organ gives the appearance of a structure that is substantially more uniform in its micromechanical properties—and, in total, less stiff—than is the organ of Corti of mammals.

The total number of sensory hair cells in the archosaur hearing organ is comparable to that of the mammalian cochlea and varies across species (see Fig. 3.2). From two to four intergrading hair cell types have been defined on the basis of arbitrarily selected morphological criteria (for a detailed discussion see Section 2.2.1). Since the avian papilla is mostly relatively short, there is a much larger number of cells in a single cross section than seen in mammals. There can be more than 50 hair cells across its widest apical area, compared to four to six in mammals (Fig. 3.1B). Toward the basal area of the bird papilla, however, the number reduces to 6 to 10 hair cells in a cross

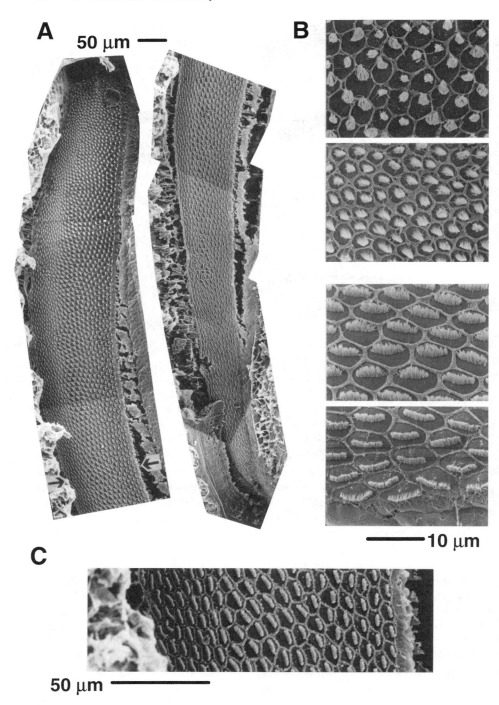

section. In *C. crocodilus*, Leake (1977) counted a total of 11,500 hair cells and 29 hair cells across the papilla at the apex.

2.1.1. The Primitive Condition

In order to understand trends over evolutionary time, we have to know which groups of birds are considered to be conservative in their structure and which specialized. The most primitive living birds are the flightless ratites (Carroll 1988). Due to the lack of detailed information on these important species, we have studied *D. novaehollandiae* (Fischer 1998; Köppl et al. 1998). Most hair cells of the basilar papilla in *D. novaehollandiae* are remarkably tall and columnar in shape, some being more than four times taller than they are wide. As in all other species, the hair cells in *D. novaehollandiae* generally become smaller and wider toward the abneural side and toward the base of the papilla. However, it is only in the very abneural, basal papilla that a small area is found whose hair cells fulfil the original criterion for short hair cells (SHC), that is, they are wider than tall. Thus, this primitive papilla consists almost entirely of tall hair cells (THC). The hair cells do not all have the sensitive axis of their stereovillar bundles oriented abneurally (i.e., with the bundle long axis parallel to the neural edge), but show rotations of up to 70° toward the apex. The maximum rotation in the papilla of *D. novaehollandiae* is at about 70% of the papillar length from the basal end and roughly midway across the papilla. From this position, the rotational angle of hair cell bundles in *D. novaehollandiae* decreases to zero in all directions. Thus even at this relatively early stage of evolutionary development, the auditory organ already shows all the features regarded as being typical for avian papillae (Manley 1990).

FIGURE 3.3. Scanning electron microscopic views of the avian basilar-papilla surface. (**A**) Low-power reconstruction of the papilla of *Serinus canarius*. The left column shows the apical half (apical is to the top and neural to the left) and the right column the basal half of the papilla. At the extreme basal end there are remnants of the tectorial membrane, most of which has been removed to expose the hair cell surface. The bright dots represent the stereovillar bundles of individual hair cells. (**B**) High-power views of small groups of hair cells from *Dromaius novaehollandiae*. The upper pair of pictures is from an apical and the bottom pair from a basal location of the papilla. In each pair of pictures, the upper panel shows an example from the abneural side, the lower panel an example from the neural side. All pictures are given with the same magnification. Hair cells, with their dark surfaces, are separated by the microvilli of the supporting cells. The four examples illustrate the variation of hair cell morphology with the position on the papilla. (**C**) Reconstruction of a strip across the papilla of *Serinus canarius* at higher magnification and at a position indicated by the arrows in **A**. Note that the long axis of the bundles is parallel to the edge of the papilla on the neural (left) and abneural (right) side and that bundles in the middle of the papilla show a systematic rotation.

2.2. General Patterns in Hair Cell Morphology in the Avian Papilla

2.2.1. The Continuum in Hair Cell Shape

In *C. crocodilus*, there are two anatomical types of hair cells, a fact to which Retzius (1884) alluded. Baird (1974) and later authors describe a sharp transition between these hair cell types (von Düring et al. 1974; Leake 1977), unlike the continuum found in birds (see below). As in birds, a substantial proportion of the hair cells in *C. crocodilus* does not lie on the free basilar membrane. There are in total about 3,000 THC in *C. crocodilus* situated over the neural limbus and 8,500 bowl-shaped SHC that mostly lie over the free basilar membrane.

In contrast to the situation in *Caiman*, Jahnke et al. (1969) correctly observed that there are no structural grounds for distinguishing discreet hair cell types in the avian papilla, even though hair cell shape changes systematically along and across the basilar papilla. The hair cell height is roughly correlated with response frequency (Fischer et al. 1992; Fischer 1994b), being generally shorter in birds hearing higher frequencies, but all papillae show a reduction in hair cell height both toward the base (i.e., toward higher frequencies) and *across* the papilla toward the abneural side. The gradients in hair cell shape are species specific (Fig. 3.4) and show various degrees of specialization (e.g., *T. alba*; Köppl 1993; Fischer 1994b). In spite of the obvious continuum in hair cell shape, the terms "tall" (THC) and "short" (SHC) were introduced by Takasaka and Smith (1971) and have been widely used to describe the extremes of the structural configurations based on a simple arbitrary morphological criterion; THC are taller than they are wide and vice versa. Except near the apical end, THC are found predominantly supported by the superior cartilaginous plate or neural limbus (below which lie the ganglion cells of the auditory nerve, Fig. 3.1). They are the less specialized of the two types and are most similar to the typical hair cell of more primitive groups of vertebrates (Takasaka and Smith 1971; Chandler 1984). The SHC typically occupy most of the space over the free basilar membrane. Intermediate hair cells (INHC) were described as being intermediate in both shape and position, with, however, no specific limits to their form parameters (Tanaka and Smith 1978). A few hair cells at the basal end of the papilla in *G. gallus domesticus* and many hair cells of the basal 3 mm of the *T. alba* papilla have been called lenticular hair cells (Smith 1985). They are also flattened like SHC, but their stereovillar bundles lie eccentrically on the cell's neural surface.

Recent detailed anatomical studies have confirmed that avian hair cells do in fact form a structural continuum along and across the papilla (Fischer et al. 1988; Gleich and Manley 1988; Fischer 1992, 1994a, 1994b; Fischer et al. 1992; Manley et al. 1993; Brix et al. 1994; Gleich et al. 1994; Manley

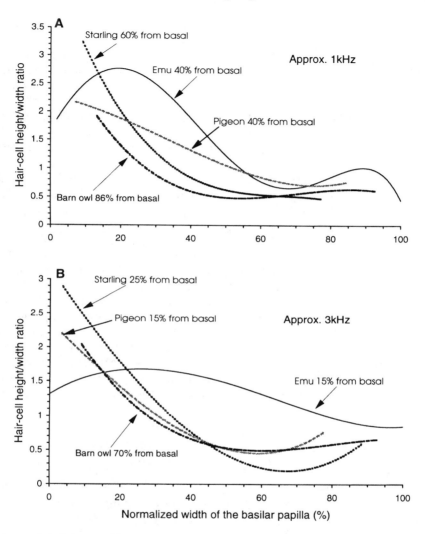

FIGURE 3.4. Diagram illustrating the changes in hair cell dimensions (in this case the hair cell height/width ratio) as a function of the normalized position *across* the papilla in four bird species. Shown are the data from *Sturnus vulgaris, Columba livia, Dromaius novaehollandiae*, and *Tyto alba* for a position along the papilla that responds to (**A**) 1 kHz and (**B**) 3 kHz. The lines are third-order polynomial fits to the data points available. For each line, the absolute position of the frequency represented on the individual species' papilla is shown. For example, 3 kHz in *Columba livia* is found at a position 15% of the papilla's length from the basal end. The change of the hair cell height/width ratio across the papilla is generally more rapid in the higher-frequency region.

et al. 1996). The gradients observed are similar, but species specific. The terms THC and SHC represent the extreme forms, but additional terms for hair cells, such as "intermediate" and "lenticular," imply the existence of definable subgroups that do not exist, and thus these terms should be abandoned. Indeed, because not only hair cell shape, but other parameters (e.g., of the hair-cell bundle) also change independently of each other along and across the epithelium, it can be argued that *every hair cell in the avian papilla* is in fact unique!

Detailed transmission electron microscopic (TEM) studies of the papillae of *Sturnus vulgaris* (starling), *G. gallus domesticus*, *T. alba*, and *D. novaehollandiae* (Fischer 1992, 1994b, 1998; Fischer et al. 1992) revealed the existence of hair cells in an abneural, basal area of the basilar papilla that were devoid of afferent contacts (Fig. 3.5). These cells are not only found in an "extremely abneural" position, as suggested by Smolders et al. (1995), but can in fact, depending on the species, cover half the width of the papillar area at its base (Fischer 1994b). Fischer's finding offers—for the first time—a functionally definable basis for distinguishing populations of avian hair cells and he thus suggested redefining the terms THC and SHC using their innervation pattern: THC are those hair cells that are afferently innervated, SHC are those that have no afferent innervation. This suggestion has been adopted here. In several species (but not in *D. novaehollandiae*; Fischer 1998), there is a rough correspondence between the papillar areas defined by this and those defined by the earlier "cell shape" criterion. A large population of hair cells without an afferent innervation presents a very interesting challenge to those studying the function of the avian papilla.

2.2.2. The Pattern of Hair Cell Bundle Rotation

Birds are also unique among tetrapods in a different respect, for their hearing organ shows a systematic change in hair cell bundle orientation patterns across and along the epithelium. Hair cells lying along both neural and abneural edges of the papilla have their stereovillar bundles all oriented nearly perpendicularly (±20°) to the edge of the papilla (defined as an "abneural" orientation), similar to the orientation seen in mammalian hair cells. Cells in the center of the papilla, however, especially in the apical half, tend to have their bundles rotated toward the apex, and the orientation gradually changes in any cross section from either edge toward the middle of the papilla (Fig. 3.6). Although this tendency is hardly noticeable at the papillar base, the rotational angle increases toward the apex. A study of the details of bundle structure showed that the axis of sensitivity of the bundles is also rotated (Pickles et al. 1989). Bundle rotation has also been observed in the most primitive bird ears studied, those of *D. novaehollandiae* (Köppl et al. 1998) and the nandu (*Rhea americana*; Jørgensen and Christensen 1989).

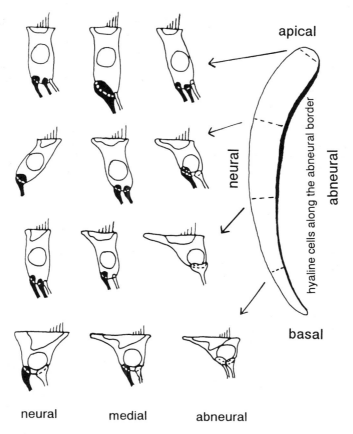

FIGURE 3.5. Schematic drawing to illustrate the shape of hair cells and the innervation pattern for different positions along and across the avian papilla. In this case, the data are from *Gallus gallus domesticus* (kindly provided by F.P. Fischer). On the right is shown a schematic outline of the papilla, with the apex at the top and the neural side to the left. The location of hyaline cells along the abneural edge is shown as a black strip. For each of four cross-sectional positions, three hair cells are shown together with their innervating fibers, one hair cell each from the neural third (leftmost column), the middle third (middle column), and the abneural third of the hair cell field (right column of hair cells). Afferent fibers are shown filled in black, efferent fibers are shown as open outlines.

The exact position of this area of rotated hair cell bundles varies with species and appears to be correlated with the hearing range. Thus in species whose hearing range is biased toward higher frequencies (e.g., *S. canarius* and *T. guttata*; Gleich et al. 1994), the most strongly rotated bundles (here only 40° rotation) are found at the extreme apical end of the papilla. In *S. vulgaris* and *C. livia*, the strongest rotations (70°–90°) are found at a posi-

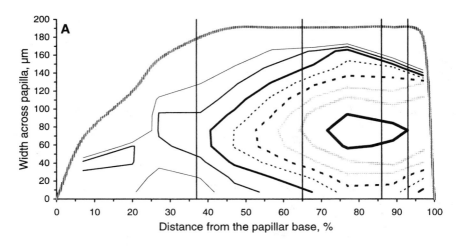

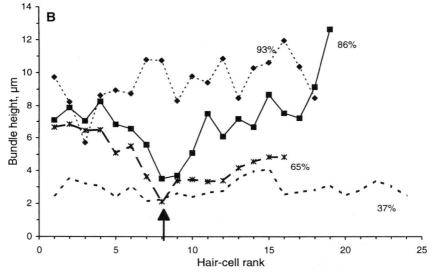

FIGURE 3.6. Diagram illustrating the pattern of rotation of hair bundles and the correlation with bundle height in the papilla of *Sturnus vulgaris*. (**A**) Schematic drawing of iso-orientation contours on the papilla of *S. vulgaris* as a function of position along and across the papilla. The neural edge of the papilla is considered to be straight and forms the abscissa. The thick gray line shows the width of the papilla (i.e., the distance to the abneural edge) as a function of the position along the papilla's length from the basal end (i.e., the apex is to the right). The various other lines are the bundle iso-orientation contours drawn in steps of 10°, from 0° (thin, continuous line near the abneural edge) to 70° (thickest, continuous line). Thus the largest rotations are found for hair cell bundles in the middle of the papilla between about 73% and 93% from the base. The four vertical lines show the position along the papilla of the data series shown in the bottom diagram. (**B**) Stereovillar bundle height in μm as a function of the position of the hair cell across the papilla, for four cross sections of the papilla at 37%, 65%, 86%, and 93% of the distance from the base of the papilla. The cross sections at 65% and 86% show changes in bundle height (thick arrow on the abscissa) correlated with the orientation changes shown in the top diagram.

tion centered 20% from the apex (Gleich and Manley 1988). In *D. novaehollandiae*, which has very good low-frequency hearing, they are nearer 30% from the apex (Köppl et al. 1998). In contrast to other bird species, hair bundle rotation in *T. alba* does not change gradually across the papilla, but suddenly (Fischer et al. 1988). In this species, the sudden change in hair cell orientation from neural to abneural is accompanied by a sharp change in the number of stereovilli per hair cell bundle.

In *C. crocodilus*, there is no detailed information about hair cell bundle rotation. Although Leake (1977) reported that "all hair cells have the same polarization, with the kinocilium situated laterally on the cell surface," the intent of this statement was to contrast the situation in *C. crocodilus* to that of lizards, where hair cell groups exist that show 180° opposite bundle orientation (Manley, Chapter 4). Scrutiny of the figures in Leake (1977) reveals that hair cell bundles do show some rotation, but this question needs to be studied in detail before a comparison to birds can be made.

2.2.3. The Shape of the Bundle, Stereovillar Numbers, and Height

The hair cell bundles in birds are each covered and surrounded by tectorial membrane, which forms a connection to the tallest stereovilli but otherwise leaves a dome-shaped cavity around the bundle (Smith 1985; Runhaar 1989). The tectorial membrane attaches to supporting-cell microvilli all around the bundle. Thus in birds, there is essentially no fluid space beneath the tectorial membrane, and neighboring hair cell bundles will be strongly coupled through the tectorial membrane.

The number of stereovilli per hair cell bundle typically rises from near 50 at the apical end to 200 or more at the basal end. Parallel to this increase, the shape of the bundle as seen from above is generally rounded in apical hair cells, and, since the stereovilli are also relatively tall, has the general appearance of a shaving brush. In the base of the papilla, the bundles are elongated and short and look more like the bristles of a toothbrush (Fig. 3.3). The actual bundle shape is, however, likely to be more determined by the cell's response frequency than the absolute position along the papilla. In high-frequency species such as *S. canarius* and *T. guttata*, even bundles in the apex are elongated (Gleich et al. 1994). Depending on the species and the animal's age, a true kinocilium may be present in the center of the abneural edge of the bundle; if it is absent, a basal body is still visible in sectioned material.

The height of the hair cell stereovillar bundles varies systematically along the papilla, but the height at any given absolute position varies between species. In the budgerigar (or parakeet, *Melopsittacus undulatus*), for example, the maximal bundle height at the low-frequency, apical end of the papilla is only about 4 μm (Manley et al. 1993), compared to more than 12 μm at the equivalent position in the papilla of *C. livia*, which is an acknowledged low-frequency specialist (Gleich and Manley 1988). The height of the tallest stereovilli in any given bundle (fixed, embedded) varies in *S. vulgaris* from about 2.7μm basally to 9.4μm apically and in *C. livia* from 4.0μm to 12.7μm (Gleich and Manley 1988). The increase in height is not necessarily monotonic, and in *S. vulgaris*, *C. livia*, and *T. alba* it is much faster in the apical third (Fischer et al. 1988; Gleich and Manley 1988). Within the specialized auditory-foveal region of *T. alba* (see Section 3.4.1.2), the bundle height hardly changes, being about 1.4μm along the entire basal half of the papilla (Fischer et al. 1988; Fischer 1994a).

There can be a consistent difference between the height of the neural and abneural bundles in each transect across the papilla. In *D. novaehollandiae*, *C. livia*, and *S. vulgaris*, hair cell bundles tend to be shortest on medially lying cells (Gleich and Manley 1988; Köppl et al. 1998). This trend is very strong in the apical third of the papilla, in the region where the bundles are rotated. At 0.4mm from the apex in *S. vulgaris*, a transect of the papilla shows that neurally, bundles are 7–8μm tall. In the region of maximal bundle rotation they are, however, only 3.5μm tall. Their height increases again toward the abneural side, where they are even taller (10μm) than on neural hair cells (Fig. 3.6; Gleich and Manley 1988). This tendency for the bundles of medial hair cells to be shorter (and therefore stiffer) is thus strongest exactly in the region of bundle rotation, being most obvious in this species between 0.4 and 1mm from the apex. It is not seen at 0.2mm from the apex, and has disappeared by 1.4mm, or 45% of the papillar length from the apex, where the maximal bundle rotation in the medial region of the papilla has reduced to about 25°. A very similar pattern of stereovillar bundle height is also seen in *D. novaehollandiae* (Köppl et al. 1998). It thus seems likely that the function of this very large reduction in bundle height is to stiffen the hair cell bundles in the center of the papilla (all other things being equal, a bundle of half the height is four times stiffer). Similar tendencies were also seen in *C. livia* (Gleich and Manley 1988) and in *M. undulatus* (Manley et al. 1993).

In *R. americana*, Jørgensen and Christensen (1989) reported that the bundles of the tall hair cells were, for the same position along the length of the papilla, generally taller than those of the short hair cells. At the apex, the neural bundles were 8.7μm high, the abneural ones only 7.3μm. Near the base, the relevant values were 4.3μm and 3.1μm. In *C. crocodilus*, the bundles of the tall hair cells were also taller than those of the short hair cells in any cross section of the papilla, but no details are available (Leake 1977). Data from these species suggest that differences in stereovillar bundle height *across* the epithelium are an evolutionarily old feature.

In the most primitive bird studied in detail, *D. novaehollandiae*, the numbers of stereovilli on neurally lying and abneurally lying hair cells are similar, with the exception of the most basal 20% of the papilla. Here, the neurally lying cells have 10–30% more stereovilli (Köppl et al. 1998). In *G. gallus domesticus*, *C. livia*, *S. vulgaris*, and *M. undulatus*, neurally lying hair cells of the basal 40–60% of the papilla have up to 60% higher numbers of stereovilli than abneurally lying cells. When plotted as a function of the frequency map of each species' papilla, it can be seen that the largest differences in stereovillar numbers between neural and abneural hair cells tend to be found near the highest response frequencies (Fig. 3.7). In the apical region of the papilla, in contrast, the numbers of stereovilli in hair

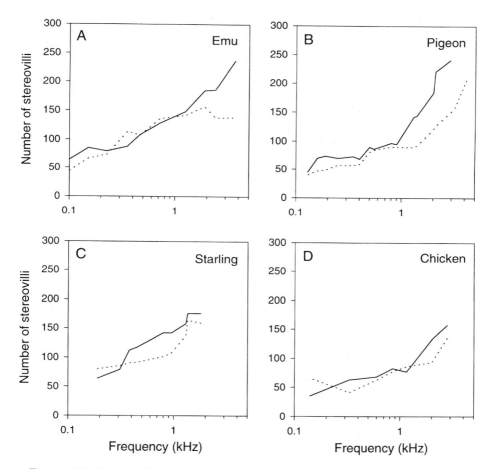

FIGURE 3.7. Diagram illustrating the changes in the number of stereovilli per hair cell bundle as a function of the response frequency of the hair cell region from which the anatomical data were collected, in four species of bird. For each species, data from neural cells are shown as a continuous line, those for abneural cells as a broken line. Whereas in *Dromaius novaehollandiae*, the number of stereovilli per hair cell bundle only differs for very high-frequency cells, the differences in *Columba livia* are clear above 1 kHz. In both *Gallus gallus domesticus* and *Sturnus vulgaris*, the differences are either absent or smaller. Larger differences in stereovillar numbers are known from other species for which there are no frequency maps available (e.g., *Serinus canarius* and *Taeniopygia guttata*).

cells across the papilla hardly differ. In the tufted duck (*A. fuligula*), the difference in stereovillar numbers extends along 80% of the papilla toward the apex. The extreme cases are found in *S. canarius* and *T. guttata*, where neurally lying hair cells along the entire length of the papilla have up to twice as many stereovilli than abneurally lying cells. An increase in stereovillar numbers would, of course, also increase bundle stiffness, but not as radically as a change in bundle height (the relationship between bundle stiffness and stereovillar numbers is approximately linear, but stiffness changes with the square of the bundle height). An increase in stereovillar numbers would also increase the number of transduction channels, and thus perhaps sensitivity.

The papilla of *T. alba* does not show great differences in stereovillar bundle height in any given hair cell region (Fischer 1994a), but there are variations (up to 100%) in the *number* of stereovilli in each bundle. These changes occur at the positions where the bundles change their orientation (see Section 2.2.2). The above data thus suggest that there can be substantial and systematic changes in the stiffness of hair cell bundles across the avian papilla, especially in the apical half.

2.3. General Patterns in Hair Cell Innervation

The afferent component of the cochlear nerve contains mostly myelinated fibers. In a study of the cochlear ganglion in *G. gallus domesticus*, Fischer et al. (1994) demonstrated that only about 3% of the afferent fibers are unmyelinated. In that study, the proportion of unmyelinated lagenar macula afferents (17.5%) and of unmyelinated efferent fibers (57%) was very much higher. Cochlear afferent fibers had diameters between 0.5 and 3.0 μm. In *T. alba*, Köppl (1997b) found over 31,000 afferent fibers to the basilar papilla, virtually all of them being myelinated. Their diameter was on average 2.75 μm, but the diameter was frequency-dependent. Whereas low-frequency fibers only had diameters of about 2 μm, the average diameter rose to between 3 and 5.5 μm in the 5.4- to 7-kHz region of the nerve. At the basal, highest-frequency region, the diameter was again smaller (2.0 μm; Köppl 1997b).

The ratio of hair cells to nerve fibers falls from about 1.7 in *D. novae-hollandiae* (Köppl and Wegscheider 1998) to 0.5 in *S. canarius* (Gleich et al. 1998). In *T. alba* it is also 0.5, although the number of hair cells is very high (Köppl 1997b). In *G. gallus domesticus* and *S. vulgaris* this ratio is 0.73 and 0.58, respectively (Köppl and Wegscheider 1998). Thus advanced species show a denser afferent innervation, especially on neural hair cells. In the auditory fovea of *T. alba*, the most neural hair cell is the only cell that is relatively tall (Köppl 1993) and these cells each receive up to 20 afferent fibers, in contrast to typically one to two afferents in other species (Fischer 1994a, 1994b). In *C. livia*, THC at a position about 20% from the neural edge receive the most dense afferent innervation (Smolders et al.

1995), a conclusion based on the frequency of contacts from stained, physiologically characterized afferents.

As noted briefly above, the innervation of hair cell types differs (Takasaka and Smith 1971; Chandler 1984; von Düring et al. 1985; Smith 1985; Fischer et al. 1992). In general, THC have large afferent and small efferent contacts (Fischer 1992), but these differences reverse toward the abneural side of the papilla. According to the new definition (Fischer 1994b; see also Section 2.2.1), SHC only have large efferent contacts. In a study of serial ultrastructural sections in one apical area of the papilla in *S. vulgaris*, Fischer et al. (1992) found that neurally lying hair cells received up to three afferents. More medially on the papilla, each hair cell was contacted by one to two afferents. The efferent innervation of neural and medial hair cells in the apical area was weak. Of three SHC, all synapsed only with one to two large efferent fibers. In more basal regions, abneurally lying hair cells have much larger efferent synaptic areas than do THC (Fig. 3.5).

2.3.1. Afferent Innervation

The numbers of afferent fibers contacting hair cells correlates well with the total afferent synaptic contact area (Fischer 1992, 1994b). Afferent fibers tend to innervate exclusively one hair cell; fibers that innervate more than two hair cells were rarely seen in ultrastructural material (Fischer 1992). Similarly, branching was not common in stained, physiologically characterized afferents, and fibers innervating more than one hair cell tended to be found only in the abneural-apical cochlear region (Köppl and Manley 1997; Gleich 1989; Smolders et al. 1995).

The densest synaptic innervation (as indicated by the number and area of the synaptic contacts) is found on THC that lie at a short distance from the neural edge (Fischer 1992, 1994b). There is some suggestion that the innervation density correlates with the "importance" of different frequencies, for the position of the largest contact areas of the afferent terminals is related to the most sensitive hearing range in different species (Fischer 1994b). In *T. alba*, for example, the afferent terminals of hair cells in the foveal region occupy much greater areas of the hair cells than in the papillar apex. In this specialized region, the most neural hair cell receives up to 20 afferent fibers, whose diameter at the level of the body of the auditory nerve (3–5.5 μm) is large compared to those of afferent fibers from other regions (Köppl 1997b).

2.3.2. Efferent Innervation

Efferent neurons, that enable the central nervous system to modulate the sensory organ, also exist in birds, but as yet we have only little knowledge of their functional significance. As the afferent and efferent innervation of the principal hair cell groups in birds parallels that of mammals (Manley et al. 1989), the avian hearing organ is a promising object for studying the role of the efferent system.

2.3.2.1. Efferents in the Basilar Papilla

Efferent fibers in birds only form axosomatic contacts, axodendritic contacts such as those on the afferents below mammalian inner hair cells are absent or extremely rare (Takasaka and Smith 1971; Firbas and Müller 1983; Chandler 1984; Fischer 1992). In contrast to mammals, with their separate populations of the lateral and medial olivo-cochlear efferent systems (Warr 1992), there is a gradual change in the size and density of efferent innervation across the avian basilar papilla. The neurally lying tall hair cells are innervated by thin axosomatic contacts from efferent fibers, whereas the abneurally lying short hair cells receive large efferent synapses (Fischer 1992, 1994b, 1998; Fischer et al. 1992). Thus the functional specialization of the abneural part of the sensory epithelium with a small or no afferent and a strong efferent component is very pronounced in birds. The function of SHC is thus restricted to the hearing organ itself and is under the control of the efferent system. In contrast to the mammalian efferent system (Guinan 1996), only one efferent transmitter—acetylcholine—has been identified in birds (Takasaka and Smith 1971; Firbas and Müller 1983; Cohen 1987), although only about 70% of the cochlear efferent neurons are cholinergic (Code and Carr 1994). Thus, it seems that the modulatory activity of the efferent system is predominantly mediated by the same substance that is the main transmitter in the mammalian efferent system. The type of acetylcholine receptor on the abneurally lying hair cells in birds and mammals also seems to be very similar. As measured in isolated outer and short hair cells, it is a novel type that has both nicotinic and muscarinic properties and shows responses to some atypical cholinergic antagonists such as strychnine and bicucculine (Fuchs and Murrow 1992a, 1992b). Activation by acetylcholine triggers a small inward current (mainly calcium) that opens calcium-activated potassium channels. The resulting potassium outward current leads to a hyperpolarization of the hair cell within 50–100 msec. There are no studies of THC in this regard.

In addition to the sensory hair cells, abneurally located and nonsensory hyaline cells in the basilar papilla of birds and *Caiman* are contacted by a network of efferent fibers (Drenckhahn et al. 1991; Keppler et al. 1994; Ofsie and Cotanche 1996). Since hyaline cells contain contractile proteins (Drenckhahn et al. 1991; Cotanche et al. 1992) their efferent innervation may be involved in regulating the mechanical properties of the basilar papilla.

2.3.2.2. Efferents in the Brain Stem

In mammals, at least two systems of efferent fibers originate in the ventral auditory brain stem from the general regions of the lateral and the medial superior olive. In birds, which possess only a unitary superior olive, this clear separation is not present. In *G. gallus* several authors have identified the origin of cochlear and lagenar efferents in two overlapping cell groups medial to the dorsal and ventral part of the nucleus facialis (Schwarz et al. 1981; Whitehead and Morest 1981; Strutz and Schmidt 1982; Cole and Gummer 1990; Schwarz et al. 1992; Kaiser 1993; Kaiser and Manley 1994). In general, a total of 300–400 retrogradely labeled efferents were counted for one ear. The fibers and cell bodies of neurons innervating the lagenar macula and the basilar papilla, however, are in such close proximity in the periphery that tracer injections into the hearing organ always result in labeling of efferent cell bodies from both sense organs. Vestibular efferents traced from the lagenar macula alone (Code 1995; Kaiser and Manley 1996) belonged to the dorsal cell group. Whether the papillar efferents stem from both cell groups or whether their site of origin is limited to the ventral cell group is still unclear. In *C. crocodilus* the brain stem location of efferent cell bodies closely resembles the distribution in *G. gallus domesticus* (Strutz 1981).

2.4. Trends in Papillar Structure Discernible in More Modern Groups

In their general anatomical features, bird groups show different levels of development and specialization that are correlated with their evolutionary status. The primitive ratites are followed by the water birds such as ducks, then primitive land bird groups such as chickens and pigeons, and finally advanced land birds such as song birds and owls (Feduccia 1980; Carroll 1988). The comparative data on hearing-organ structure in the discussion below are from the following papers: *D. novaehollandiae* (Fischer 1994b, 1998; Köppl et al. 1998), *R. americana* (Jørgensen and Christensen 1989), *C. livia* (Takasaka and Smith 1971; Gleich and Manley 1988), *G. gallus domesticus* (Tanaka and Smith 1978; Tilney and Saunders 1983; Tilney et al. 1987; Brix et al. 1994; Manley et al. 1996), *T. alba* (Smith et al. 1985; Fischer et al. 1988; Köppl 1993; Köppl et al. 1993; Fischer 1994a, 1994b), domestic duck *Anas platyrhynchos* (Chandler 1984), tufted duck *Aythya fuligula* (Manley et al. 1996), seagulls *Larus argentatus* and *Larus marinus* (Counter and Tsao 1986), *M. undulatus* (Manley et al. 1993), and several species of song birds, *S. vulgaris* (Gleich and Manley 1988; Fischer et al. 1992), *S. canarius* and *T. guttata* (Gleich et al. 1994).

The papillar morphology of every species is unique and in some species shows obvious anatomical specializations. *C. livia* and *G. gallus domesticus*, for example, show (different) specializations for low-frequency reception. *S. canarius* and *T. guttata*, in contrast, have an excellent high-frequency sensitivity at the expense of the low frequencies. The most specialized bird studied to date is *T. alba*, showing an auditory fovea at high frequencies. It should be noted that the "typical" features of advanced avian papillae were not all achieved simultaneously but developed to some extent independently of each other. Many of them are also correlated with the increased emphasis on sensitive high-frequency hearing in more advanced birds. The trends observed can be regarded as continuations of the following evolutionary developments whose beginnings are already visible in primitive species such as *D. novaehollandiae.*

- The hair cell height decreases in more advanced birds in parallel with an increase of the high-frequency hearing range and of the proportion of SHC.
- The area of true SHC, without afferent innervation, increases in more advanced birds, reaching 60% across the width in the base of the papilla in *T. alba.* Whereas in *D. novaehollandiae*, most hair cells are afferently innervated, in the basal half of the papilla in more advanced birds, only hair cells over 40–80% of the width of the papilla from the neural side are afferently innervated, the proportion depending on the species and the actual papillar position. During evolution, the population of hair cells whose function is confined to the papilla itself—the SHC—became larger, and this was correlated with the acquisition of responses to higher frequencies.
- The pattern of hair cell bundle rotation seen in primitive papillae is maintained and shifts with the known or presumed frequency map of the cochlea.
- In general, the total number of hair cells (see Section 2.1 for examples) is reduced in parallel with a decrease of the papillar length (except in *T. alba*) and an increase in the number of hair cells with a large surface area. Body size limits the papillar dimensions at least in small song birds.
- The number of afferent nerve fibers is not reduced proportionately to the number of hair cells, resulting in an increase of the innervation density from *D. novaehollandiae* to more advanced species, especially on neural hair cells (see Section 2.3).
- The differences in hair cell innervation are accompanied by changes in the ionic channels of the hair cell types. Murrow and Fuchs' (1990) study of membrane channels showed changes in the size of the inactivating potassium current in hair cells of *G. gallus domesticus* that correlated very well with the synaptic area occupied by the efferent fibers on hair cells from the same papillar position (Fischer 1992; Manley 1995).
- The hair cell bundles specialize in terms of the number and height of stereovilli (see Section 2.2.3). The micromechanics of the papilla's response to sound may thus have become more complex compared to the situation in primitive birds.

2.5. Structural Specializations in Individual Species

There are a large number of structural similarities between the different avian papillae. Here, we shall only refer to data that show significant deviations from the basic pattern. Possible functional correlates of these specializations are treated in Section 3.5.3.

1. In *C. livia*, the basilar papilla does not increase steadily in width from basal to apical but disproportionately widens in the apical third. This widening is accompanied by an increase in hair cell density and large increases of the cross-sectional dimensions of the basilar and tectorial membranes (Gleich and Manley 1988).
2. In *G. gallus domesticus*, the anatomy of the most apical region of the papilla differs from that seen in other birds. Lavigne-Rebillard et al. (1985) describe a crescent-shaped, apical region they termed the "very distal part," in which the hair cells show more resemblance to vestibular than to auditory receptors. The hair cell bundle rotation in this area is reversed compared to the rest of the apex (Brix et al. 1994).
3. In *S. canarius* and the *T. guttata*, with their very short basilar papillae, typical "basal" high-frequency features are more pronounced than typical "apical" low-frequency features (Gleich et al. 1994). Not only is the maximum width and number of hair cells across the papilla smaller than in other birds, the discrepancy in stereovillar number between neural and abneural hair cells is very pronounced. Apical hair cells have a rather large number of stereovilli and the maximum rotation of the stereovillar bundles is reached at the extreme apical end.
4. The basilar papilla of *T. alba* is the longest so far described in birds (Fig. 3.2). The apical half of the papilla in *T. alba* shows structural patterns that resemble those of the entire papillae of other birds. The basal half of the papilla in *T. alba* is, however, exceptional (Smith et al. 1985; Fischer et al. 1988; Köppl 1993). Smith et al. (1985) found that in *T. alba*, the basal end, where the height of the stereovilli is nearly constant ($1.4\,\mu$m; Fischer et al. 1988), has a marked thickening of the basilar membrane, a feature that has been found in specialized areas of some mammalian (e.g., bat) cochleae. As the basal area also differs in some other respects from that of other birds (Fischer et al. 1988), it can be regarded as a specialization for high-frequency hearing.

3. Function of the Basilar Papilla

The detection and analysis of airborne sound was an important factor during avian evolution and the hearing of birds specialized for several purposes including communication, detection of prey, or even echolocation. The inner ear transforms the incoming acoustic signal into trains of action potentials in auditory nerve fibers. These spike trains are the information that is available for subsequent processing in the brain. Only those features of the sound that are transformed by the inner ear and coded in the neural activity can be perceived and used by the bird. Physically, sounds can be described in spectral, temporal, and intensity terms and we will address the capability of the avian auditory system to extract and code these features.

3.1. The Avian Basilar Papilla: Mechanisms of Frequency Selectivity

It has long been suggested that there is more than one mechanism of frequency analysis in the vertebrate inner ear (Klinke 1979; Manley 1979, 1986). More recently, fundamentally different mechanisms of frequency selectivity, which often coexist, have been recognized in terrestrial vertebrates (Manley 1986, 1990). Although it is not yet possible in any individual case to cleanly separate the different mechanisms (which are not mutually exclusive), for descriptive purposes they will be treated separately. The two fundamental mechanisms of frequency-selectivity are described in the next two sections.

3.1.1. Frequency Selectivity Through Electrical Characteristics of the Hair Cell Membrane

Electrical tuning is a primordial property of hair cells (Crawford and Fettiplace 1981; Hudspeth 1986; Manley 1986, 1990). A voltage-sensitive Ca^{2+} conductance and a Ca^{2+}-sensitive K^+ conductance in hair cells can, under stimulation, produce a resonance whose properties can explain a good deal of the frequency selectivity of hair cells. In the frog sacculus and the turtle basilar papilla, the variations in the number and the kinetics of these channels are thought to be responsible for the different best, or characteristic frequencies (CFs) of hair cells (Crawford and Fettiplace 1981; Ashmore and Attwell 1985). Such channels have also been described in low-CF hair cells of *G. gallus domesticus* (Fuchs et al. 1988).

Three other factors, (1) the presence of preferred intervals in the spontaneous activity of primary auditory nerve fibers in several species (see Section 3.3.1); (2) the finding of specific deviations from expected phase responses (Gleich 1987); and (3) a strong temperature sensitivity in the frequency tuning (Schermuly and Klinke 1985), together suggest that birds have retained electrical tuning in their hair cells, at least at low CF (for a more exhaustive review see Manley and Gleich 1992). Thus a major part of the tuning selectivity of low-CF avian THC seems to reside in the properties of individual cell membranes of the hair cells. Klinke and Smolders (1993) regard electrical tuning as an "adequate candidate" for "the active contribution to frequency tuning in the avian ear."

The effect of electrical tuning will diminish toward higher frequencies, making it important to take into account the frequency range in question when comparing tuning mechanisms. We do not yet know how high in frequency this mechanism can operate at the body temperatures of birds, but Wu et al. (1995) have shown that using reasonable assumptions with regard to the temperature dependence of the channels involved, the upper frequency limit of electrical tuning might extend to above 4 kHz. The highest limits found for indicators of electrical tuning in preferred intervals in the spontaneous activity of primary auditory nerve fibers in birds are between 1.5 kHz in *S. vulgaris* (Manley et al. 1985), 2.5 kHz in *C. livia* (Temchin 1988), and 5 kHz in *T. alba* (Köppl 1997a). Above the limit of function of electrical tuning, it would be necessary for higher CFs to be analyzed using other mechanisms, presumably predominantly macro- and micromechanical in nature.

3.1.2. Frequency Selectivity Resulting from Micro Mechanical Factors

A variety of structures can interact in complex ways to produce mechanical tuning, but it is not yet possible to quantify the contribution of individual parameters. In archosaurs, these could be, for example, the structural characteristics of the hair cell stereovillar bundles (e.g., number and height of stereovilli), the mass of the tectorial membrane, and active movement processes in hair cells. Many structures show strong morphological gradients in all birds, and especially the stereovillar bundles can be expected to play an important role in the frequency response. In archosaur hearing organs, the hair cells are rather firmly connected to the tectorial membrane, making it likely that large numbers of hair cells and their area of tectorial membrane would form some sort of coupled resonant unit. A recent model of tuning in a lizard, when applied to birds, suggested that these assumptions are reasonable (Authier and Manley 1995). That *non-monotonic* changes in the various single morphological gradients along the basilar papilla can together result in a monotonic frequency map has been demonstrated for the bobtail lizard (*Tiliqua rugosa*; see Manley, Chapter 4).

That hair cells can generate active motions under certain conditions results in the generation of otoacoustic emissions, that is, sounds that emerge from the cochlea and are sometimes present spontaneously (see Section 3.7). The presence of these emissions in birds suggests that active motion of hair cells or their bundles also play a role in mechanical tuning in these species.

3.2. Hair Cell Physiology

3.2.1. In Vivo Hair Cell Responses

The only report of successful in vivo recording of avian hair cell responses is that of Patuzzi and Bull (1991) from abneural, basal hair cells (mainly SHC) of the cochlea in *G. gallus domesticus*. Here, SHC showed a frequency selective alternating current (AC) response upon stimulation with sound. The most sensitive frequencies of these hair cells were between 0.6 and 2.0 kHz, consistent with a relatively basal recording site within the basilar papilla (see frequency maps below). The frequency selectivity appeared to lie below that of avian auditory nerve fibers (that innervate THC) in a similar frequency range and corresponded to data obtained by mechanical measurements of the basilar membrane in *C. livia* (Gummer et al. 1987).

3.2.2. In Vitro Hair Cell Responses

A number of studies have been performed on isolated hair cells from the basilar papilla of *G. gallus domesticus*. An analysis of isolated tall hair cells was carried out under tight-seal, whole-cell recording conditions by Fuchs et al. (1988). Differential distributions of voltage-gated calcium channels, calcium-activated potassium channels, and additional potassium channels along (Fuchs and Evans 1990; Fuchs et al. 1990) and across (Murrow and Fuchs 1990; Murrow 1994) the basilar papilla were observed. These differences in ionic channels of individual hair cells probably contribute to the functional gradients present in the avian basilar papilla (Manley 1995).

The mechanical properties of isolated hair cells from the basilar papilla of *G. gallus domesticus* gave no evidence for frequency-selective behavior upon mechanical stimulation up to 1kHz with a water jet stimulator (Brix and Manley 1994). Only apparently damaged bundles showed broadly tuned, low-frequency resonances. A possible explanation for the lack of a detectable micromechanical frequency selectivity in isolated hair cells is the disruption of coupling and mass load of the tectorial membrane to the stereovillar bundles. Isolated hair cells also displayed bundle movements or shape changes upon current injection (Brix and Manley 1994). However, due to the relatively small size of avian hair cells (compared to typical mammalian hair cells), it was not possible to investigate whether hair cells from *G. gallus domesticus* are capable of fast movements at auditory frequencies.

3.3. Afferent Nerve Fiber Responses

Single-fiber physiological data from the auditory nerve are available for *D. novaehollandiae* (Manley et al. 1997), *C. livia* (Smolders et al. 1995), redwing blackbird (*Agelaius phoeniceus*; Sachs et al. 1980), *G. gallus domesticus* (Manley et al. 1991), *S. vulgaris* (Manley et al. 1985), *T. alba* (Köppl 1997a, 1997c), and *C. crocodilus* (Klinke and Pause 1980). Data from cochlear efferents in birds are only available for *G. gallus domesticus* (Kaiser and Manley 1994).

3.3.1. Spontaneous Afferent Nerve Fiber Activity in Birds

The vast majority of auditory nerve fibers in birds produces action potentials in the absence of external acoustic stimulation, and the rates of spontaneous activity have a monomodal distribution. In contrast to mammalian data, bird data do not show a significant population of fibers with spontaneous discharge rates below five spikes per second; neither do they show a notch in the distribution between 15 and 20 spikes per second. However, the upper limit of spontaneous rates and the mean spontaneous rates vary considerably between studies, even within one species. This is best documented for *C. livia*, where the mean values of spontaneous activity in different studies vary by almost 300%, ranging from 35/sec to 91/sec (e.g., Hill et al. 1989a; Smolders et al. 1995). Differences in the anesthetic regimes do not obviously account for the variable outcome of these studies and maturational changes may only be a possible explanation in cases where animals of different ages are involved (Richter et al. 1996; Manley et al. 1997). Overall, these rate variations across studies have yet to be adequately explained and it is not clear whether differences in spontaneous discharge rates between bird species really exist.

Spontaneous activity that results from stochastic release of transmitter packets at the afferent synapse should result in a discharge pattern in which the intervals between action potentials show a Poisson-like distribution, as generally seen in mammalian primary afferents. Avian auditory nerve fibers often show significant deviations from this pattern. There are more short intervals than expected from a Poisson distribution (Gummer 1991b) and many cells produce an unexpectedly large proportion of interspike intervals whose period is related to their most sensitive frequency (Fig. 3.8). The latter results in peaks (preferred intervals) and valleys in inter spike interval histograms. The presence of avian preferred intervals was first described in *S. vulgaris* (Manley 1979; Manley and Gleich 1984; Manley et al. 1985) and subsequently observed in *C. livia* (Temchin 1982, 1988; Gummer 1991a; Klinke et al. 1994), *G. gallus domesticus* (Manley et al. 1991; Salvi et al. 1992) and *T. alba* (Köppl 1997a).

We have suggested that these preferred intervals reflect spontaneous oscillations of the hair cell potential related to an electrical hair cell tuning mechanism (Manley 1979; Manley and Gleich 1984; Manley et al. 1985). Ruggero (1973) in a mammal and Klinke et al. (1994) for a bird showed that externally added acoustic noise near the thresholds of single fibers can induce evoked preferred intervals (in this case phase-locking to the preferred frequency component of the noise), that are related to the CF of the respective neuron. Klinke et al. (1994) thus argued that preferred intervals in *C. livia* are traceable to external or internal animal noise. In agreement with this, some studies of spontaneous activity that show a predominance of preferred intervals also show very high mean rates of spontaneous activity that may be artifactual (Salvi et al. 1992). However, in other studies, clipping the columella or obstructing the external ear canal did not affect the characteristics of the spontaneous discharges, although these procedures reduced the sensitivity to acoustic stimulation by 40 to 50 dB (Sachs et al. 1980; Temchin 1988). An acoustic drive could also be derived from normal body noises such as breathing and heart beat (Lewis and Henry 1992; Manley, Köppl, and Yates 1997; Richter et al. 1996). However, Crawford and

Tabelle1

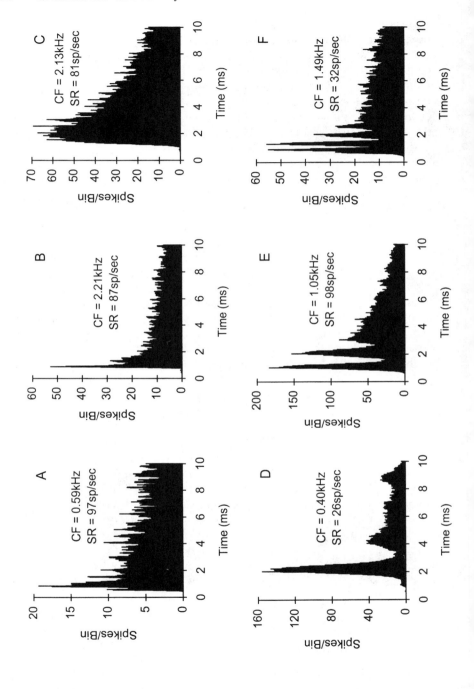

Fettiplace (1980) found significant spontaneous activity and preferred intervals in the isolated half-head preparation of the red-eared turtle. Body noise might explain preferred intervals only in the most sensitive units, but the thresholds of cells with spontaneous preferred intervals varied over a 30-dB range in *C. livia* (Klinke et al. 1994) and we found preferred intervals in auditory nerve fibers of *S. vulgaris* that had thresholds as high as 50–80 dB SPL. Köppl (1997a) also observed no correlation between the fiber threshold and the presence of preferred intervals in *T. alba*. In addition, if body noise is inducing preferred intervals in birds, it is difficult to understand why the same thing is not seen in mammals, although body noise does affect the spontaneous discharge in mammals in other ways (Lewis and Henry 1992).

As indicated above, membrane voltage noise generated by the individual hair cell (e.g., ionic channel activity that is not related to mechano-electrical transduction of acoustic input) contributes to spontaneous discharges in primary afferents. In mammals, this hair cell noise is generated beyond the micromechanical cochlear filter stage and is only lowpass filtered by the hair cell membrane. This could explain the lack of preferred intervals in spontaneous activity in mammals, while allowing the possibility that external acoustic stimuli passing through the micromechanical filter complex can induce evoked preferred intervals. In birds, however, spontaneous preferred intervals of auditory afferents indicate that the hair cell noise is subjected to a narrow *bandpass* filter operating at the level of the individual hair cell. Studies on isolated avian tall hair cells have directly confirmed the presence of an electrical tuning mechanism analogous to that found in the turtle (Fuchs, Nagai, and Evans 1988).

3.3.2. Spontaneous Afferent Nerve Fiber Activity in *Caiman crocodilus*

Unlike birds, and despite the very similar cochlear anatomy, spontaneous activity rates in *C. crocodilus* are bimodally distributed, as in mammals, with approximately 30% of the fibers having discharge rates below 20/sec (Klinke and Pause 1980; Smolders and Klinke 1984, 1986). The population of mammal and *C. crocodilus* afferent units that have spontaneous rates near zero does not exist in birds. The origin of this discrepancy between the different archosaur groups is not yet understood. The highest spontaneous

---

FIGURE 3.8. Time-interval histograms of spontaneous activity from six auditory nerve fibers in *Sturnus vulgaris*. All histograms show the predominance of short inter-spike intervals that is typical for auditory afferents. Some cells show no obvious preferred intervals (**A–C**). Other cells show prominent peaks and valleys in the distribution of the inter-spike intervals (**D–F**), where the time between successive peaks or successive valleys correlates with the reciprocal of the acoustic CF of the respective cell. For each cell, the characteristic frequency (CF) and the spontaneous discharge rate (SR) are indicated.

discharge rates in *C. crocodilus* (80/sec; Klinke and Pause 1980) are below the highest found in birds and mammals. The lower body temperature may at least partially explain this difference. No reports about preferred intervals in the spontaneous activity of *C. crocodilus* auditory nerve fibers have been published.

3.3.3. Responses to Simple Tonal Stimuli

Archosaur auditory nerve fibers respond only to a restricted frequency range with a modulation of their discharge rate, that is, they are frequency selective. The most frequent response pattern to an acoustic stimulation is an increase of the discharge rate (Fig. 3.9). Data from *S. vulgaris* and *C. livia* indicate that the maximum discharge rates in birds are on average substantially higher than those of mammals; some of this difference may be due to the higher body temperature of birds (Sachs et al. 1980; Manley et al. 1985). Many avian auditory nerve fibers, however, also respond with a *decrease* of their discharge rate (i.e., to below the spontaneous activity rate) to stimuli of frequencies lower and higher than those of their excitatory response area, a phenomenon known as single-tone, or primary, suppression (Fig. 3.10; Gross and Anderson 1976; Manley et al. 1985, 1991; Temchin 1988; Hill et al. 1989a; Gummer 1991a).

In *C. crocodilus*, maximal discharge rates to tones are lower than in birds or mammals (average 120 spikes/sec at 27°C; Klinke and Pause 1980). Rate-level functions in *C. crocodilus* were mostly nonmonotonic, decreasing strongly at levels above saturation.

Rate-level (I/O)-functions in birds have been studied in some detail (Manley et al. 1985; Richter et al. 1995; Köppl et al. 1997; Fig. 3.10). In *G. gallus domesticus*, the slope of the I/O functions increases with age after hatching—this mostly reflects age-dependent changes in the maximum discharge rate (Manley et al. 1991). Between posthatching days 2 and 21, the I/O function slopes in *G. gallus domesticus* almost double; the maximum rates rise by about 40% and the dynamic ranges in decibels fall by 20%. These changes occur although most other features of the activity of the auditory nerve fibers (e.g., the characteristics of the frequency tuning curves) did not change over the same developmental time period (Manley et al. 1991). In developing *D. novaehollandiae* (Manley et al. 1997) and *C. livia* (Richter et al. 1996) also, tuning curve characteristics are constant during early posthatching development, but the spontaneous rate increases over time.

Three types of I/O-functions (saturating, sloping, and straight, at CF) have recently been characterized in mammals (Winter et al. 1990) and subsequently also in relatively low-CF fibers of adult *C. livia* (Richter et al. 1995). In higher-CF fibers from *D. novaehollandiae* (i.e., out of the range of cycle-to-cycle phase locking), however, Köppl et al. (1997) found that almost all fibers were of the sloping-saturation type (Fig. 3.10). They suggested that the fact that the break-points on the sloping saturation in each

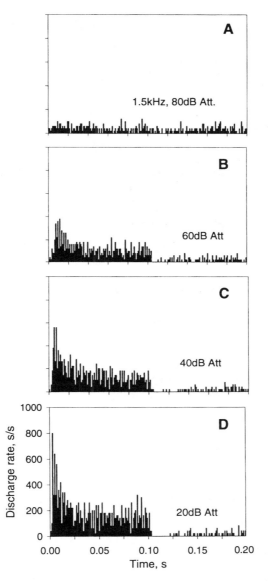

FIGURE 3.9. Peri-stimulus-time-histograms from an auditory-nerve fiber from *Dromaius novaehollandiae* on stimulation with 100 ms CF-tone pips at four different levels. Stimulus frequency was 1.5 kHz and the attenuation of the stimulus is given in each histogram; high numbers indicate soft, low numbers loud stimuli. (**A**) Near threshold for an increase in discharge rate. (**B, C, D**) Sound pressure was increased by 20 dB for each histogram. There is a strong phasic response at stimulus onset, especially at high stimulus levels, followed by a sustained tonic increased discharge rate during the stimulus. At the higher levels, the spontaneous discharge rate is reduced after the end of the stimulus and did not recover over the 200 ms period shown in the histograms.

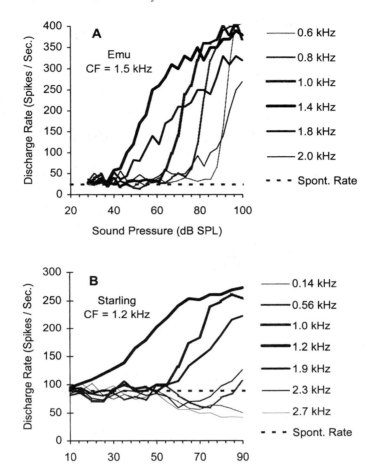

FIGURE 3.10. Input/output functions of avian auditory nerve fibers. (**A**) I/O functions from an auditory afferent of *Dromaius novaehollandiae* measured at six different frequencies, below and above CF. The function near CF (1.4 kHz) shows a very steep slope between 40 and 60 dB SPL and a more gradual increase of discharge rate at higher stimulus levels, typical for the "sloping saturation" type. (**B**) Rate-intensity functions measured at seven different frequencies in an auditory afferent from *Sturnus vulgaris*. Gray lines are below CF, black lines at and above CF. This fiber showed prominent primary suppression below and above CF. At stimulus levels above 60 dB, the discharge rate is below spontaneous rate for the low-frequency (0.14 kHz) and high-frequency (2.7 kHz) stimuli. At slightly higher and lower frequencies (0.56 kHz and 2.3 kHz), there is primary suppression at intermediate stimulus levels, and the fiber only showed an increase of the rate above spontaneous level at very high stimulus intensities (above 80 dB).

fiber correlated very well with the fiber's sensitivity indicates that the cochlear amplifier in birds is local in nature and not global as in mammals (Köppl et al. 1997). This is consistent with the fact that the sharp tuning of neural elements in birds is not reflected in the tuning of the basilar membrane (Gummer et al. 1987), and suggests that the mechanical output of the cochlear amplifier in birds is not fed into the motion of the basilar membrane on a large scale as it is in mammals (see Section 3.5.1).

3.3.4. Suppression of Discharge Rate

The phenomenon of single-tone or primary suppression of the spontaneous rate in a substantial number of primary auditory fibers in birds has already been introduced above. It is much easier to detect in cells with high spontaneous activity and thus statistics on its occurrence have to be treated with care. It was found in approximately 20% of fibers in *S. vulgaris* (Gleich, unpublished data) and approximately 30% in *C. livia* (Temchin 1988; Gummer 1991a; Figs. 3.10 and 3.11). Many data indicate that single-tone suppression is a general phenomenon in nonmammalian species (birds: Manley et al. 1985, 1991; Temchin 1988; Hill et al. 1989a; Gummer 1991a; tokay gecko: Eatock et al. 1981; frog: Sachs 1964; Christensen-Dalsgaard and Jorgensen 1996; fish: Fay 1990) but that it is absent in the cat (Kiang et al. 1965; Sachs and Kiang 1968). However, the presence of single-tone suppression has also been reported for low-frequency auditory nerve fibers in the gerbil (Schmiedt et al. 1980; Henry and Lewis 1992). The origin of single-tone suppression is not well understood. We and others proposed that the firm coupling between the tectorial membrane and the stereovillar bundle may play a crucial role in producing single-tone suppression in nonmammalian species (Manley et al. 1985; Temchin 1988; Gummer 1991a). Alternatively Hill et al. (1989a, 1989b) proposed that integration of bioelectric signals at the primary afferent dendrite may contribute to suppression of spike discharges. In their model, the presence of primary suppression in birds and its absence in most mammals might be related to smaller space constants and the higher number of hair cells across the papilla in the avian cochlea.

A phenomenon present in both birds and mammals is two-tone rate suppression, where the excitation caused by a tone within the excitatory response area is suppressed by a second tone (Sachs and Kiang 1968; Sachs et al. 1974). Systematic investigations in mammals showed that two-tone suppression can be induced in all fibers by using tones below and above CF and that the frequency-intensity suppressive areas show a consistent asymmetry (Sachs and Kiang 1968). In birds, two-tone rate suppression appears more variable among neurons than it is in mammals (Manley et al. 1991; Chen et al. 1996). In *G. gallus domesticus*, 50% of the neurons showed two-tone rate suppression on both sides of the tuning curve, 38% only on the high side, 2% only on the low-frequency side; it was not detectable in 10% of the neurons studied (Chen et al. 1996).

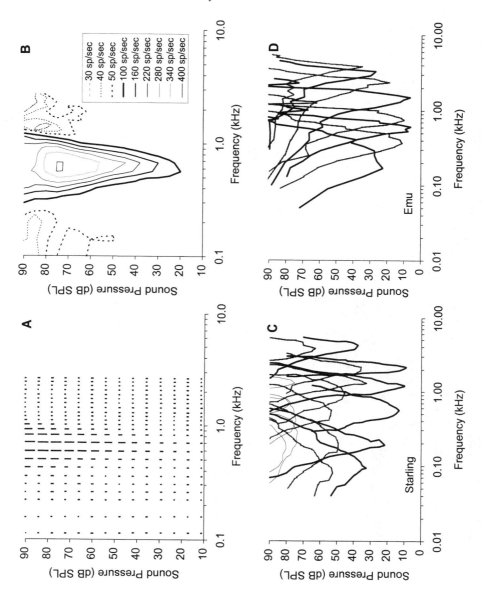

fig.3.11

3.3.5. Frequency Selectivity

The response map of auditory units in the frequency-intensity domain becomes obvious when a matrix of frequencies and stimulus levels is presented and the respective discharge rates are represented by the height of a bar (Fig. 3.11A). The excitatory response area can be seen as an area of elevation of discharge rate. At low stimulus levels, only stimuli within a small frequency range elicit an increase in rate. At higher levels, the width of the response area is greater; nevertheless, the units only respond over a restricted frequency range even at the highest levels tested.

The response characteristics in the frequency-intensity domain can also be illustrated by calculating iso-response contours from such response matrices (Fig. 3.11B). Curves defined using a criterion just above the spontaneous rate and thus close to the response threshold represent the so-called *tuning curve* and are used to characterize the CF (tip of the tuning curve), the threshold at CF and various measures of frequency selectivity (Q_{10dB}, high- and low-frequency slopes) of the respective neuron. The dashed curves in Fig. 3.11 characterize the areas where the neural responses were suppressed below the spontaneous rate by the test tone (i.e., areas of primary suppression).

The examples from *S. vulgaris* (Fig. 3.11C) illustrate that avian excitatory tuning curves are generally V-shaped and on average symmetrical around the CF (on a logarithmic frequency scale; Manley et al. 1985). Several measures related to tuning curves have been quantified in a number of avian species (e.g., *C. livia*: Sachs et al. 1974; *S. vulgaris*: Manley et al. 1985; *G. gallus domesticus*: Manley et al. 1991; *D. novaehollandiae*: Manley et al.

◀ ───

FIGURE 3.11. The response of auditory nerve fibers in the frequency-intensity domain. (**A**) Bar graph illustrating the discharge rate of an auditory nerve fiber from *Sturnus vulgaris* as the height of a bar for a matrix of test stimuli covering a range of frequencies and sound pressure levels. Excitation (increased discharge rate) occurs only over a narrow frequency range, even at high sound pressures. (**B**) Iso-rate contours calculated from the response matrix presented in (**A**). The continuous lines show excitatory iso-rate functions calculated using different rate criteria (see key), while the dotted lines indicate areas where the stimulus reduced the discharge rate below the level of the spontaneous rate (primary suppression). This unit responded near its CF (0.6–0.7 kHz) with a fairly linear increase of the discharge rate for sound pressures between 20 and 60 dB, reaching a maximal rate near 75 dB. At even higher stimulus levels, the rate declined, resulting in the closed iso-rate function. (**C**) Typical examples of near-threshold tuning curves from *S. vulgaris*. The curves are on average symmetrical with a narrow, frequency-selective tip, and there is no indication of a low-frequency tail that is typical for mammalian tuning curves. (**D**) Typical examples of near-threshold tuning curves from *Dromaius novaehollandiae*. Note the more prominent tails on the low-frequency flanks of some tuning curves.

1997; *T. alba*: Köppl 1997a). In general, the tuning curve symmetry changes gradually across the hearing range. In *G. gallus domesticus* and *S. vulgaris* low-CF cells (below 1–1.5 kHz) tend to have steeper low-frequency flanks than high-frequency flanks, and high-CF cells show the reverse behavior (Manley et al. 1985, 1991). In *D. novaehollandiae* this transition appears at lower frequencies, so that the tuning curves show a more pronounced asymmetry (Manley et al. 1997; Fig. 3.11D). In addition, high CF tuning curves in *D. novaehollandiae* show low-frequency tails similar to those found in mammals.

3.3.5.1. Tuning Curve Thresholds

The distribution of neural thresholds at any given CF in birds can exceed 50 dB (Fig. 3.12A). This threshold variability has been described in several studies and in different species (Sachs et al. 1974; Manley et al. 1985; Gummer 1991a; Salvi et al. 1992; Richter et al. 1995; Smolders et al. 1995; Manley et al. 1997). It is present in birds in good physiological condition, as shown, for example, by the fact that high- and low-threshold units of the same CF are often encountered successively in a single recording penetration (Manley et al. 1985; Gummer 1991a). Threshold variation, however, may be reduced at the higher CFs, as has been shown for *D. novaehollandiae* (Manley et al. 1997) and *T. alba* (Köppl 1997a). The wide range of thresholds between individual fibers may contribute to the large dynamic range of the behavioral performance in the detection of intensity differences in birds (Klump and Baur 1990), despite the fact that most individual cells have only a smaller dynamic range. The most sensitive thresholds at a given CF are related to the respective threshold of the behavioral audiogram (Sachs et al. 1974; Manley et al. 1985; Gleich 1994; Köppl 1997a). In most species, the lowest thresholds are found between 1 and 3 kHz and increase toward both lower and higher frequencies (Fig. 3.12A).

3.3.5.2. Frequency Selectivity of Tuning Curves

The frequency selectivity of a unit is quantified using the frequency width of the tuning curve; narrow curves indicate a high selectivity. For comparison of physiological data, the tuning curve width is usually determined 10 dB and/or 40 dB above the respective threshold and a quality factor (Q_{10dB} or Q_{40dB}) is calculated (CF/width$_{xdB}$) and characterizes the frequency selectivity of the unit independently of its CF. Despite a considerable variability of Q_{10dB} at a given CF, the average frequency selectivity clearly increases with the CF (Fig. 3.12B; Sachs et al. 1974; Manley et al. 1995; Salvi et al. 1992; Köppl 1997a).

The frequency selectivity of avian auditory fibers near the tip of the tuning curve (Q_{10dB}) tends on average to be higher than in mammals (Sachs et al. 1974; Manley et al. 1985). This difference in tuning curve width is even

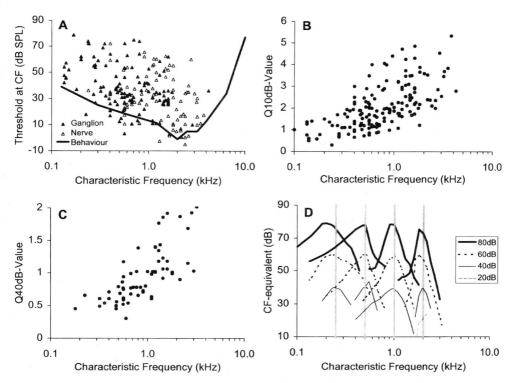

FIGURE 3.12. (**A**) The distribution of threshold at CF for a sample of cochlear ganglion cells (filled triangles) and auditory nerve fibers (open triangles) from *Sturnus vulgaris*. Inaccessibility of high frequencies with the ganglion approach (Manley et al. 1985) and lower incidence of low CFs with the nerve approach (Klump and Gleich 1991) are responsible for the different CF ranges encountered. In the region of overlap, however, there were no obvious differences between both recording sites. The heavy line shows a behavioral audiogram for *S. vulgaris* (Kuhn et al. 1982). (**B**) Q_{10dB} of individual primary neural elements of *S. vulgaris* as a function of the CF. Despite large variability, there is a clear trend for an increase of Q_{10dB} values with increasing CF. (**C**) Q_{40dB} as a function of the CF. The values obtained in *S. vulgaris* are on average almost twice those of guinea pig auditory nerve units in a comparable frequency range (Gleich, unpublished data). (**D**) Excitation patterns derived from the response of a large number of auditory units in the auditory nerve of *S. vulgaris* to test tones at 0.25, 0.5, 1, and 2 kHz (gray vertical lines) at stimulus levels increasing from 20 to 80 dB SPL (Gleich 1994). The examples illustrate that the high-frequency slopes of the excitation patterns are on average steeper on the high-frequency side and that there is no pronounced change in the shape of the excitation pattern comparable to the high-frequency spread of excitation found in mammals at high stimulus levels.

more obvious when comparing the selectivity 40 dB above threshold. Avian tuning curves are on average only half as wide, and thus much more frequency selective, 40 dB above threshold than those of mammals (in a comparable frequency range). This is especially striking in *T. alba*, a bird where a large number of tuning curves of unusually high CF (up to 9 kHz) is available (Köppl 1997a). The poorer frequency selectivity of auditory nerve fibers in mammals is at least partly due to the shallow slopes of the low-frequency flank of their tuning curves. With the exception of higher-frequency afferent fibers of *D. novaehollandiae* (Fig. 3.11D), the typical low-frequency tails of mammalian tuning curves have not been found in birds, but even where they are found, the frequency selectivity remains higher than in mammals (Manley et al. 1997).

The tuning curves in *C. crocodilus* also did not show the low-frequency tails reported for mammals, however, and, in contrast to the situation in most birds, the slopes of the high-frequency tuning curve flanks (30–180 dB/oct) were almost all steeper than the slopes of the low-frequency flanks (Klinke and Pause 1980). In other respects, the data from *C. crocodilus* were similar to those from birds, for example, there was a considerable variation of thresholds (50 dB) at any given CF, and the sharpness of tuning increased with CF and decreased with increasing threshold.

Unlike in mammals, the tuning characteristics of avian auditory nerve fibers are temperature dependent. The tuning curve shifts to lower frequencies with a decrease of body temperature and to higher frequencies with an increase. On average, there is a 0.07 octave shift in CF for a 1°C temperature shift between 36°C and 38°C (Schermuly and Klinke 1985). These temperature-dependent shifts in birds are similar to those found in other nonmammalian species (gecko, *Gekko gecko*: Eatock and Manley 1981; *C. crocodilus*: Smolders and Klinke 1984; frog: Stiebler and Narins 1990) and contrast to neural data from mammals (e.g., Gummer and Klinke 1983). Spontaneous otoacoustic emissions in mammals are, however, temperature sensitive (Ohyama et al. 1992), an effect that seems to be over-ridden at a processing stage between hair-cell transduction and the cochlear nucleus. It has been suggested that the temperature dependence of tuning in nonmammalian vertebrates (over and above any effects on fluid viscosity) is related to the temperature dependence of the ion-channel kinetics in the hair cell membrane (Eatock and Manley 1981). Differences in temperature dependency of tuning in nonmammals and mammals may thus reflect differences in the respective tuning mechanisms.

3.3.5.3. Population Responses—Excitation Patterns

Instead of studying the response characteristics of single fibers to different stimuli, one can investigate the response of a large sample of cells to a single test stimulus and reconstruct the pattern of excitation along the auditory papilla. Models based on such excitation patterns in mammals have been used to explain psychophysical phenomena, such as intensity discrimination in humans. Since many psychophysical data have become available for *S. vulgaris* (e.g., Klump and Baur 1990; Klump and Okanoya 1991), it was desirable to describe its cochlear excitation pattern. Excitation patterns for a range of frequencies (0.125–2 kHz) and stimulus levels (10–90 dB SPL) were derived from the responses of a large population of auditory afferents (Fig. 3.12D; Gleich 1994). In contrast to mammals, the so-called "spread of excitation" with increasing stimulus levels was not seen. Instead, the symmetry of the excitation patterns in *S. vulgaris* proved to be independent of the stimulus level, and the high-frequency flank of the excitation pattern was steeper than the low-frequency flank. The difference between mammalian and avian excitation patterns is most likely related to the different shapes of the respective single-neuron tuning curves. Based on these neural excitation patterns and a critical band scale derived from the cochlear frequency map (from Gleich 1989), Buus et al. (1995) developed a model of excitation patterns for *S. vulgaris*. The model indicates that the minimum detectable frequency differences for different frequency ranges correspond to equal distances along the basilar papilla and predicts very well the behaviorally determined level discrimination in *S. vulgaris*, except near threshold.

3.3.6. Temporal Resolution

3.3.6.1. *Phase Locking to Tones*

A further measure of the neuronal response to tonal stimulation, besides the rate of the spike discharges, is the degree of synchronization of the spikes to the phase of stimulus cycles, known as phase locking (Fig. 3.13A). Phase locking is prominent at low frequencies (generally below 2 kHz) in auditory nerve fibers of birds and deteriorates toward higher frequencies (Fig. 3.13B; Sachs et al. 1974; Gleich and Narins 1988; Hill et al. 1989; Manley et al. 1997). In the response of each cell, the phase lag increases systematically with stimulus frequency, as indicated by phase-versus-frequency plots (Fig. 3.13C). Some of this phase lag can be attributed to a middle ear delay (about 200 μs), but most of the delay is within the inner ear (travel time within the basilar papilla/tectorial membrane complex, synaptic delay, and action-potential conduction delays; Gleich and Narins 1988).

In *S. vulgaris*, the phase-versus-frequency functions of auditory afferents showed systematic deviations from a simple constant delay that would result in a straight line (Fig. 3.13C). The phase-versus-frequency functions could be mathematically separated into a putative constant delay plus the phase response of a simple LRC-type filter (Gleich 1987) analogous to that used by Crawford and Fettiplace (1981) to describe the phase response of turtle hair cells. The resulting best-fit functions explained very well the residual nonlinearity in the phase-versus-frequency functions of the auditory nerve fibers in *S. vulgaris* (Fig. 3.13D). This nonlinear component in the phase-versus-frequency response of auditory afferents in *S. vulgaris* can

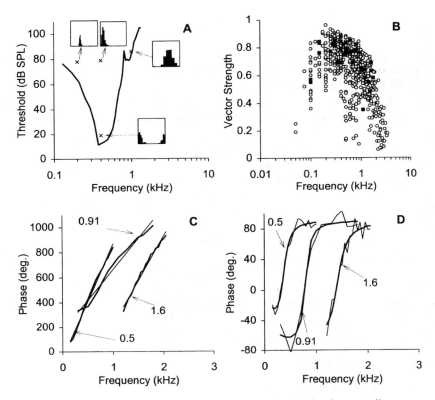

Figure 3.13. (**A**) Diagram illustrating phase locking behavior in an auditory nerve fiber of *Dromaius novaehollandiae* with a CF of about 0.4 kHz. The continuous line represents the rate-threshold tuning curve for this fiber. The Xs mark the frequency-level combinations that were used to collect the data of the phase diagrams corresponding to each X (arrows). Each of the squares shows a histogram, the abscissa of which covers 360° of phase (i.e., one cycle of the stimulus phase) at that frequency. The black columns show the number of action potentials that fell in each time bin within the cycle. In each case, the action potentials only occurred within a limited portion of each cycle: This proportion gets larger at higher frequencies (i.e., the box at top right). (Courtesy of C. Köppl.) (**B**) Vector strength, a measure of the degree of phase locking as a function of test frequency. Data were collected in response to 90 dB SPL stimuli within the excitatory tuning curve (open symbols) or at CF (filled symbols) in auditory afferents of *Sturnus vulgaris*. (**C**) Plots of the preferred response phase (shown as a phase lag) versus the test frequency of three auditory afferents in *S. vulgaris* (heavy lines). The cells show a prominent phase accumulation with increasing test frequency; however, there are systematic deviations from the straight line that would represent a constant delay (thin lines are linear regressions through the neural phase-versus-frequency functions). The numbers indicate the CF of the respective cells in kHz. (**D**) The neural phase response was separated into a constant delay and the phase contribution of a simple LRC-filter (as that proposed by Crawford and Fettiplace (1981) to describe the filter properties of turtle auditory hair cells), by using an iterative best-fit procedure that varied the resonant frequency and the frequency selectivity. Shown are the phase response of the best-fit filter (thick line) and the difference between the neural phase response and the calculated constant delay component (thin line). The nonlinear component of the phase response could be modeled very well by the phase response of the simple LRC-filter.

be explained as the contribution of a filter process. The filter frequencies and the acoustic CFs of the individual fibers showed a highly significant correlation, but the calculated filter frequencies were, on average, slightly lower than the measured acoustic CFs. The difference in the resonance frequencies of the filters and the corresponding fiber's CF resembles the difference in the equivalent frequency of the preferred intervals in spontaneous activity and their CF in this species (Manley et al. 1985; Gleich 1987).

The mean threshold for phase locking in *S. vulgaris* and, in a comparable frequency range, in *T. alba* was on average 10 dB below the criterion threshold for an increase in discharge rate (Gleich and Narins 1988; Köppl 1997c). In other words, auditory nerve fibers rearrange their spontaneous discharges in the time domain to follow the cycles of the stimulus tone without significantly raising the discharge rate, thus, this occurs below the rate threshold. The difference between the two thresholds diminished at higher frequencies in both the starling and the owl. In *C. livia* Schermuly and Klinke (1990a) described infrasound sensitive neurons that did not respond to normal auditory stimuli. However, for stimulus frequencies below 20 Hz these units showed a prominent phase locked modulation of their discharges over a range of sound pressures. Despite a deep modulation of the neural activity, these units showed on average only a 15% rate increase above spontaneous activity at the highest stimulus level (90 dB SPL). In the normal auditory frequency range and with the exception of *T. alba*, the ability to phase lock was highest around 0.4 kHz and decreased steeply for frequencies above 1 to 1.5 kHz (Fig. 3.13B; Gleich and Narins 1988; Hill et al. 1989; Manley et al. 1997). The upper limit of significant phase locking is species dependent: 4 kHz in *S. vulgaris* (Gleich and Narins 1988) and *D. novaehollandiae* (Manley et al. 1997); 5–6 kHz in *A. phoeniceus* (Woolf and Sachs 1977); and 10 kHz in *T. alba* (Sullivan and Konishi 1984; Köppl 1997c). In other species, this has been shown to be related to the decline of the AC-component of the hair-cell response with increasing frequency (Russell and Palmer 1986; Weiss and Rose 1988). Thus, phase locking is generally only prominent in the low-frequency range. However, despite various proposed models, it is presently not clear to what degree the rate- and the phase-locking properties contribute to the physiological basis of the behavioral performance in the recognition of complex signals. The physiological basis of the higher-frequency phase locking of *T. alba*, while obviously a specialization for the accurate processing of bilateral time cues for sound localization, has also not yet been established.

3.3.6.2. Gap Detection and Amplitude Modulation

Two experimental paradigms have been widely used to investigate the temporal resolution of the auditory system in several species, including humans. One is gap detection, that is, measuring the ability to detect an abrupt change in a continuous signal. The other paradigm concerns the detection of sinusoidal amplitude (envelope) modulation of a continuous signal. Both have been investigated at the level of primary auditory nerve fibers in *S. vulgaris* (Klump and Gleich 1991; Gleich and Klump 1995).

The minimum detectable gap in noise for a population of auditory fibers in *S. vulgaris* showed a monomodal distribution with a median of 12.8 ms and minimum values of 1.6 to 3.2 ms (Klump and Gleich 1991). This distribution is very similar to data collected in the auditory forebrain of *S. vulgaris* (Buchfellner et al. 1989), indicating that the ability of the auditory system to detect gaps is determined in the periphery and preserved in the ascending auditory pathway. Thus the most sensitive cells in the auditory periphery are sufficient to explain the behavioral minimum-detectable gap in *S. vulgaris* (1.8 ms; Klump and Maier 1989).

Temporal modulation transfer functions (TMTFs) of sinusoidally modulated noise were measured for auditory afferents in *S. vulgaris* by determining the modulation threshold (i.e., the depth of modulation necessary to elicit a significant phase locking to the modulation frequency) for a range of modulation frequencies. This procedure permitted a direct comparison of the neuronal data with psychoacoustic functions obtained with similar stimuli in the same species (Klump and Okanoya 1991). The shapes of the average neural TMTF and the behavioral function are very similar (Fig. 3.14) and even reflect response differences due to varying stimulus

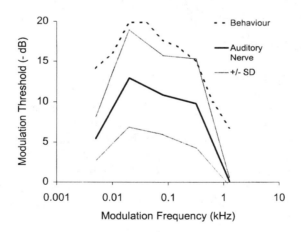

FIGURE 3.14. Modulation threshold as a function of the modulation frequency of a pulsed broadband noise stimulus (Gleich and Klump 1995). The continuous black line shows the mean through data obtained from 38 auditory afferents in *Sturnus vulgaris*; the gray lines indicate ± one standard deviation. The dotted line is the behavioral function determined in *S. vulgaris* using the same stimuli (from Klump and Okanoya 1991). The shape of the behavioral and neural functions are similar, both showing a bandpass characteristic. In addition, the responses of the most sensitive cochlear fibers appear sufficient to explain the behavioral performance.

paradigms (Gleich and Klump 1995). These data, as well as those from the gap-detection studies, suggest that the temporal resolution of the stimulus envelope in the periphery is neither degraded nor improved in the ascending auditory pathway.

3.4. Functional Organization of the Basilar Papilla

3.4.1. Frequency Representation in the Basilar Papilla

3.4.1.1. Methods

There have been five different approaches to determine the tonotopic arrangement of the frequency representation in the basilar papilla: (1) direct, stroboscopic-light observation, (2) traumatization, (3) the Mössbauer effect, (4) capacitative probe, and (5) single-fiber tracing studies. Since traumatization employs unphysiologically high sound pressure levels and because it is difficult to differentiate between frequency- and level-specific effects (Cotanche et al. 1987), studies using noise-damage methods will not be discussed here (for a detailed review see Manley 1996).

The first demonstration of a tonotopic frequency representation along the basilar membrane of *G. gallus domesticus* was obtained by von Békésy (1960) by direct stroboscopic observation of the basilar membrane motion upon stimulation with different frequencies. However, these measurements, though similar to some recent measurements, were obtained at unphysiologically high sound pressures and in cadaver ears.

The frequency-response characteristics of the basilar membrane of *C. crocodilus* were reported by Wilson et al. (1985). Using a capacitative probe, the displacement amplitude of the basilar membrane for different frequencies was measured at points between about 0.5 and 2 mm from the basal end. The more basal locations had higher cut-off frequencies, indicating the presence of a tonotopic organization. Unlike the more sharply tuned responses of primary auditory nerve fibers (see Section 3.3.7), however, the mechanical vibration of the basilar membrane was unaffected by a change of temperature. Both the tuning characteristics and the phase responses of the basilar membrane indicated the presence of a traveling wave, with, however, very little phase shift when compared to that of the mammalian basilar membrane. Since the tuning of single fibers is sharper and their phase shift is much greater than that measured at the basilar membrane (see also Section 3.3.5.1), one or more filter mechanisms must intervene between the observed basilar membrane motion and the nerve fiber responses.

In *C. livia*, the Mössbauer technique was used to determine the frequency response of the basilar membrane in the most basal 1.3 mm of the cochlea (Gummer et al. 1987). In specimens in good physiological condition, the basilar membrane was shown to be tonotopically organized, with a high-frequency limit of 6 kHz and a mapping constant of 0.9 mm per octave. The mapping constant of the papilla derived from neural data is, however, 0.63 mm (Smolders et al. 1995). In addition to this difference, Gummer et al. (1987) found that the frequency selectivity of the basilar membrane was worse than that of single auditory nerve fibers of similar CF in this species (Sachs et al. 1974), indicating that an additional filter stage exists subsequent to the motion of the basilar membrane. However, the data of Gummer et al. (1987) also indicate a dependence on the physiological condition of their preparation. As with the other methods discussed above, the basilar membrane data do not necessarily provide a map for that part of the papilla that lies over the neural limbus and that, at least in this frequency range, receives most of the afferent innervation.

To obtain the frequency map at the level of those hair cells that have most of the afferent synapses, tracer labeling of auditory afferents has been used. With these techniques, it became feasible to determine the location of the hair cell(s) contacted by physiologically characterized fibers (Manley et al. 1987; Gleich 1989; Chen et al. 1994; Jones and Jones 1995; Smolders et al. 1995; Köppl and Manley 1997). This approach enabled not only the study of basal-to-apical frequency gradients, it also allowed the correlation of other physiological measures with position (e.g., neural versus abneural). In *T. alba*, focal HRP injections were made in the cochlear nuclei at locations of known CF and stained fibers traced to the papilla (Köppl et al. 1993).

3.4.1.2. Frequency Maps in Different Species

Since the approach to tracing functionally characterized fibers to their innervated hair cells has proven to be so precise, only these data will be considered in the discussion below. To compare frequency maps across species, it is necessary to fit adequate functions to the scatter plots from individual species and treat the data sets in an identical way, for the choice of the function can strongly affect the derived map (Gleich 1989; Köppl et al. 1993; Chen et al. 1994; Buus et al. 1995). We digitized the published data that were unavailable to us directly and fitted a slightly modified version of the function originally proposed by Greenwood (1961) to all data sets (Buus et al. 1995). As an additional constraint on the fitting procedure, we set the frequency at the apical end to be <0 Hz. For *T. alba* (Köppl et al. 1993), *C. livia* (Smolders et al. 1995) and *D. novaehollandiae* (Köppl and Manley 1997), available data extend to within the most basal 5% of the papilla. In *S. vulgaris* and *G. gallus domesticus*, however, no data are available for the basal quarter of the papilla. Thus the upper limit of the frequencies represented on the papilla in these two species was estimated and fixed at 6.3 kHz in *S. vulgaris* and 4.7 kHz in *G. gallus domesticus* (the highest frequency with a threshold 30 dB above the best threshold of the audiogram; Gleich 1989; Saunders and Salvi 1993; Buus et al. 1995; Dooling et al., Chapter 7). The resulting maps provided an adequate representation of all data sets except for the adult *G. gallus domesticus*, where the requirement to map only positive frequency values onto the papilla could not be adequately satisfied. However, combining all data from *G. gallus domesticus* into one fitting procedure resulted in a map almost indistinguishable from that calculated only for the younger specimens. Thus for the species comparison, we use the combined map for *G. gallus domesticus*. A systematic deviation of the data points in *T. alba* at low frequencies from the "Greenwood" map has been observed previously (Köppl et al. 1993) and may be a result of the enormous overrepresentation of the 5–10 kHz octave in the auditory foveal region.

The resulting best-fit functions illustrate the species-specific differences (Fig. 3.15). The frequency distributions of the basilar papillae of *D. novaehollandiae* and *C. livia* lie to the right of all other functions, indicating that

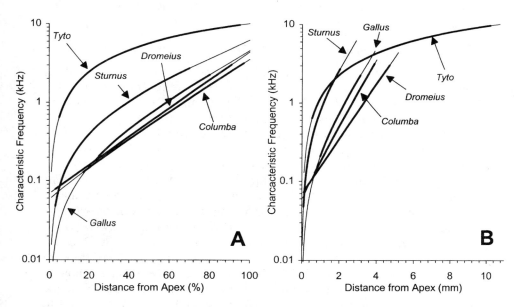

FIGURE 3.15. Cochlear frequency maps of several avian species based on the innervation sites within the papilla of stained and physiologically characterized fibers. The functions are best fits through the data points, obtained using a slightly modified Greenwood function (for details see Section 3.4.1.2; Buus et al. 1995). The heavy lines indicate the measurement range, thin lines show projections extending beyond the range of available data. For a given location on the papilla, *Columba livia* and *Dromaius novaehollandiae* respond to the lowest frequency (except perhaps at very low frequencies), while *Gallus gallus domesticus*, *Sturnus vulgaris*, and *Tyto alba* respond to increasingly higher frequencies. (**A**) Frequency representation represented as a percentage of total papillar length. (**B**) Frequency representation as absolute distance from the papillar apex.

at a given distance from the apex (Fig. 3.15A; in percentage of total papilla length), these papillae respond to the lowest frequency. The maps of *G. gallus domesticus*, *S. vulgaris*, and *T. alba* are shifted to the left to successively higher frequencies. All maps show a variation of the slope of their frequency map across frequency, with a steeper slope at low frequencies and a shallower slope at high frequencies, but the degree of the change of slope is very small in primitive species (*D. novaehollandiae, C. livia*) and progressively higher in other species. Similar differences at low frequencies have been described for mammals (e.g., Greenwood 1990). The map of *T. alba* lies at the extreme; this species has unusually good high-frequency hearing and a great over representation of the 5–10 kHz octave (6 mm per octave), indicated by the very shallow slope over the basal half of the papilla (Köppl et al. 1993). A plot of the maps in terms of absolute basilar papilla dimensions illustrates that the frequency distribution in the apical half of the *T. alba* papilla is in fact fairly similar to that of *S. vulgaris*, and emphasizes the high-frequency specialization (Fig. 3.15B).

3.4.1.3. Developmental Changes in the Frequency Representation

The first published frequency maps based on the labeling of physiologically characterized cochlear ganglion cells in birds were from 2-day- and 3-week-old *G. gallus domesticus* (Manley et al. 1987). As seen in the map provided by von Békésy (1960), the horseradish peroxidase (HRP) injections revealed a similar low- to high-frequency gradient from the apex toward the base in both age groups of *G. gallus domesticus*. However, in contrast to the assumptions made from the sound trauma data of Rubel and Ryals (1983), the HRP-based maps for 2-day- and 21-day-old *G. gallus domesticus* were not statistically different, indicating that no posthatch shift of the frequency representation in the neural hair cell area occurs between 2 and 21 days of age, at least for frequencies up to 2 kHz. These findings have recently been extended to the E19 embryos of *G. gallus domesticus* by Jones and Jones (1995), who found no significant difference to the frequency distribution of the 2-day-old *G. gallus domesticus*, indicating that the frequency distribution on the papilla is apparently stable from prehatching to three weeks of age. These and other data relevant to the development of frequency representation have been discussed in detail in a recent review (Manley 1996) and will therefore not be dealt with further.

3.4.2. Changes in Function Across the Papilla's Width

The avian basilar papilla has many hair cells across its width (see Section 2.1). Also, the proportion of these hair cells that are afferently innervated is species dependent. Whereas apically, hair cells across the entire width are innervated by afferent fibers, in the basal half of the papilla of all species, a more-or-less large proportion of the hair cells (up to >50% of the papilla's width) is not afferently innervated (Fig. 3.5; Fischer 1994b). Nonetheless, the avian papilla offers the opportunity of examining data from stained fibers for correlations with the position of the innervated hair cell(s) across that part of the width of the epithelium that is afferently innervated.

In *S. vulgaris*, Gleich (1989) first showed that afferent fibers contacting hair cells at different distances from the neural edge differ in their physiological properties. In the CF-range between 0.6 and 1.8 kHz, the audiogram of *S. vulgaris* is nearly flat and thus does not contribute to threshold variation. In this region, there was a linear correlation between the threshold and the position of the innervated hair cell, such that fibers contacting hair cells near the neural edge were the most sensitive. Fibers contacting hair cells progressively nearer the abneural edge were less sensitive to the extent of 6 dB per hair cell. Threshold differences have also been reported across the papilla of *C. livia* (Smolders et al. 1995), but there was a great deal of scatter. These data, which were pooled across frequencies after correction for the audiogram, showed a complex change of threshold across the epithelium. There, the most sensitive cells on average were found at a position about 20% of the distance from the neural edge. In *D. novaehollandiae* also (Köppl and Manley 1997), no linear gradient was found, and afferent fibers at any position across the width of the papilla had comparatively low thresholds—in other words, in *D. novaehollandiae* there was no threshold trend across the papilla. In *C. livia*, the threshold changes were accompanied by variation in the Q_{10dB}, with maximal values on average at the place of minimal threshold (Smolders et al. 1995).

The data thus suggest that gradients in sensitivity and frequency selectivity of afferent fibers innervating hair cells at different positions across the basilar papilla are found in birds, but that their pattern and prominence is species specific. It may also be that such gradients are different at different frequency positions along the papilla, and that pooling data for different positions obscures trends. In this respect, it should be noted that the frequency areas sampled in the three studies mentioned above differed. In *S. vulgaris* and *C. livia*, the sample of labeled fibers was skewed toward low frequencies (Gleich 1989; Smolders et al. 1995), whereas in *D. novaehollandiae*, it was more homogeneous across frequencies (Köppl and Manley 1997). In the data of *D. novaehollandiae*, above 2 kHz the range of thresholds was much smaller than found at lower frequencies in all species (Manley et al. 1997). Gradients across the papilla may thus be expected to be more prominent in the low- to mid-frequency range. These frequencies lie in the apical to middle area of the papilla, where there are obvious anatomical gradients across the papilla, such as hair cell bundle rotation angle and stereovillar height and number (see also Sections 2.2 and 3.5.1), which may all influence the sensitivity.

3.5. Functional Implications of Variations in Morphology

Our understanding of the function of the avian basilar papilla is still relatively poor. It is thus difficult to realize the full implications of the anatomical variability discussed above. Since each of the bird papillae investigated to date is unique, however, we would expect there to be species specificities in the excitation patterns and the responses of the hair cells and their associated nerve fibers.

3.5.1. The Micromechanics of Stimulation of the Avian Papilla

In birds, it is apparent that the mechanical stimulus for most THC must come through the tectorial membrane. The structural differences between the THC and the SHC in birds are substantial and comparable to the differences between inner and outer hair cells of mammals. The evolutionary origin of separate hair cell populations in mammals apparently derived from the selection pressure for specialization—the outer hair cells for active motion in response to sound, and the inner hair cells for sensory transformation. We have put forward the hypothesis that the abilities of primitive vertebrate hair cells (both sensory transduction and active motility; Crawford and Fettiplace 1985) predispose large arrays of hair cells to this kind of specialization. It is conceivable that the selection pressures acting on a large uniform population of hair cells in both birds and mammals could produce—independently and convergently—a similar "division of labor" between hair cell groups (Manley et al. 1989; Manley and Köppl 1998).

Of course, there is now substantial evidence that the division of labor in the mammalian cochlea is based on the involvement of the outer hair cells (OHC)—through active and fast length changes—in the so-called cochlear amplifier (Patuzzi 1996). Is there any evidence for active movement in bird cochleae? We have recently demonstrated the presence of high-frequency (up to 11 kHz) spontaneous otoacoustic emissions (SOAE) in *T. alba* (see Section 3.7.2). The presence of such SOAE is generally accepted as evidence for an active process in the hearing organ. The mechanisms thought to be producing SOAE in mammals (i.e., the active changes in hair cell length) are, however, not appropriate for the avian situation (Manley 1995). Instead, active bundle movements, as demonstrated in turtle (Crawford and Fettiplace 1985) and in sacculus hair cell bundles (Benser et al. 1996) are more likely to be involved. Active movement of hair cell bundles would feed energy into the tectorial membrane. Laterally coupled mechanical energy in the plane of the tectorial membrane would be appropriate for an active mechanism in birds (Manley 1995). Our working hypothesis (based on lizard data: Manley 1995; Manley and Gallo 1997) is that the adaptational myosin motors of hair cell bundles (Hudspeth and Gillespie 1994) are fast enough to feed energy into the motion cycle-by-cycle if the myosins are stretch-activated. At present, the functions of the SHC are insufficiently understood to do more than speculate. Nonetheless, a finite-element model of the avian papilla constructed assuming that SHC feed mechanical energy into the tectorial membrane produced a maximum stimulus for hair cells that lie near the neural edge of the papilla (Steele 1996), consistent with the neural data.

There is evidence in some non-avian species that the tectorial membrane contributes mass to a resonance system (bobtail lizard, see *Tiliqua rugosa:* Manley et al. 1988; Manley et al. 1989; gecko, *G. gecko*: Authier and Manley 1995; guinea pig, *Cavia procellus*: Zwislocki and Cefaratti 1989). If the stimulus to the hair cells depends mainly on a resonance of the tectorial membrane and its associated hair cell bundles and less or only indirectly on the vibrational amplitude of the basilar membrane, hair cells over the basilar membrane would not necessarily be more sensitive. In *G. gallus domesticus*, the pattern of hair cell bundle orientations led Tilney et al. (1987) to suggest that the direction of hair cell stimulation is not radial as in mammals. Similar morphological patterns have since been found in all avian species. If the stimulation pattern is radial, the change of hair cell bundle orientation across the papilla would certainly reduce the effectiveness of stimulation on rotated cells. The maximal threshold effect in *S. vulgaris* would, however, be less than 1 dB per hair cell (Gleich 1989), which alone is not sufficient to explain the range of sensitivity differences found in that species (see also Section 3.4.2).

An interesting correlation to note in this regard is that, taken together, the sensitivity differences between single auditory nerve fibers in the species investigated so far (*G. gallus domesticus, S. vulgaris, C. livia, D. novaehollandiae,* and *T. alba*) tend to be greatest at lower CF. Toward high CFs, which, due to technical problems, are not well represented in recordings of some species, the variation in threshold is much smaller. Whereas 50–60 dB differences are common for units below 1 kHz (Manley 1990), above 2 kHz the variation in *D. novaehollandiae* is only 25–30 dB (Manley et al. 1997). In *T. alba* it is even less (Köppl 1997a.). As the afferent nerve fibers innervate more of the width of the papilla apically and only about half or less of the width basally, this difference is perhaps not unexpected.

As noted in Section 2.2.3, it is possible that the higher numbers of stereovilli on the neural hair cells in most species also make them more sensitive, since this will be correlated with an equivalent increase in the number of tip links per hair cell (Pickles et al. 1989) and the number of transduction channels connected to these tip links. These potential changes in sensitivity would of course add to any changes resulting from stereovillar bundle rotation angle. An increase in numbers of stereovilli, all else being equal, would increase the stiffness of the bundle, but this could be compensated for by a reduction in stiffness via other means. Here, there are three obvious possibilities.

One possibility is a reduction in the thickness of the short axis of the whole bundle (bundle width) while increasing the long axis (bundle length), which would decrease the number of connections in any column of stere-ovilli. This was seen, for example, in *A. fuligula* (Manley et al. 1996). The addition of stereovilli in such cases is achieved by adding columns to the ends of the bundle (making it longer), and not by making the bundle wider. The second possibility of reducing stiffness is a reduction in stereovillar diameter. In *A. fuligula* and other species (e.g., *S. canarius* and *T. guttata*: Gleich et al. 1994; *M. undulatus*: Manley et al. 1993), the stereovilli of the bundles of neural cells in the basal two-thirds of the papilla are thinner than those on abneural cells and thus less stiff. In *C. livia* and *S. vulgaris*, however, no difference in average stereovillar diameter was found between neural and abneural hair cells (Gleich and Manley 1988). The third possibility of reducing stiffness is an increase in bundle height, and different species show different patterns of bundle height distribution (see Section 2.2.3). In general, neurally situated bundles are in fact taller, and, therefore, if their stiffness were only determined by the height of the bundle, less stiff. Together these differences between hair cell bundles might produce a more sensitive bundle on the neural side that has the same overall stiffness, com-patible with the same frequency responses on the neural and abneural sides of papilla. Of course, if stiffness is not fully compensated for, it is possible that hair cells across the papilla do not have the same response frequency. Unfortunately, the available mapping data give insufficient information on this.

3.5.2. A Hypothesis on the Mechanism of an Active Process in the Avian Papilla

At present, the following scenario can be constructed for the excitation of hair cells in the avian papilla. An incoming stimulus near threshold would activate first those hair cells over the basilar membrane, that is, the SHC. These cells (that themselves have no afferent innervation) respond to the stimulus by producing active movements of their bundles (see Section 3.7) that are, at least at their characteristic frequency, in phase with the sound stimulus. This additional acoustic energy is fed into the radial movement of the tectorial membrane. Steele's (1996) model suggests that acoustic energy traveling across the tectorial membrane in a radial direction toward the neural limbus would be reflected by the harder material of the limbus and interact with waves in the tectorial membrane in such a way that a maximum of shear movement occurs at a position near, but not at, the neural edge. This maximum of stimulation would, at threshold, mainly stimulate the small population of (tall) hair cells in this location. These THC would thus have the lowest thresholds.

The exact pattern of stimulus *across* the papilla would depend on the position *along* the papilla (and thus the frequency used) and the specific details of the morphological patterns of the hair cells in that area (i.e., bundle rotation and height patterns, stereovillar numbers, etc.) and thus on the species. Due to the complexity of this situation, it will not be an easy task to examine this hypothesis experimentally. An examination of the rate-level functions of auditory nerve fibers in *D. novaehollandiae* suggested that the cochlear amplification process is confined to localized groups of hair cells in that species, and does not globally affect all cells across the epithelium as in mammals (Köppl et al. 1997).

3.5.3. Species-Specific Physiology

Major differences are not seen in the hearing ranges (measured behaviorally or as the range of CFs of the tuning curves) of most avian species studied so far (100 Hz to about 6–8 kHz; Konishi 1970). *T. alba*, however, (an "advanced land bird") has exceptional high-frequency hearing and a specialized basal papillar area. Sullivan and Konishi (1984), Köppl et al. (1993), and Köppl (1997a) report data that, together with the behavioral audiogram (Konishi 1973; Dyson et al. 1998), indicate that CFs in the auditory nerve of *T. alba* exceed 10 kHz. The anatomical features are correlates of a specific adaptation for processing of the high frequencies (5–10 kHz) used by *T. alba* for sound localization (Konishi 1973). These frequencies are processed in an auditory fovea (Köppl et al. 1993). Also in *T. alba*, the abruptness of the change in the orientation of hair cell bundles across the papilla is striking and correlates in position both with a change in the number of stereovilli on the hair cells and, at least in the neural half of the papilla, with the edge of the neural limbus (Fischer et al. 1988). That is, hair cells on the two sides of the outer edge of the neural limbus have quite different stereovillar bundle orientation and number.

Although hair cell morphological parameters can differ substantially between species at any absolute position along the epithelium, these parameters correlate much better for the different species when plotted relative to the response frequencies of the different papillar positions (Fischer 1994b).

C. livia and *G. gallus domesticus* ("primitive land birds") have differently specialized apical papillar areas sensitive to very low sound frequencies. The infrasound sensitivity of *C. livia* also disappears upon removal of the cochlear duct (Kreithen and Quine 1979), proving that the receptors lie in the basilar papilla or the lagena macula. Klinke and Schermuly (1986) and Schermuly and Klinke (1990a) report finding extremely low-frequency, phase locked responses in auditory nerve fibers of *C. livia*. When stained, these fibers were found to innervate hair cells in the specialized abneural area of the apical part of the basilar papilla (see Section 2.5; Schermuly and Klinke 1990b). In *G. gallus domesticus*, Warchol and Dallos (1989a) report finding cochlear nucleus cells that responded to low-frequency sound (10–500 Hz). These authors (Warchol and Dallos 1989b) also traced the low-CF fibers to the apical papillar area, suggesting that the "very distal part" (as described by Lavigne-Rebillard et al. 1985) might be specialized in a functionally similar way to the broad, abneural, apical region in *C. livia*. As the anatomical patterns are very different, however, it is probable that *C. livia* and *G. gallus domesticus* have developed a low-frequency specialization quite independently of each other. In most previous experiments with other species, however, the sound systems were often unable to stimulate adequately below 100 Hz and phase locking responses were neglected, so it is difficult to estimate the relative occurrence of very low frequency responses in other avian species. However, in *D. novaehollandiae*, certainly a primitive species, a specific search for such low-frequency specializations turned up only negative results (Manley et al. 1997).

Although no physiological data are available for *S. canarius* and *T. guttata*, there is some evidence from behavioral studies that these birds with their very short papillae have a good sensitivity at high frequencies, similar to that of other passerine birds with substantially longer basilar papillae, while their thresholds are substantially higher at low frequencies (Dooling 1992). This can be correlated with the predominance throughout the papilla of anatomical features typically more pronounced in the basal half of the basilar papilla of other species (Gleich et al. 1994). Thus, in these small songbirds with a very short basilar papilla, there may have been a trade-off such that retaining a good sensitivity at high frequencies resulted in a compromised low-frequency sensitivity.

3.6. Efferent Fibers

In contrast to mammals, where quite a substantial number of compound action potential (CAP), cochlear microphonic (CM), single-neuron, and otoacoustic emission measurements are available (Guinan 1996), only very little is known about the physiological properties and effects of papillar efferents in birds. Recently, Kaiser and Manley (1994) developed an experimental approach that allowed single-cell recordings from putative cochlear efferents in the auditory brainstem of *G. gallus domesticus* using stereotactic coordinates obtained from tracer experiments. These fibers differed from fibers of the afferent pathway by their low spontaneous activity, broad frequency tuning, long latency responses, long modes of their time-interval histograms of spontaneous activity, and by their temporal response characteristics to sound. There are two principally different response types. One type responds to sound with excitation (chopper- or primary-like pattern), resembling mammalian efferents (medial olivocochlear bundle). A second and novel efferent type having a higher spontaneous activity responded with complete suppression of discharge activity during tonal stimulation (Fig. 3.16). The effects of these two types within the hearing organ are as yet unknown. They could, however, be opposite; the excitation type, as in mammals, could lead to an inhibition of the auditory response, whereas the suppression type could, in silence, provide the hearing organ with a steady input that is reduced during a specific

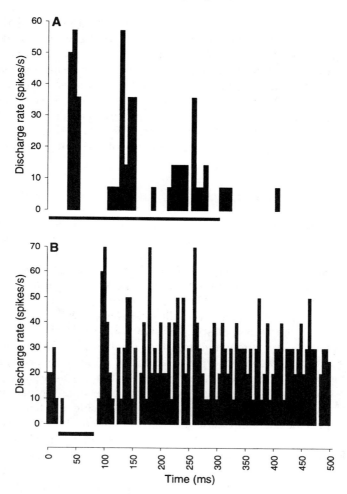

FIGURE 3.16. Peri-stimulus-time histograms of the response of two different efferent fibers from *Gallus gallus domesticus* upon stimulation with a tone pip (bars below the histograms; figure kindly provided by A. Kaiser). (**A**) A chopper-type response similar to that found in mammalian medial olivocochlear fibers innervating outer hair cells (e.g., Robertson and Gummer 1985). (**B**) A suppression-type response. The neural discharge of this type of unit was completely suppressed during the stimulus. A prominent off response, occasionally consisting of several peaks, followed the stimulus offset.

sound. In other octavolateralis systems, such as the vestibular system of monkeys and toads or the lateral-line system of fish, efferent stimulation is known to produce both inhibitory and facilitatory effects on afferent neurons (Goldberg and Fernández 1980; Highstein and Baker 1985; Sugai et al. 1991).

Our knowledge of the effect of the efferent system in birds is minimal. Desmedt and Delwaide (1963, 1965) measured the effect of electrical stimulation of the efferent bundle on cochlear potentials. As in mammalian studies, the CAP amplitude decreased and the CM potentiated markedly during efferent activation; this could be inhibited by strychnine. In *T. alba* (Taschenberger 1995; Taschenberger and Manley 1996), stimulation of the contralateral ear with sound can suppress, but also facilitate, the amplitude of distortion-product emissions. These opposing effects could be related to activation of the different types of efferent fibers described above.

3.7. Otoacoustic Emissions

Since their discovery by Kemp (1978), otoacoustic emissions have created great interest, for they provide a noninvasive method of investigating the function of the inner ear. In one form or another, they have been reported from representatives of all classes of terrestrial vertebrates (Köppl 1995; Whitehead et al. 1996). This not only suggests that the generation of emissions is a primitive property of vertebrate auditory hair cell arrays, but also, in view of the morphological variability between the hearing organs of different vertebrate classes, that a comparative investigation of emissions could contribute to understanding the mechanisms underlying their generation. These mechanism(s) may, of course, differ between species or between groups of vertebrate animals, but from an evolutionary point of view, one would initially assume that their presence in different vertebrate classes is based on a common, primitive underlying mechanism. While self-sustaining mechanical oscillations of the hair-cell body are generally thought to be the origin of spontaneous emissions in mammals, there is as yet no direct evidence to support this assumption in non mammals. Avian SHC have no subsurface cisternae and are often extremely short (a few microns high, e.g., Fischer 1994b) with little active cytoplasm, suggesting that if they influence papillar micromechanics, they could only do so via stiffness changes of the stereovillar bundle or by bundle movement (Manley 1995; see also Section 3.5.2).

Spontaneous otoacoustic emissions have been reported from mammals, anuran amphibians, and lizards (for review, see Köppl 1995). In birds, simultaneous-evoked emissions occur in *S. vulgaris* (Manley et al. 1987), distortion-product emissions have been reported from *S. vulgaris* and *G. gallus domesticus* (Kettembeil et al. 1995) and from *T. alba* (Taschenberger and Manley 1998), but SOAE were only found in *T. alba* (Manley and Taschenberger 1993; Taschenberger 1995; Taschenberger and Manley 1997). Spontaneous otoacoustic emissions in the other species may be so small as to be technically very difficult to detect. In addition, anesthetic agents depress emissions in birds (Kettembeil et al. 1995).

3.7.1. Simultaneous-Evoked Emissions in *Caiman crocodilus* and *Sturnus vulgaris*

Simultaneous-evoked emissions have been reported both from *C. crocodilus* (Klinke and Smolders 1984) and *S. vulgaris* (Manley et al. 1987). Simultaneous otoacoustic emissions in the ear canal of *C. crocodilus* appeared as ripples in the waveform of sound pressure level of a low-level swept tone in the outer ear canal. The emissions were fairly broadband, non linear in their growth with increasing SPL and suppressible by second tones. The most effective suppressor tones were near the emission frequency. Very similar data were described from a study of simultaneous-evoked otoacoustic emissions in *S. vulgaris* (Manley et al. 1987). In *S. vulgaris*, such emissions were measurable mostly in the upper half of the species' hearing range, and varied in level between −30 and +2 dB SPL. Suppression tuning curves were measured by adding second tones and searching for a criterion amount of suppression. Their shape was strongly reminiscent of neural tuning curves in the same species (Manley et al. 1985).

3.7.2. Spontaneous Otoacoustic Emissions (SOAE) in *Tyto alba*

So far, SOAE from birds have only been reported from *T. alba*, where 65% of ears showed SOAE (at an average of 2.1 per ear; Manley and Taschenberger 1993; Taschenberger 1995; Taschenberger and Manley 1997). *T. alba* requires very little anesthetic, and, since distortion product otoaconstic emissions (DPOAE) have been shown to be anesthesia sensitive (Kettembeil et al. 1995), this may be the reason why so far it has not been possible to record SOAE from birds other than *T. alba*. The SOAE had center frequencies between 2.3 and 10.5 kHz, but almost all were higher than 7.5 kHz (Fig. 3.17A). Their peak amplitudes lay between −5.8 and 10.3 dB SPL. Like the DPOAE discussed below, the SOAE were unstable in amplitude (and here even in frequency) during and between recording sessions. The center frequency of the emissions was temperature dependent, shifting with an average rate of 0.039 octaves/°C (Taschenberger 1995).

In the presence of ipsilateral external tones, SOAE could be suppressed, and suppression tuning curves (STC) were V-shaped, with the sensitive tip (sensitivity down to about 0 dB SPL) near the center frequency of the SOAE (Fig. 3.17C). The Q_{10dB} sharpness coefficient was similar to the tuning selectivity of DPOAE (see below) and of auditory nerve fibers in *T. alba* (Köppl 1997a). With the exception of SOAE reported at very high frequencies from a species of constant frequency bat (Kössl and Vater 1985) and reports of occasional high-frequency emissions in human neonates (Burns et al. 1992), *T. alba* shows on average the highest-frequency SOAE ever reported from undamaged cochleae, certainly of nonmammals. When compared to the behavioral audiogram (Konishi 1973; Dyson et al. 1998) and the cochlear map (Köppl et al. 1993), it is obvious that SOAE in this

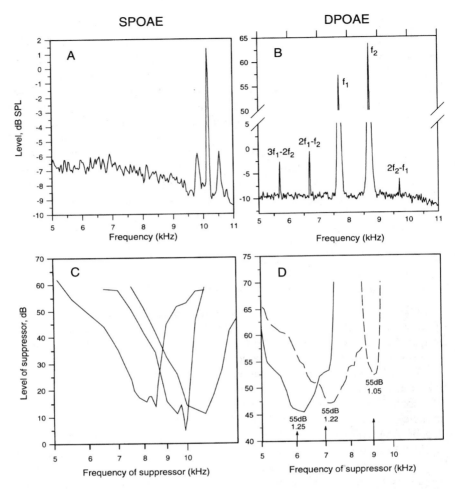

FIGURE 3.17. Spontaneous and distortion-product otoacoustic emissions in *Tyto alba*. The two top diagrams show ear canal spectra for (**A**) the spontaneous otoacoustic emissions (SOAE) and (**B**) distortion product otoacoustic emissions (DPOAE). The respective diagrams below illustrate the sensitivity and shape of suppression tuning curves (STC) for (**C**) SOAE and (**D**) DPOAE. SOAE almost all had center frequencies above 7.5 kHz. Their STC were sensitive and sharply tuned (**C**). The DPOAE could be generated at any frequency, but for comparison with the SOAE, we show three STC for the suppression of the product 2f1–f2 at higher frequencies (**D**). Below each STC is given the primary tone level and primary tone frequency ratio used in each case. The arrows indicate f₁. The width of the STC depends (also) on the ratio of the primary tones, being narrower when closely spaced primary tones are used. (Courtesy of G. Taschenberger.)

species originate primarily in the upper quarter of the animal's hearing range, and mostly from the specialized area previously described as being part of an auditory fovea. Indeed, most SOAE originate in the basal-most 3.5 mm of the papilla in *T. alba*.

3.7.3. Distortion-Product OAE in *Gallus gallus domesticus*, *Sturnus vulgaris*, and *Tyto alba*

When the inner ear is stimulated by two tones of different frequency (f1 and f2, where f2 > f1), various combination-tone or distortion-product spectral peaks are generated that are emitted through the middle ear and can be measured in the external meatus. The distortion product (DP) 2f1–f2 and, in some cases, 2f2–f1 have been measured in the ear canal of both conscious and anesthetized *S. vulgaris* and *G. gallus domesticus* and in the lightly anaesthetized *T. alba* (Fig. 3.17B). The effect of a third suppressive tone and the behavior of the DP under deeper anesthesia were also studied (Kettembeil et al. 1995).

In general, the DP characteristics in birds resembled those of lizards and mammals, but first appeared at somewhat higher primary-tone levels. The best frequencies of third tones suppressing 2f1–f2, as measured from suppression tuning curves in *G. gallus domesticus* and *S. vulgaris*, lay near the first primary tone (f1). As in lizards, facilitation via a third tone was also seen for the DP in both species, often at levels below those eliciting suppression. Also in these two species, the DP 2f1–f2 disappeared completely at the onset of deep anesthesia and recovered to its original magnitude when the anesthesia was lightened, sometimes with a considerable delay. The compound action potential (CAP) was somewhat more sensitive to anesthesia than the DP. Control experiments showed that this anesthesia effect was not a result of hypoxia (Kettembeil et al. 1995). It is also extremely rapid and thus unlikely to be due to changes in middle ear air pressure affecting the transfer characteristic of the middle ear (Larsen et al. 1996).

Although *T. alba* is known to possess an auditory fovea (Köppl et al. 1993) and prominent morphological specializations in the auditory periphery (Smith et al. 1985; Fischer et al. 1988), many features of DPOAE in this bird are very similar to those described for *G. gallus domesticus* and *S. vulgaris*, and for reptiles and mammals (Taschenberger 1995; Taschenberger and Manley 1998). Interestingly, the highest DPOAE output levels and best thresholds were found both at the lowest f1 (the lower primary frequency of the two-tone stimulus) frequency investigated (1 kHz) and for f1 frequencies of 5 to 7 kHz, that lie within the auditory fovea. In these f1 frequency ranges also, the optimal primary-tone frequency ratio was smaller than at the other frequencies. In some cases, the threshold DPOAE sound pressures were only 43 dB below the primary-tone levels. The amplitude of emissions, however, tended to drift over time.

Suppression tuning curves of DPOAE of *T. alba* were V-shaped and had their sensitive tips near the frequency of f1, suggesting DPOAE generation near the place in the papilla where the frequency of f1 is encoded (Fig. 3.17D; Taschenberger and Manley 1998). The same is true for such STC in other nonmammals (Köppl and Manley 1993, Köppl 1995). As a measure of frequency selectivity, the Q_{10dB}-value of the STC increased as a function of frequency up to a value of 8.1 and thus strongly resembled the equivalent increase in frequency selectivity with CF of auditory nerve fibers in *T. alba* (Köppl 1997a).

In the lightly anesthetized *T. alba*, Taschenberger (1995) found that both SOAE and DPOAE varied spontaneously in amplitude with time. One possible cause of this instability in SOAE and DPOAE amplitudes could be efferent influences on hair cells that may vary as a function of the momentary anesthetic state and/or of the animal's state of arousal. Taschenberger (1995) and Taschenberger and Manley (1996) therefore examined whether contralateral sound stimulation (noise and pure tones) could—via efferent fibers—influence SOAE and DPOAE in *T. alba*. Contralateral tones and noise of frequencies above 7 kHz were used, which are attenuated >70 dB across the interaural canal and thus do not directly affect the ear under investigation. Effects were seen with very low levels of contralateral stimulation and after the middle ear muscles had been severed. In contrast to the generally suppressive efferent effects described in mammals, both reductions in level and increases in amplitude of SOAE and DPOAE were found as a result of applying contralateral noise (Taschenberger and Manley 1996). Such increases in DPOAE level are especially interesting in view of the existence of avian auditory efferents whose activity is suppressed in the presence of sounds (Kaiser and Manley 1994; see also Section 3.6). If the spontaneous activity of these efferents normally inhibits hair cells, then the suppression of such efferent fibers during tonal stimulation should lead to a disinhibition at the hair cell level. It is possible that the DPOAE level increases observed are attributable to such a mechanism.

Noise stimulation produced both inhibitory and facilitatory effects. In general, contralateral noise was accompanied by an increase in amplitude of small DPOAE but a decrease in amplitude of large DPOAE, and caused a strong reduction in the drift of amplitude. This may indicate that the extremes of the drift seen were due to changes in the relative levels of activity in the two kinds of efferents. Contralateral *tonal* stimulation, on the other hand, was generally associated with inhibitory effects only. The effect was strongly frequency selective and tuned to the same contralateral frequency range as the ipsilateral primary frequencies. This suggests that tonal suppression is mainly mediated by sharply tuned efferents and that their contralateral projection is tonotopic, with corresponding afferent and efferent cochlear maps, as in mammals. Thus clear similarities, but also differences, exist at peripheral levels in efferent effects in birds and mammals, and these need further examination.

3.8. Trauma and Regeneration

Over the past 10 years, the effects of damage to the avian cochlea have been studied in much detail since it was first shown that birds (in contrast to mammals) can replace damaged or lost auditory hair cells (Cotanche 1987). Thus birds serve as a model for studying hair cell regeneration and cochlear repair processes with the ultimate aim of improving and developing a therapy for human deafness that is associated with hair cell loss. The processes associated with damage and regeneration of auditory hair cells after acoustic and aminoglycosid-induced trauma have recently been reviewed by Cotanche et al. (1994) and will be the subject of a forthcoming volume in this series.

While many strains of various mammalian species with genetic abnormalities of their inner ear are known and characterized (e.g., Steel 1995) there is only one documented example of a nonmammalian species with a hereditary hearing deficit. This is a special strain of *S. canarius*, the Belgian Waterslager canary, that shows a genetically determined hearing loss (Okanoya and Dooling 1985, 1987; Okanoya et al. 1990; Gleich and Dooling 1995; Gleich et al. 1995) with associated pathologies of the basilar papilla and hair cells (Gleich et al. 1994). Although these birds continuously produce new hair cells (Gleich et al. 1997) they are unable to completely repair their inner ear pathologies.

4. Summary

The avian auditory papilla shows an extremely high degree of structural and functional organization. There is evidence for a functional separation of tall from short hair cells, in that the tall hair cells tend to lie neurally and be placed at or near the position of maximal stimulation. Only these hair cells are afferently innervated, and some of the characteristics of their associated afferent fibers correlate with the position of the innervated hair cell. The short hair cells are not afferently innervated, and these hair cells may play a role in the mechanics of the local papillar area, in that their stereovillar bundles may actively move in phase with the stimulus and thus add energy to the system—in other words act as a cochlear amplifier. The presence of spontaneous otoacoustic emissions is one manifestation of this active process.

There is considerable evidence that the specialization of the avian cochlea in terms of its hair cell types developed gradually during the evolution of the birds, for the size of the SHC population increased greatly from the early representatives (represented here by *D. novaehollandiae*) to later ones (such as song birds and owls). Individual morphological characteristics underlying the differentiation of hair cell types did not necessarily evolve simultaneously and in parallel. Rather each species shows a unique mosaic of characteristics. In some cases, clear physiological correlates of these specializations have been described. Too little is known about the papilla of Crocodilia to make similar comments.

In general, it can be said that the avian auditory organ is a highly specialized assemblage of hair cells and that a broad survey of species in the various avian groups suggest that during their evolution, selective pressures existed for increased differentiation into functional types. The end result is morphologically quite different to the cochlea of mammals. Nonetheless physiological studies have shown that the avian papilla, at least over the range of frequencies it responds to, matches or exceeds the mammal hearing organ in its performance.

Acknowledgments. Supported by a grant to GAM from the Deutsche Forschungsgemeinschaft (SFB 204 "Gehör"). We thank F.P. Fischer for providing Figure 3.5, A. Kaiser for his contribution on the sections on efferents (Sections 2.3.2 and 3.6) and for providing Figure 3.16, C. Köppl for providing emu data shown in Figures 3.9, 3.10, 3.11, and 3.13 and for comments on a previous version of the manuscript, J. Strutz for comments and encouragement, and G Taschenberger for the data shown in Figure 3.17.

References

Ashmore JF, Attwell D (1985) Models for electrical tuning in hair cells. Proc Roy Soc B 226:325–344.
Authier S, Manley GA (1995) A model of frequency tuning in the basilar papilla of the Tokay gecko, *Gekko gecko*. Hear Res 82:1–13.
Baird IL (1974) Anatomical features of the inner ear in submammalian vertebrates. In: Keidel WD, Neff WD (eds) Handbook of Sensory Physiology Vol. V/1. Berlin, New York: Springer-Verlag, pp. 159–212.
Benser ME, Marquis RE, Hudspeth AJ (1996) Rapid, active hair-bundle movements in hair cells from the bullfrog's sacculus. J Neurosci 16:5629–5643.
Brix J, Manley GA (1994) Mechanical and electromechanical properties of the stereovillar bundles of isolated and cultured hair cells of the chicken. Hear Res 76:147–157.
Brix J, Fischer FP, Manley GA (1994) The cuticular plate of the hair cell in relation to morphological gradients of the chicken basilar papilla. Hear Res 75:244–256.
Buchfellner E, Leppelsack H-J, Klump GM, Häusler U (1989) Gap detection in the starling (*Sturnus vulgaris*): II. Coding of gaps by forebrain neurons. J Comp Physiol A 164:539–549.
Burns EM, Arehart KH, Campbell SL (1992) Prevalence of spontaneous otoacoustic emissions in neonates. J Acoust Soc Am 91:1571–1575.

Buus S, Klump GM, Gleich O, Langemann U (1995) An excitation pattern model for the starling (*Sturnus vulgaris*). J Acoust Soc Am 98:112–124.

Carroll RL (1988) Vertebrate Palaeontology and Evolution. New York: Freeman.

Chandler JP (1984) Light and electron microscopic studies of the basilar papilla in the duck, *Anas platyrhynchos*: I. The hatchling. J Comp Neurol 222:506–522.

Chen L, Salvi R, Shero M (1994) Cochlear frequency-place map in adult chickens: intracellular biocytin labeling. Hear Res 81:130–136.

Chen L, Salvi RJ, Trautwein PG, Powers N (1996) Two-tone rate suppression boundaries of cochlear ganglion neurons in normal chickens. J Acoust Soc Am 100:442–450.

Christensen-Dalsgaard J, Jorgensen MB (1996) One-tone suppression in the frog auditory nerve. J Acoust Soc Am 100:451–457.

Code RA (1995) Efferent neurons to the macula lagena in the embryonic chick. Hear Res 82:26–30.

Code RA, Carr CE (1994) Choline acetyltransferase-immunoreactive cochlear efferent neurons in the chick auditory brainstem. J Comp Neurol 340:161–173.

Cohen GM (1987) Acetylcholinesterase activity in the embryonic chick's inner ear. Hear Res 28:57–63.

Cole KS, Gummer AW (1990) A double-label study of efferent projections to the cochlea of the chicken, *Gallus domesticus*. Exp Brain Res 82:585–588.

Cotanche DA (1987) Regeneration of hair cell stereociliary bundles in the chick cochlea following severe acoustic trauma. Hear Res 30:181–196.

Cotanche DA, Saunders JC, Tilney LG (1987) Hair cell damage produced by acoustic trauma in the chick cochlea. Hear Res 25:267–286.

Cotanche DA, Henson MM, Henson OW Jr (1992) Contractile proteins in the hyaline cells of the chicken cochlea. J Comp Neurol 324:353–364.

Cotanche DA, Lee KH, Stone JS, Picard DA (1994) Hair cell regeneration in the bird cochlea following noise damage or ototoxic drug damage. Anat Embryol 189:1–18.

Counter SA, Tsao P (1986) Morphology of the seagull's inner ear. Acta Otolaryngol 101:34–42.

Crawford AC, Fettiplace R (1980) The frequency selectivity of auditory nerve fibers and hair cells in the cochlea of the turtle. J Physiol Lond 306:79–125.

Crawford AC, Fettiplace R (1981) An electrical tuning mechanism in turtle cochlear hair cells. J Physiol 312:377–412.

Crawford AC, Fettiplace R (1985) The mechanical properties of ciliary bundles of turtle cochlear hair cells. J Physiol 364:359–379.

Desmedt JE, Delwaide PJ (1963) Neural inhibition in a bird: effect of strychnine and picrotoxin. Nature 200:583–585.

Desmedt JE, Delwaide PJ (1965) Functional properties of the efferent cochlear bundle of the pigeon revealed by stereotaxic stimulation. Exp Neurol 11:1–26.

Dooling RJ (1992) Hearing in birds. In: Fay RR, Popper AN, Webster DB (eds) The Evolutionary Biology of Hearing. Heidelberg, New York: Springer-Verlag, pp. 545–559.

Drenckhahn D, Merte C, von Düring M, Smolders J, Klinke R (1991) Actin, myosin and alpha-actinin containing filament bundles in hyaline cells of the caiman cochlea. Hear Res 54:29–38.

Dyson ML, Klump GM, Gauger B (1998) Absolute hearing thresholds and critical masking ratios in the European barn owl: a comparison with other owls. J Comp Physiol A 182:695–702.

Eatock RA, Manley GA (1981) Auditory nerve fibre activity in the tokay gecko: II, temperature effect on tuning. J Comp Physiol A 142:219–226.

Eatock RA, Manley GA, Pawson L (1981) Auditory nerve fibre activity in the tokay gecko: I, implications for cochlear processing. J Comp Physiol A 142:203–218.

Fay RR (1990) Suppression and excitation in auditory nerve fibers of the goldfish, *Carassius auratus*. Hear Res 48:93–110.

Fay RR (1992) Structure and function in sound discrimination among vertebrates. In: Fay RR, Popper AN, Webster DB (eds) The Evolutionary Biology of Hearing, Heidelberg, New York: Springer-Verlag, pp. 229–263.

Feduccia A (1980) The Age of Birds. Cambridge: Harvard University Press.

Fettiplace R (1987) Electrical tuning of hair cells in the inner ear. Trends Neurosci 10:421–425.

Firbas W, Müller G (1983) The efferent innervation of the avian cochlea. Hear Res 10:109–116.

Fischer FP (1992) Quantitative analysis of the innervation of the chicken basilar papilla. Hear Res 61:167–178.

Fischer FP (1994a) Quantitative TEM analysis of the barn owl basilar papilla. Hear Res 73:1–15.

Fischer FP (1994b) General pattern and morphological specializations of the avian cochlea. Scanning Microsc 8:351–364.

Fischer FP (1998) Hair-cell morphology and innervation in the basilar papilla of the emu (*Dromaius novaehollandiae*). Hear Res 121:112–124.

Fischer FP, Köppl C, Manley GA (1988) The basilar papilla of the barn owl *Tyto alba*: a quantitative morphological SEM analysis. Hear Res 34:87–101.

Fischer FP, Brix J, Singer I, Miltz C (1991) Contacts between hair cells in the avian cochlea. Hear Res 53:281–292.

Fischer FP, Miltz C, Singer I, Manley GA (1992) Morphological gradients in the starling basilar papilla. J Morphol 213:225–240.

Fischer FP, Eisensamer B, Manley GA (1994) Cochlear and lagenar ganglia of the chicken. J Morphol 220:71–83.

Fuchs PA, Evans MG (1990) Potassium currents in hair cells isolated from the cochlea of the chick. J Physiol Lond 429:529–521.

Fuchs PA, Murrow BW (1992a) Cholinergic inhibition of short (outer) hair cells of the chick's cochlea. J Neurosci 12:800–809.

Fuchs PA, Murrow BW (1992b) A novel cholinergic receptor mediates inhibition of chick cochlear hair cells. Proc R Soc London B 248:35–40.

Fuchs PA, Nagai T, Evans MG (1988) Electrical tuning in hair cells isolated from the chick cochlea. J Neurosci 8:2460–2467.

Fuchs PA, Evans MG, Murrow BW (1990) Calcium currents in hair cells isolated from the cochlea of the chick. J Physiol Lond 429:553–568.

Garrick LD, Lang JW, Herzog HA (1978) Social signals of adult American alligators. Bull Amer Mus Nat Hist 160:153–192.

Gleich O (1987) Evidence for electrical tuning in the starling inner ear. In: Elsner N, Creutzfeld O (eds) New frontiers in brain research. Stuttgart, New York: Georg Thieme Verlag, p. 101.

Gleich O (1989) Auditory primary afferents in the starling: correlation of function and morphology. Hear Res 37:255–268.

Gleich O (1994) Excitation patterns in the starling cochlea: a population study of primary auditory afferents. J Acoust Soc Am 95:401–409.

Gleich O, Dooling RJ (1995) Belgian Waterslager canaries: morphological and physiological studies of a mutant bird strain showing continuous hair-cell production. In: Manley GA, Klump GM, Köppl C, Fastl H, Oeckinghaus H (eds): Advances in Hearing Research, Singapore: World Scientific Publishers, pp. 40–49.

Gleich O, Klump GM (1995) Temporal modulation transfer functions in the European starling (*Sturnus vulgaris*): II. Responses of auditory-nerve fibres. Hear Res 82:81–92.

Gleich O, Manley GA (1988) Quantitative morphological analysis of the sensory epithelium of the starling and pigeon basilar papilla. Hear Res 34:69–86.

Gleich O, Narins PM (1988) The phase response of primary auditory afferents in a songbird (*Sturnus vulgaris* L.). Hear Res 32:81–91.

Gleich O, Dooling RJ, Manley GA (1994) Inner-ear abnormalities and their functional consequences in Belgian Waterslager canaries (*Serinus canarius*). Hear Res 79:123–136.

Gleich O, Manley GA, Mandl A, Dooling R (1994) The basilar papilla of the canary and the zebra finch: a quantitative scanning electron microscopic description. J Morphol 221:1–24.

Gleich O, Klump GM, Dooling RJ (1995) Peripheral basis for the auditory deficit in Belgian Waterslager canaries (*Serinus canarius*). Hear Res 82:100–108.

Gleich O, Dooling RJ, Presson JC (1997) Evidence for supporting cell proliferation and hair cell differentiation in the basilar papilla of adult Belgian Waterslager canaries (*Serinus canarius*). J Comp Neurol 377:5–15.

Gleich O, Ryals BM, Dooling RJ (1998) The number of auditory nerve fibers in normal canaries and Belgian Waterslager canaries. 21st Midwinter Res. Mtg. Assoc. Res. Otolaryngol., Abstr. Nr. 788.

Goldberg JM, Fernández C (1980) Efferent vestibular system in the squirrel monkey: anatomical location and influence on afferent activity. J Neurophysiol 43:986–1025.

Greenwood DD (1961) Critical bandwidth and the frequency coordinates of the basilar membrane. J Acoust Soc Am 33:1344–1356.

Greenwood DD (1990) A cochlear frequency-position function for several species—29 years later. J Acoust Soc Am 87:2592–2605.

Gross NB, Anderson DJ (1976) Single unit responses recorded from the first order neuron of the pigeon auditory system. Brain Res 101:209–222.

Guinan JJ (1996) Physiology of olivocochlear efferents. In: Dallos P, Popper AN, Fay RR (eds) The Cochlea. New York: Springer-Verlag, pp. 435–502.

Gummer AW (1991a) First order temporal properties of spontaneous and tone-evoked activity of auditory afferent neurones in the cochlear ganglion of the pigeon. Hear Res 55:143–166.

Gummer AW (1991b) Postsynaptic inhibition can explain the concentration of short interspike intervals in avian auditory nerve fibres. Hear Res 55:231–243.

Gummer AW, Klinke R (1983) Influence of temperature on tuning of primary-like units in the guinea pig cochlear nucleus. Hear Res 12:367–380.

Gummer AW, Smolders JWT, Klinke R (1987) Basilar membrane motion in the pigeon measured with the Mössbauer technique. Hear Res 29:63–92.

Hennig W (1983) Stammesgeschichte der Chordaten. Hamburg: Paul Parey Verlag.

Henry KR, Lewis ER (1992) One-tone suppression in the cochlear nerve of the gerbil. Hear Res 63:1–6.

Highstein SM, Baker R (1985) Action of the efferent vestibular system on primary afferents in the toadfish, *Opsanus tau*. J Neurophysiol 54:370–384.

Hill KG, Mo J, Stange G (1989a) Excitation and suppression of primary auditory fibres in the pigeon. Hear Res 39:37–48.

Hill KG, Mo J, Stange G (1989b) Induced suppression in spike responses to tone-on-noise stimuli in the auditory nerve of the pigeon. Hear Res 39:49–62.

Hill KG, Stange G, Mo J (1989) Temporal synchronization in the primary auditory response in the pigeon. Hear Res 39:63–74.

Hudspeth AJ (1986) The ionic channels of a vertebrate hair cell. Hear Res 22:37.

Hudspeth AJ, Gillespie PG (1994) Pulling strings to tune transduction—adaptation by hair cells. Neuron 12:1–9.

Jahnke V, Lundquist PG, Wersäll J (1969) Some morphological aspects of sound perception in birds. Acta Otolaryngol 67:583–601.

Jones SM, Jones TA (1995) The tonotopic map of the embryonic chick cochlea. Hear Res 82:149–157.

Jørgensen JM, Christensen JT (1989) The inner ear of the common rhea *(Rhea americana L.)*. Brain Behav Evol 34:273–280.

Kaiser A (1993) Das efferente Bündel des Hörorgans beim Huhn—Ursprung, Projektionen und Physiologie. Doctoral Dissertation, Institut für Zoologie der Technischen Universität München.

Kaiser A, Manley GA (1994) Physiology of putative single cochlear efferents in the chicken. J Neurophysiol 72:2966–2979.

Kaiser A, Manley GA (1996) Brainstem projections of the Macula lagenae in the chicken. J Comp Neurol 374:108–117.

Katayama A, Corwin JT (1989) Cell production in the chicken cochlea. J Comp Neurol 281:129–135.

Kemp DT (1978) Stimulated acoustic emissions from the human auditory system. J Acoust Soc Am 64:1386–1391.

Keppler C, Schermuly L, Klinke R (1994) The course and morphology of efferent nerve fibres in the papilla basilaris of the pigeon (*Columba livia*). Hear Res 74:259–264.

Kettembeil S, Manley GA, Siegl E (1995) Distortion-product otoacoustic emissions and their anaesthesia sensitivity in the European Starling and the Chicken. Hear Res 86:47–62.

Ketten DR (1992) The marine mammal ear: specializations for aquatic audition and echolocation. In: Fay RR, Popper AN, Webster DB (eds) The Evolutionary Biology of Hearing, Heidelberg, New York: Springer-Verlag, pp. 717–750.

Kiang NYS, Watanabe T, Thomas EC, Clark LF (1965) Discharge Patterns of Single Fibers in the Cat's Auditory Nerve. Cambridge, Mass., MIT Press.

Klinke R (1979) Comparative physiology of primary auditory neurons. In: Hoke M, de Boer E (eds) Models of the Auditory System and Related Signal Processing Techniques. Scand Audiol Suppl 9:49–61.

Klinke R, Pause M (1980) Discharge properties of primary auditory fibres in *Caiman crocodilus*; comparisons and contrasts to the mammalian auditory nerve. Exp Brain Res 38:137–150.

Klinke R, Schermuly L (1986) Inner ear mechanics of the cr
basilar papillae in comparison to neuronal data. Hear Res

Klinke R, Smolders JWT (1984) Hearing mechanisms in ᴜ.
Bolis L, Keynes RD, Maddrell SHP (eds) Comparative Phy.
Systems. Cambridge: Harvard University Press, pp. 195–211.

Klinke R, Smolders JWT (1993) Performance of the avian inner ear. Hᵢ
Res 97:31–43.

Klinke R, Müller M, Richter CP, Smolders J (1994) Preferred intervals in birds aᵢ.
mammals: a filter response to noise? Hear Res 74:238–246.

Klump GM, Baur A (1990) Intensity discrimination in the European starling
(*Sturnus vulgaris*). Naturwiss 77:545–548.

Klump GM, Gleich O (1991) Gap detection in the European starling (*Sturnus
vulgaris*) III. Processing in the peripheral auditory system. J Comp Physiol
A. 168:469–476.

Klump GM, Maier EH (1989) Gap detection in the starling (*Sturnus vulgaris*): I.
Psychophysical thresholds. J Comp Physiol A 164:531–539.

Klump GM, Okanoya K (1991) Temporal modulation transfer functions in the
European starling (*Sturnus vulgaris*): I. Psychophysical modulation detection
thresholds. Hear Res 52:1–12.

Konishi M (1970) Comparative neurophysiological studies of hearing and vocalizations in songbirds. Z vergl Physiol A 157:687–697.

Konishi M (1973) How the owl tracks its prey. Amer Sci 61:414–424.

Konishi M, Knudsen EI (1979) The oilbird: hearing and echolocation. Science
204:425–427.

Köppl C (1993) Hair-cell specializations and the auditory fovea in the barn owl
cochlea. In: Duifhuis H, Horst JW, van Dijk P, van Netten S (eds) Biophysics of
Hair-Cell Sensory Systems. London: World Scientific Publishers, pp. 216–222.

Köppl C (1995) Otoacoustic emissions as an indicator for active cochlear mechan-
ics: a primitive property of vertebrate auditory organs. In: Manley GA, Klump
GM, Köppl C, Fastl H, Oeckinghaus H (eds) Advances in Hearing Research.
Singapore: World Scientific Publishers, pp. 207–216.

Köppl C (1997a) Frequency tuning and spontaneous activity in the auditory nerve
and cochlear nucleus magnocellularis of the barn owl *Tyto alba*. J Neurophysiol
77:364–377.

Köppl C (1997b) Number and axon calibres of cochlear afferents in the barn owl.
Auditory Neurosci 3:313–334.

Köppl C (1997c) Phase locking to high frequencies in the auditory nerve and
cochlear nucleus magnocellularis of the barn owl, *Tyto alba*. J Neurosci
17:3312–3321.

Köppl C, Manley GA (1993) Distortion-product otoacoustic emissions in the bobtail
lizard, II: Suppression tuning characteristics. J Acoust Soc Am 93:2834–2844.

Köppl C, Manley GA (1997) Frequency representation in the emu basilar papilla.
J Acoust Soc Am 101:1574–1584.

Köppl C, Wegscheider A (1998) Axon numbers and diameters in the avian auditory
nerve. 21st Midwinter Res. Mtg. Assoc. Res. Otolaryngol., Abstr. Nr. 401.

Köppl C, Gleich O, Manley GA (1993) An auditory fovea in the barn owl cochlea.
J Comp Physiol 171:695–704.

Köppl C, Yates GK, Manley GA (1997) The mechanics of the avian cochlea: rate-
intensity functions of auditory-nerve fibres in the emu. In: Lewis E, Long G, Leake

P, Narins P, Steele C (eds) Diversity in Auditory Mechanics. Singapore: World Scientific, pp. 77–82.

Köppl C, Gleich O, Schwabedissen G, Siegl E, Manley GA (1998) Fine structure of the basilar papilla of the emu: implications for the evolution of avian hair-cell types. Hear Res 126:99–112.

Kössl M, Vater M (1985) Evoked acoustic emissions and cochlear microphonics in the mustache bat, *Pteronotus parnellii*. Hear Res 19:157–170.

Kreithen ML, Quine DB (1979) Infrasound detection by the homing pigeon: a behavioural audiogram. J Comp Physiol 129:1–4.

Ladhams A, Pickles JO (1996) Morphology of the monotreme organ of Corti and macula lagena. J Comp Neurol 366:335–347.

Larsen ON, Dooling RJ, Ryals BM (1996) Roles of intracranial pressure in bird audition. In: Lewis E, Long G, Leake P, Narins P, Steele C (eds) Diversity in Auditory Mechanics. Singapore: World Scientific, pp. 11–17.

Lavigne-Rebillard M, Cousillas H, Pujol R (1985) The very distal part of the basilar papilla in the chicken: a morphological approach. J Comp Neurol 238:340–347.

Leake PA (1977) SEM observations of the cochlear duct in *Caiman crocodilus*. Scanning Electron Micoscopy, II, 437–444.

Lewis ER, Henry KR (1992) Modulation of cochlear nerve spike rate by cardiac activity in the gerbil. Hear Res 63:7–11.

Manley GA (1979) Preferred intervals in the spontaneous activity of primary auditory neurones. Naturwiss 66:582.

Manley GA (1986) The evolution of the mechanisms of frequency selectivity in vertebrates. In: Moore BCJ, Patterson RD (eds) Auditory Frequency Selectivity. New York: Plenum, pp. 63–72.

Manley GA (1990) Peripheral Hearing Mechanisms in Reptiles and Birds. Heidelberg: Springer-Verlag.

Manley GA (1995) The avian hearing organ: a status report. In: Manley, GA, Klump GM, Köppl C, Fastl H, Oeckinghaus H (eds) Advances in Hearing Research. Singapore: World Scientific Publishers, pp. 219–229.

Manley GA (1996) Ontogeny of frequency mapping in the peripheral auditory system of birds and mammals: a critical review. Aud Neurosci 3:199–214.

Manley GA, Gallo L (1997) Otoacoustic emissions, hair cells and myosin motors. J Acoust Soc Amer 102:1049–1055.

Manley GA, Gleich O (1984) Avian primary auditory neurones: the relationship between characteristic frequency and preferred intervals. Naturwiss 71:592–594.

Manley GA, Gleich O (1992) Evolution and specialization of function in the avian auditory periphery. In: Fay RR, Popper AN, Webster DB (eds) The Evolutionary Biology of Hearing. Heidelberg, New York: Springer-Verlag, pp. 561–580.

Manley GA, Köppl C (1998) Phylogenetic development of the cochlea and its innervation. Curr Opinion Neurobiol 8:468–474.

Manley GA, Taschenberger G (1993) Spontaneous otoacoustic emissions from a bird: a preliminary report. In: Duifhuis H, Horst JW, van Dijk P, van Netten S (eds) Biophysics of Hair-Cell Sensory Systems. London: World Scientific Publ. Co., pp. 33–39.

Manley GA, Gleich O, Leppelsack HJ, Oeckinghaus H (1985) Activity patterns of cochlear ganglion neurones in the starling. J Comp Physiol A 157:161–181.

Manley GA, Brix J, Kaiser A (1987) Developmental stability of the tonotopic organization of the chick's basilar papilla. Science 237:655–656.

Manley GA, Schulze M, Oeckinghaus H (1987) Otoacoustic emissions in a song bird. Hear Res 26:257–266.

Manley GA, Yates G, Köppl C (1988) Auditory peripheral tuning: evidence for a simple resonance phenomenon in the lizard *Tiliqua*. Hear Res 33:181–190.

Manley GA, Gleich O, Kaiser A, Brix J (1989) Functional differentiation of sensory cells in the avian auditory periphery. J Comp Physiol 164:289–296.

Manley GA, Köppl C, Yates GK (1989) Micromechanical basis of high-frequency tuning in the bobtail lizard. In: Wilson JP, Kemp D (eds) Mechanics of Hearing. New York: Plenum Press, pp. 143–150.

Manley GA, Haeseler C, Brix J (1991) Innervation patterns and spontaneous activity of afferent fibres to the lagenar macula and apical basilar papilla of the chick's cochlea. Hear Res 56:211–226.

Manley GA, Kaiser A, Brix J, Gleich O (1991) Activity patterns of primary auditory-nerve fibres in chickens: development of fundamental properties. Hear Res 57:1–15.

Manley GA, Schwabedissen G, Gleich O (1993) Morphology of the basilar papilla of the budgerigar *Melopsittacus undulatus*. J Morphol 218:153–165.

Manley GA, Meyer B, Fischer FP, Schwabedissen G, Gleich O (1996) Surface morphology of the basilar papilla of the tufted duck *Aythya fuligula* and the domestic chicken *Gallus gallus domesticus*. J Morphol 227:197–212.

Manley GA, Köppl C, Yates GK (1997) Activity of primary auditory neurones in the cochlear ganglion of the emu *Dromaius novaehollandiae*: spontaneous discharge, frequency tuning and phase locking. J Acoust Soc Am 101:1560–1573.

Murrow BW (1994) Position-dependent expression of potassium currents by chick cochlear hair cells. J Physiol Lond 480:247–259.

Murrow BW, Fuchs PA (1990) Preferential expression of transient current (I_A) by short hair cells of the chick's cochlea. Proc Roy Soc Lond (Biol) 242:189–195.

Ofsie MS, Cotanche DA (1996) Distribution of nerve fibers in the basilar papilla of normal and sound-damaged chick cochleae. J Comp Neurol 370:281–294.

Ohyama K, Sato T, Wada H, Takasaka T (1992) Frequency instability of the spontaneous otoacoustic emissions in the guinea pig. Abstr. 15. Mtg, Assoc. Research in Otolaryngol. p. 150.

Okanoya K, Dooling RJ (1985) Colony differences in auditory thresholds in the canary (*Serinus canarius*). J Acoust Soc Am 78:1170–1176.

Okanoya K, Dooling RJ (1987) Strain differences in auditory thresholds in the canary (*Serinus canarius*). J Comp Psychol 101:213–215.

Okanoya K, Dooling RJ, Downing RD (1990) Hearing and vocalizations in hybrid Waterslager-Roller canaries (*Serinus canarius*). Hear Res 46:271–276.

Padian K, Chiappe LM (1998) The origin and early evolution of birds. Biol Rev 73:1–42.

Patuzzi RP (1996) Cochlear micromechanics and macromechanics. In: Dallos P, Popper AN, Fay RR (eds) The Cochlea. New York: Springer-Verlag, pp. 186–257.

Patuzzi RP, Bull CL (1991) Electrical responses from the chicken basilar papilla. Hear Res 53:57–77.

Pickles JO, Brix J, Comis SD, Gleich O, Köppl C, Manley GA, Osborne MP (1989) The organization of tip links and stereocilia on hair cells of bird and lizard basilar papillae. Hear Res 41:31–42.

Retzius G (1884) Das Gehörorgan der Wirbelthiere II Das Gehörorgan der Reptilien, der Vögel und der Säugethiere. Stockholm: Samson and Wallin.

Richter CP, Heynert S, Klinke R (1995) Rate-intensity functions of pigeon auditory-nerve afferents. Hear Res 83:19–25.

Richter CP, Sauer G, Hoidis S, Klinke R (1996) Development of activity patterns in auditory nerve fibres of pigeons. Hear Res 95:77–86.

Rubel EW, Ryals BM (1983) Development of the place principle: acoustic trauma. Science 219:512–514.

Ruggero MA (1973) Response to noise of auditory nerve fibres in the squirrel monkey. J Neurosci 36:569–587.

Runhaar G (1989) The surface morphology of the avian tectorial membrane. Hear Res 37:179–187.

Russell I, Palmer AR (1986) Filtering due to inner hair-cell membrane properties and its relation to the phase-locking limit in cochlear nerve fibers. In: Moore BCJ, Patterson RR (eds) Auditory Frequency Selectivity. New York, London: Plenum Press, pp. 198–207.

Sachs MB (1964) Responses to acoustic stimuli from single units in the eighth nerve of the green frog. J Acoust Soc Am 36:1956–1958.

Sachs MB, Kiang NYS (1968) Two-tone inhibition in auditory-nerve fibres. J Acoust Soc Am 43:1120–1128.

Sachs MB, Lewis RH, Young ED (1974) Discharge patterns of single fibers in the pigeon auditory nerve. Brain Res 70:431–447.

Sachs MB, Woolf NK, Sinnott JM (1980) Response properties of neurons in the avian auditory system: comparisons with mammalian homologues and consideration of the neural encoding of complex stimuli. In: Popper AN, Fay RR (eds) Comparative Studies of Hearing in Vertebrates. New York: Springer-Verlag, pp. 323–353.

Salvi RJ, Saunders SS, Powers NL, Boettcher FA (1992) Discharge patterns of cochlear ganglion neurons in the chicken. J Comp Physiol 170:227–241.

Saunders SS, Salvi RJ (1993) Psychoacoustics of normal adult chickens: thresholds and temporal integration. J Acoust Soc Am 94:83–90.

Schermuly L, Klinke R (1985) Change of characteristic frequencies of pigeon primary auditory afferents with temperature. J Comp Physiol A 156:209–211.

Schermuly L, Klinke R (1990a) Infrasound sensitive neurones in the pigeon's cochlear ganglion. J Comp Physiol A 166:355–363.

Schermuly L, Klinke R (1990b) Origin of infrasound sensitive neurones in the papilla basilaris of the pigeon: a HRP study. Hear Res 48:69–78.

Schmiedt RA, Zwislocki J, Hamernik RP (1980) Effects of hair-cell lesions on responses of cochlear nerve fibres. I. Lesions, tuning curves, two-tone inhibition and responses to trapezoidal-wave patterns. J Neurophysiol 43:1367–1389.

Schwartzkopff J, Winter P (1960) Zur Anatomie der Vogel-Cochlea unter natürlichen Bedingungen. Biol Zentralblatt 79:607–625.

Schwarz IE, Schwarz DWF, Frederickson JM, Landolt JP (1981) Efferent vestibular neurons: a study employing retrograde tracer methods in the pigeon (*Columba livia*). J Comp Neurol 196:1–12.

Schwarz DWF, Schwarz IE, Dezsoe A (1992) Cochlear efferent neurons projecting to both ears in the chicken, *Gallus domesticus*. Hear Res 60:110–114.

Smith CA (1985) Inner ear. In: King AS, McLeland J (eds) Form and Function in Birds, Vol 3. London: Academic Press, pp. 273–310.

Smith CA, Konishi M, Schull N (1985) Structure of the barn owl's (*Tyto alba*) inner ear. Hear Res 17:237–247.

Smolders JWT, Klinke R (1984) Effects of temperature on the properties of primary auditory fibres of the spectacled caiman, *Caiman crocodilus* (L.). J Comp Physiol 155:19–30.

Smolders JWT, Klinke R (1986) Synchronized responses of primary auditory fibre populations in *Caiman crocodilus* (L.) to single tones and clicks. Hear Res 24:89–103.

Smolders JWT, Ding-Pfennigdorff D, Klinke R (1995) A functional map of the pigeon basilar papilla: correlation of the properties of single auditory nerve fibres and their peripheral origin. Hear Res 92:151–169.

Steel KP (1995) Inherited hearing defects in mice. Annu Rev Genet 29:675–701.

Steele CR (1996) Three-dimensional mechanical modeling of the cochlea. In: Lewis E, Long G, Lyon RF, Narins P, Steele CR, Hecht-Poinar E (eds) Diversity in Auditory Mechanics. Singapore: World Scientific, pp. 455–461.

Stiebler IB, Narins PM (1990) Temperature-dependence of auditory nerve response properties in the frog. Hear Res 46:63–82.

Strutz J (1981) The origin of centrifugal fibers to the inner ear in *Caiman crocodilus*. A horseradish peroxidase study. Neurosci Lett 27:95–100.

Strutz J, Schmidt CL (1982) Acoustic and vestibular efferent neurons in the chicken (*Gallus domesticus*). Acta Otolaryngol 94:45–51.

Sugai T, Sugitani M, Ooyama H (1991) Effects of activation of the divergent efferent fibers on the spontaneous activity of vestibular afferent fibers in the toad. Jpn J Physiol 41:217–232.

Sullivian WE, Konishi M (1984) Segregation of stimulus phase and intensity coding in the cochlear nucleus of the barn owl. J Neurosci 4:1787–1799.

Takasaka T, Smith CA (1971) The structure and innervation of the pigeon's basilar papilla. J Ultrastruct Res 35:20–65.

Tanaka K, Smith CA (1978) Structure of the chicken's inner ear. Am J Anat 153:251–271.

Taschenberger G (1995) Spontane otoakustische Emissionen und Verzerrungsprodukt-Emissionen der Schleiereule, *Tyto alba guttata*. Doctoral Dissertation, Institut für Zoologie der Technischen Universität München.

Taschenberger G, Manley GA (1996) Influence of contralateral acoustic stimulation on distortion-product otoacoustic emissions in the barn owl. 19th Midwinter Res Mtg Assoc Res Otolaryngol., Abstr. Nr. 87.

Taschenberger G, Manley GA (1997) Spontaneous otoacoustic emissions in the barn owl. Hear Res 110:61–76.

Taschenberger G, Manley GA (1998) General characteristics and suppression tuning properties of the distortion-product otoacoustic emission 2f1–f2 in the barn owl. Hear Res 123:183–200.

Temchin AN (1982) Acoustical reception in birds. In: Ilyichev VD, Gavrilov VM (eds) Acta XVII Congressus Internat Ornithologicus, Moscow, August 1982, pp. 275–282.

Temchin AN (1988) Discharge patterns of single fibres in the pigeon's auditory nerve. J Comp Physiol A 163:99–115.

Tilney LG, Saunders JC (1983) Actin filaments, stereocilia, and hair cells of the bird cochlea. I. Length, number, width, and distribution of stereocilia of each hair cell are related to the position of the hair cell on the cochlea. J Cell Biol 96:807–821.

Tilney MS, Tilney LG, DeRosier DJ (1987) The distribution of hair cell bundle lengths and orientations suggests an unexpected pattern of hair cell stimulation in the chick cochlea. Hear Res 25:141–151.

von Békésy G (1960) Experiments in Hearing. (Wever EG, trans) New York: McGraw-Hill.

von Düring M, Karduck A, Richter HG (1974) The fine structure and the inner ear in *Caiman crocodilus*. Z Anat Entwickl-Gesch 145:41–65.

von Düring M, Andres KH, Simon K (1985) The comparative anatomy of the basilar papillae in birds. Fortsch d Zool 30:681–685.

Warchol ME, Dallos P (1989a) Neural response to very low-frequency sound in the avian cochlear nucleus. J Comp Physiol A 166:83–95.

Warchol ME, Dallos P (1989b) Localization of responsiveness to very low frequency sound on the avian basilar papilla. Abstracts 12th Mtg. Assoc Res Otolaryngol, p. 125.

Warr WB (1992) Organization of olivocochlear efferent systems in mammals. In: Webster DB, Popper AN, and Fay RR (eds) The Mammalian Auditory Pathway: Neuroanatomy. New York: Springer-Verlag, pp. 411–448.

Weiss TF, Rose C (1988) Stages of degradation of timing information in the cochlea: a comparison of hair-cell and nerve-fibre responses in the alligator lizard. Hear Res 33:167–174.

Werner CF (1938) Das Innenohr des Helmkasuars und anderer, 'Ratiten'. Zool Anz 124:67–74.

Whitehead MC, Morest DK (1981) Dual populations of efferent and afferent cochlear axons in the chicken. Neurosci 6:2351–2365.

Whitehead ML, Lonsbury-Martin B, Martin GK, McCloy MJ (1996) Otoacoustic emissions: animal models and clinical observations. In: Van De Water TR, Popper AN, Fay RR (eds) Clinical Aspects of Hearing. New York: Springer Verlag, pp. 199–257.

Wilson JP, Smolders JWT, Klinke R (1985) Mechanics of the basilar membrane in *Caiman crocodilus*. Hear Res 18:1–24.

Winter IM, Robertson D, Yates GM (1990) Diversity of characteristic frequency rate-intensity functions in guinea-pig auditory nerve fibres. Hear Res 45:191–202.

Woolf NK, Sachs MB (1977) Phase-locking to tones in avian auditory-nerve fibers. J Acoust Soc Am 62:46.

Wu Y-C, Art JJ, Goodman MB, Fettiplace R (1995) A kinetic description of the calcium-activated potassium channel and its application to electrical tuning of hair cells. Prog Biophys Molec Biol 63:131–158.

Zwislocki J, Cefaratti L (1989) Tectorial-membrane II: Stiffness measurements in vivo. Hear Res 42:211–227.

4
The Hearing Organs of Lizards

Geoffrey A. Manley

1. Reptiles, Lizards, and the Structure of the Hearing Organ

The reader may well ask the question: Why deal with the hearing organ of lizards and not all reptiles? The answer lies in the great diversity of the group of animals placed under the old term "reptiles"; it has long been recognized that they are a highly diverse assemblage of animals. This diversity is equally great in the structure of their hearing organs, and at least three basic structures of hearing organs can be distinguished, the turtle type, the archosaur type, and the lizard type. One of the "reptilian" lines, the crocodiles, alligators, and their relatives, are more closely related to the birds than to other reptiles and they are placed with the birds in the group Archosauria. The characteristics of their hearing organs is dealt with in this volume by Gleich and Manley in Chapter 3. The turtles or Chelonia are difficult to classify, and the only certain thing is that they have been a separate group since the Triassic period. Since my earlier review (Manley 1990), much work has been carried out on the ion channels of their hair cells (e.g., Ricci and Fettiplace 1998), but to cover all this work would exceed the limits of the present review.

Another evolutionary line called the Lepidosauria has as its modern representatives the lizards (Lacertilia), the snakes (Serpentes), and the "living fossil" *Sphenodon* (Rhynchocephalia; the tuatara of New Zealand). The terms "lizard" and "snake" are no longer used in scientific classification. Instead, these two groups are placed under the heading Squamata, a monophyletic group. Snakes are closely related to monitor lizards (Lee 1997). Because there has been no work on the auditory system of snakes since my earlier review (Manley 1990), they will not be discussed again here. Instead, this review discusses only the hearing organ of lizards, which among the amniotes shows by far the greatest structural variability while always remaining recognizable as such.

There are several reasons to be interested in the hearing organs of lizards. They are interesting to the comparative and evolutionary morphologist because of the remarkable degree of structural variation between the ears of lizard families, genera, and even species. A systematic comparison using only the hearing organs of the lizard families has lead to classifications of family groups (Wever 1978; Miller 1980, 1992) that generally parallel those based on other features, for example, those of the skeleton. Thus the hearing organ's structure has become one of the morphological features used to determine the systematic status of lizard groups (e.g., Baird 1970; Wever 1978; Miller 1980, 1992). From the evolutionary point of view, the hearing organ provides a very interesting basis to ask questions such as: "Were there particular selection pressures that gave rise to the structural variation that we find, or is it the result of neutral evolution or genetic drift?" The functional morphologist and physiologist finds that these hearing organs present an extremely interesting series of natural experiments for investigating the relationship between structure and function in the auditory periphery of land vertebrates. What common functional principles can be described using the variety of lizard hearing organs, and what do the results tell us about fundamental questions regarding function in all amniote ears? The fact that most lizards are not vocal, and therefore do not use their auditory system for intraspecific communication, makes questions relating to the great variety of hearing organ morphologies highly interesting.

The purpose of this chapter is not to provide an exhaustive review of structure and function of lizards' hearing organs, as detailed information is available from other sources (e.g., Wever 1978; Miller 1980, 1985, 1992; Manley 1990; Köppl and Manley 1992). Instead, this chapter will provide only a sketch of the structures and physiological data underlying our current understanding of the function of the lizard hearing organ. It will place more emphasis on the area where the most progress has been made since the last reviews were written, for example, in the area of otoacoustic emissions. These data will be discussed in the context of modern concepts of the function of hair cell sensory epithelia and of new results of modeling the papillar responses.

For the purposes of the present treatment of the lizards, the classification of Estes et al. (1988; Table 4.1) will be used, here simplified to show only the groups relevant to this chapter.

2. Morphology of the Hearing Organ

A number of detailed reviews of the anatomy of the lizard ear already exist (Baird 1970; Wever 1978; Miller 1980, 1985, 1992; Manley 1990). The present treatment will only attempt a brief overview of the general features. Table 4.2 lists the lizard species mentioned in this chapter.

As in other tetrapods, the inferior portion of the lizard otic labyrinth contains the saccule, with its saccular macula, and the cochlear duct. The cochlear duct contains anteriorly or anterio-ventrally the lagenar macula

TABLE 4.1. Phylogenetic classification of the lizards.[a]

Squamata
  Iguania
    Iguanidae (Iguanas: *Anolis sagrei, Sceloporus orcutti*)[b]
  Scleroglossa
    Gekkota
      Gekkonidae (Geckos: *Gekko gecko, Eublepharis macularius*)
    Autarchoglossa
      Scincomorpha
        Scincoidea
          Scincidae (Skinks: *Tiliqua rugosa*)
        Lacertoidea
          Lacertiformes
            Lacertidae (European lizards: *Podarcis sicula, Podarcis muralis*)
          Teiidoidea
            Teiidae (Tegus: *Callopistes maculatus, Tupinambis teguixin*)
      Anguimorpha
        Anguidae (Glass "snakes": *Gerrhonotus multicarinatus, Gerrhonotus leiocephalus*)
        Varanoidea
          Varanidae (Monitor lizards: *Varanus bengalensis, Varanus exanthematicus*)

[a] The cladistic principle of indenting subordinate groups has been followed.
[b] Species names are in italics (species often referred to are in parentheses).

TABLE 4.2. List of lizard species mentioned in this chapter

Scientific name	Common name	Family
Anolis sagrei	Bahamian brown anole	Iguanidae
Callopistes maculatus	Chile tegu lizard	Teiidae
Calotes versicolor	Indian variable lizard	Agamidae
Eublepharis macularius	Leopard gecko	Gekkonidae
Gekko gecko	Tokay gecko	Gekkonidae
Gerrhonotus leiocephalus	Texas alligator lizard	Anguidae
Gerrhonotus multicarinatus	Southern alligator lizard	Anguidae
Lacerta agilis	European sand lizard	Lacertidae
Podarcis muralis	European wall lizard	Lacertidae
Podarcis sicula	European ruin lizard	Lacertidae
Sceloporus orcutti	Granite spiny lizard	Iguanidae
Tiliqua rugosa	Bobtail skink	Scincidae
Tupinambis teguixin	Golden tegu lizard	Teiidae
Varanus bengalensis	Bengal monitor lizard	Varanidae
Varanus exanthematicus	Steppes monitor lizard	Varanidae

and, nearer the sacculo-cochlear duct, the area of sensory cells known as the basilar papilla, which forms the main hearing organ in amniote vertebrates (Fig. 4.1). In lizards, there is no division between the lagenar and basilar portions of the cochlear duct. The basilar papilla is situated on the basilar membrane, a free area of thick membrane that is supported by a connective tissue surround called the limbus and that separates the otic (endolymph-filled) from the periotic (perilymph-filled) spaces.

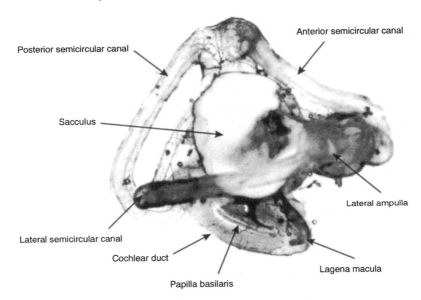

FIGURE 4.1. The right vestibular labyrinth of *Tiliqua rugosa* dissected out of its bony sheath. Dorsal is at the top, anterior to the right, lateral is toward the observer. The dark strip of cells making up the papilla basilaris can be seen lying oriented posterio-dorsally to anterio-ventrally within the cochlear duct. The apical end of the papilla is to the right, near the lagenar macula. Both the darkly stained lagenar- and basilar-papillar branch of the eighth nerve can be seen near their respective sensory epithelia. (Original photograph courtesy of C. Köppl.)

Over the course of the evolution of the lizards, the dimensions of the basilar membrane have diversified among lizard families (Fig. 4.2). The very short papilla of some families (Agamidae, Anguidae, Iguanidae, and Lacertidae), being at most a few hundred micrometers long with only a few hundred sensory hair cells, was probably shortened secondarily. In most lizard families (e.g., Teiidae, Varanidae, Scincidae, Gekkonidae), the basilar papilla length has become more than 20 times the width, and the number of hair cells varies up to about 2,000 (Table 4.3). It should be noted, however, that the length of the papilla is not necessarily an indicator of a primitive or an advanced state. Following established convention (after Miller 1980), the term *apical* will be used for the end of the cochlear duct anatomically distant from the sacculus but close to the lagenar macula and *basal* will be used for the other end (Fig. 4.1). The side of the basilar papilla where the nerve fiber bundles enter is the *neural* side, the opposite side is *abneural*.

The basilar papilla is found on the lateral surface of the basilar membrane and, in most cases, the hearing epithelium is continuous. In the varanid and lacertid families and in isolated cases in the iguanid family,

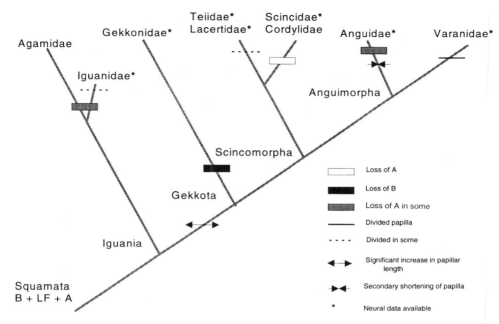

FIGURE 4.2. Putative morphological changes in the basilar papilla during the evolution of the main lizard families. It is assumed that the stem lizards (bottom left, at the origin of the Squamata) possessed a small papilla consisting of three hair cell areas (B + LF + A). These were a low-frequency area (LF), bordered both apically (A) and basally (B) by high-frequency regions presumably consisting of bidirectionally oriented hair cell areas. Following the branching off of the Iguania, a general increase in length of the papilla probably occurred, and long papillae are a feature of almost all later families. Only in the Anguimorpha, specifically here the Anguidae, was the papilla secondarily shortened. Most lizard families show some loss of the high-frequency regions, either the apical one (in skinks and some agamids, iguanids, and anguids) or the basal one (all geckos). In varanids and lacertids and some iguanids, the papilla is secondarily divided into subpapillae, an event that occurred independently at least twice.

however, the basilar papilla is almost or completely divided into two hair cell areas or subpapillae by a connective tissue bridge connecting the neural and abneural sides of the limbus (Figs. 4.2, 4.3, 4.6; Table 4.3). Although the position of this division is similar in the iguanids, lacertids, and varanids, it is likely that it was evolved independently at least twice.

All lizards have two kinds of hair cell areas (Miller 1980). Miller and Beck (1988) described cytological features and innervation patterns that define these two groups of hair cells. In one group, the hair cells have a greater basal diameter, large numbers of, and larger afferent nerve fibers, with more afferent synapses per fiber. These hair cells also have an efferent innervation. In most lizard papillae, such hair cells are located in an area where all

TABLE 4.3. Morphological characteristics of the basilar papillae of various lizard species from six families

Species (Family)	Papillar length	Sub-papillae	No. of hair cells	No. (est.) of afferent nerve fibers	Ratio of afferents to hair cells	Tectorial structure over HF hair cells
Sceloporus orcutti (*Iguanidae*)	300 m	No	85	(500)	5.9	No
Gerrhonotus multicarinatus (*Anguidae*)	400 m	No	193	890	4.6	No
Podarcis sp. (*Lacertidae*)	500 m	2	125	(540)	4.3	Continuous, unattached
Gekko gecko (*Gekkonidae*)	1,900 m	No	1,950	(1,800)	0.9	Preaxial: continuous, attached. Postaxial: 170 sallets
Tiliqua rugosa (*Scincidae*)	2,100 m	No	1,900	(900)	0.48	80 sallets
Varanus exanthematicus (*Varanidae*)	1,800 m	2	1,800	(1,600)	0.89	Continuous, attached

In most species, the number of afferent nerve fibers is not known exactly, but close estimates can be given (in parentheses)

bundles have the same (abneural) orientation. Such areas are referred to as unidirectionally oriented, thus this hair cell type was called the "unidirectional type" (UDT hair cells). Functionally, we now know that these cells are *always* responsible for low-frequency responses (below about 1 kHz; Köppl and Manley 1992). There is always only one UDT hair cell area.

The second hair cell type is the "bidirectional type" (BDT), characterized by their smaller size, the lack of an efferent innervation and the smaller and fewer afferents, that also have smaller synapses. Almost all BDT regions are also bidirectionally oriented, with groups of neurally and abneurally oriented hair cells; Figs. 4.3, 4.4, 4.8. These cells *always* respond to higher sound frequencies than about 1 kHz, with an upper frequency limit between about 4 and 7 kHz. Some BDT areas lack a tectorial membrane. There may be one or two BDT areas. If there are two, they are located apically and basally of the UDT area (e.g., as in some iguanids). If there is only one BDT area, it can be located either at the papilla's apical end (e.g., geckos) or at the basal end (others). In some cases, the bundle-orientation pattern of a hair cell area does not correspond to the hair cell type defined using the more important cytological criteria. It should be emphasized that these types of lizard hair cells are *not* homologous with the two populations of hair cells in mammals or with those in birds (Manley 1990). The arrangement of the basilar papillae of lizards strongly suggests that the two types of lizard hair cell evolved independently of those found in other tetrapod groups (Manley and Köppl 1998).

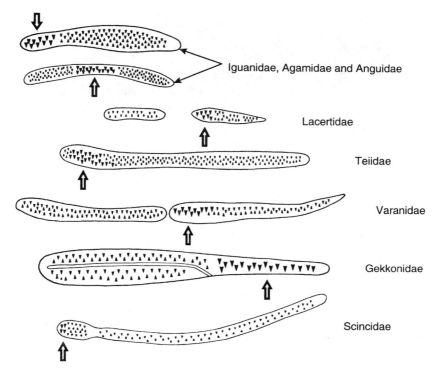

Iguanidae, Agamidae and Anguidae

Lacertidae

Teiidae

Varanidae

Gekkonidae

Scincidae

FIGURE 4.3. Typical hair cell orientation maps for different lizard families. Larger arrowheads on the epithelia indicate the areas where the hair cells all have their bundles oriented in the same direction (abneurally). The center of this area is indicated by an arrow beside the papilla. It can be seen that the location and size of the unidirectionally oriented hair cell area is family-specific. The smaller arrowheads on the epithelia indicate the bidirectionally oriented hair cell areas. (Modified after Miller 1980 and personal communication.)

The primitive condition of the squamate basilar papilla was probably like that of the tuatara *Sphenodon* sp., where all hair cell bundles are oriented abneurally (Miller 1992) and probably correspond to the UDT type. One possibility for the primitive condition in lizards is a medium-sized papilla where all hair cell areas were covered by a tectorial membrane. At that stage of evolution, there was one UDT hair cell area that was probably flanked at both ends by newly developed BDT areas in which the hair cell orientations were not strictly ordered (Fig. 4.2). In modern lizards, the arrangement of UDT and BDT hair cell areas have become family- and, to some extent, genus-specific (Figs. 4.3, 4.6).

A tectorial membrane is present in all lizard papillae, although sometimes it covers only the UDT (low-frequency) hair cell group. If present over both or all hair cell areas, its shape often differs between these areas. In some

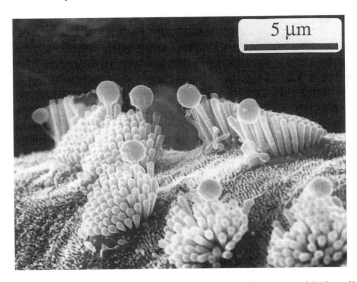

FIGURE 4.4. A scanning electron micrograph of a small group of hair cells in the basal subpapilla (BDT hair cells) of *Podarcis muralis*. The tectorial membrane has been removed to reveal the stereovillar bundles of the hair cells. The kinocilia of the bundles in this species have a prominent rounded tip, which normally contacts both the tallest of the stereovilli and the tectorial membrane. (From Manley 1990, courtesy of C. Köppl.)

papillae (e.g., in iguanids, anguids), a very fine meshwork is present instead of a tectorial structure proper and covers the BDT hair cell areas. Such hair cells have been called "free-standing" (Bagger-Sjöbäck and Wersäll 1973). In some lizard papillae, tectorial structures covering the BDT (high-frequency) hair cell areas have no connection to the limbus (e.g., skinks, part of gecko papillae) and are often subdivided along the length of the papilla into a connected chain of so-called sallets (after a fifteenth-century helmet of a similar shape; Wever 1978). Each sallet covers a certain number of hair cells—in some cases, as in geckos, a single row of hair cells across part of the papilla (Fig. 4.5). The kinocilial tip of lizard hair cells is often expanded, even bulb-like (Fig. 4.4), and appears to serve as a connecting surface between the longest stereovilli and the tectorial membrane where present.

According to the criteria of Miller (1978, 1980, 1985), and compared to primitive papillae such as those of the turtles and *Sphenodon* sp., lizard papillae are specialized. The criteria for recognizing specialization are: (a) the hair-cell density is high (300–400 per $10,000\,\mu m^2$) and the hair cells are often highly regularly arranged, (b) the papillae show two cytologically definable types of hair cell, one of which lacks an efferent innervation, (c) the number of stereovilli per bundle is relatively small (sometimes fewer

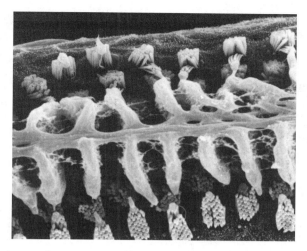

FIGURE 4.5. A scanning electron micrograph of an area of the postaxial hair cell segment of the basilar papilla of *Gekko gecko*, where the hair cells are contacted by specialized pieces of tectorial material known as sallets. Each sallet is connected to its neighbors by a strand of tectorial material and overlies a single transverse row of hair cells. Since tectorial material shrinks during fixation and drying, the lateral ends of the sallet pull on the outermost hair cells of the transverse row and either pull off or disrupt the outermost stereovillar bundles. The width of the hair-cell field across this postaxial hair cell area is about 40 µm. (Original micrograph courtesy of C. Köppl and S. Authier.)

than 40), and (d) the tectorial membrane is often highly specialized, not connected to the limbus, or absent.

Based on a comparison with other reptilian groups, Miller (1978, 1985) suggested that in lizards the primitive condition of the papilla can be recognized by large UDT hair cell areas, BDT areas without a strict organization of differently oriented hair cells, unspecialized innervation patterns, and an unspecialized tectorial membrane. Among modern lizard families, this description applies especially to the teiid and, to some extent, varanid papillae. The other types of lizard papilla, for example, the iguanid-agamid and the very similar anguid type, the gekkonid type and scincid type, are specialized. They show a stricter ordering of the bidirectional area and the specialization or loss of tectorial structures.

As noted above, hair cell orientation and cytological classification of cell type do not always correspond. Thus, although the teiid lizards have a large, unidirectionally oriented hair cell area, some of this area consists cytologically of BDT hair cells. The opposite is also found, however, in skinks, where the cytologically UDT area consists of hair cells whose bundles are not all oriented in the same direction. When two BDT areas exist, they are not necessarily mirror images of each other. Prominent differences are often seen between apical and basal BDT areas, especially where a limbic bridge divides the papilla into subpapillae.

2.1. Morphological Variety and Trends Across Families

2.1.1. Iguanid, Agamid, and Anguid Lizards

There are strong anatomical similarities between the basilar papillae of iguanid, agamid, and anguid lizards (Miller 1980). This similarity can be presumed to be partly the result of convergent evolution, because the anguid family is not closely related to the other two (Fig. 4.2). The structural variability even within one family, however, can be quite large (Fig. 4.6). Nevertheless the data from the southern alligator lizard *Gerrhonotus multicarinatus* (Anguidae) and the granite spiny lizard *Sceloporus orcutti* (Iguanidae) can be discussed together. These species generally have very small papillae (<0.5 mm) with 100–200 hair cells (Fig. 4.6, Table 4.3) and a UDT, low-frequency hair cell area covered by a dense tectorial membrane connected thinly to the neural limbic ridge. They also have one (*G. multicarinatus*) or two (*S. orcutti*) BDT, high-frequency hair cell areas that lack a tectorial structure and also occupy more space on the papilla than the low-frequency area (Miller 1973a; Mulroy 1974; Mulroy and Oblak 1985, Mulroy and Williams 1987; Turner 1987). In *Sceloporus*, the UDT hair cell area is located between the two BDT areas, that are in this case mirror images of each other. Gradients in the number and height of the stereovillar bundles are obvious both *across* the low-frequency region and *along* the high-frequency region(s).

2.1.2. Lacertid Lizards

The family of lacertid lizards is especially interesting, as they are the only family in which the basilar papilla, which is fairly small, is always completely divided into two subpapillae (Figs. 4.2, 4.3, 4.7). The lacertid lizards were the subjects of one of the first scientific attempts to behaviorally assess whether reptiles can hear (Berger 1924; see also Section 6.1). The European ruin and wall lizards *Podarcis sicula* and *Podarcis muralis* are representatives of this family (Köppl, described in Manley 1990). In these species, the subpapillae are each 160 µm long, and separated by a 70-µm bridge of limbic material. The basal subpapilla has about 60 hair cells; in the apical third, UDT-type hair cells are found and are covered by a plate-like tectorial membrane that is attached to the limbus. In the rest of this subpapilla and in all the apical subpapilla (a further 60–70 hair cells), only BDT-type hair cells are found. Both BDT areas are covered with thick, unattached, sausage-like tectorial structures. The complex patterns of the stereovillar numbers and bundle heights are consistent with the frequency responses described (see below).

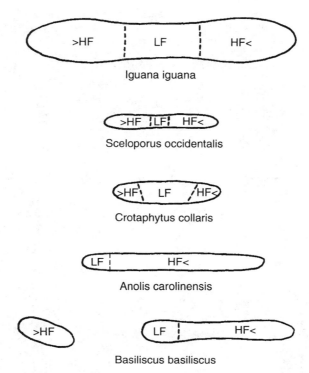

FIGURE 4.6. Schematic scale drawings of the sensory epithelial surface of the basilar papillae of five species of iguanid lizards, to illustrate the structural variation within one family. Apical is to the left in each case. Most iguanid species (e.g., the top three species) have a papilla that shows a central, low-frequency (LF) hair cell area where all hair cells have their stereovillar bundles oriented abneurally, that is flanked by two high-frequency areas that are bidirectionally oriented (HF; >< indicates that the frequency response increases toward the ends of the papilla and away from the low-frequency area). In some of these papillae, such as in species of *Sceloporus*, the high-frequency areas consist of only two hair cell rows. In some species, such as *Anolis carolensis*, the apical HF area is missing. In a few species, as illustrated for *Basiliscus basiliscus*, there is a limbic bridge between the apical and basal subpapillae. The bridge is inserted between the LF area and the apical HF area. The papillar lengths, the total number of hair cells and the percentage of LF hair cells for the five species are as follows: *Iguana iguana*: 600 μm, 310, 21%; *Sceloporus occidentalis*: 250 μm, 74, 20%; *Crotaphytus collaris*: 220 μm, 110, 50%; *Anolis carolinensis*: 400 μm, 182, 18%; *Basiliscus basiliscus*: 620 μm (including limbic bridge), 150, 20%. (Data from Miller 1973a, 1981.)

2.1.3. Monitor Lizards

The monitor lizards (of the genus *Varanus*) have large basilar papillae with a partial or complete constriction at about one-third of the distance from the apical end. A fairly thick tectorial membrane that is attached to the limbus covers all hair cells (Wever 1978). In the Bengal monitor *Varanus*

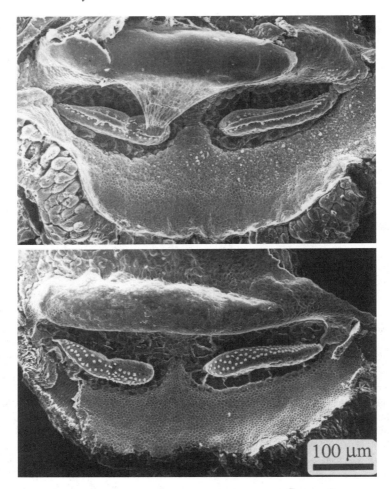

FIGURE 4.7. A scanning electron micrograph of the basilar papilla of *Podarcis muralis* with (top) and without (bottom) attached tectorial structures. The lateral wall of the cochlear duct has been opened by removal of the vestibular (Reissner's) membrane. At the top in each case is the ridge on the neural limbus, from which in the upper picture a tectorial network stretches to contact a tectorial plate over the small region of the apical papilla that is unidirectionally oriented. Both bidirectionally oriented hair cell areas are covered by thick, sausage-shaped tectorial structures that have no contact to the neural limbus. The apical (right) and basal (left) subpapillae are clearly separated by a bridge of limbic material. This papilla is about 500 μm long and has a total of about 130 hair cells. (From Manley 1990, courtesy of C. Köppl.)

bengalensis, the division of the papilla by the constriction is complete (Miller 1978). Here, the apical subpapilla has 786 hair cells, the basal sub-papilla 1,006 hair cells (Miller 1978). The apical quarter of the basal sub-papilla is unidirectionally oriented; all other hair cells are in bidirectionally oriented areas. No details are available on the patterns of hair cell bundle heights and stereovillar numbers.

2.1.4. Geckos

Geckos are most interesting, for they are the only lizards that clearly use strong vocalizations in intraspecific communication and as aggressive warning sounds (e.g., Marcellini 1977; Werner et al. 1978). This is associated with their generally nocturnal or crepuscular way of life. Wever (1978) noted that the ear structure in different gecko groups correlates with their systematic division into four subfamilies. The papilla of geckos is long (Miller 1985) and highly organized. In geckos, and (together with the related pygopods) uniquely in lizard families, the UDT hair cell area lies *basal* on the papilla (Wever 1978; Miller 1992). A thin tectorial membrane reaches down from the limbus and covers the UDT, low-frequency hair cell area. The long apical region consists of two parallel strips of highly organized, BDT hair cell areas separated by a longitudinal hiatus (Fig. 4.8). The hair cells on the neural side of the hiatus are the pre-axial group, those on the abneural side the post-axial group (Miller 1973b). The post-axial hair cells are arranged in well-organized columns across the papilla, and each column is covered by one sallet of the chain of about 170 sallets (Fig. 4.5). A "normal" tectorial membrane covers the pre-axial BDT area. In the Tokay gecko *Gekko gecko*, the papilla is 1.9 mm long and contains about 2,000 hair cells (Miller 1973b; Wever 1978; Köppl and Authier 1995). The arrangement of bundle height and stereovillar numbers on the *G. gecko* papilla is complex (Köppl and Authier 1995).

2.1.5. Skinks

Skinks also possess very well-developed basilar papillae. The papilla of the Australian bobtail skink *Tiliqua rugosa* (or *Trachydosaurus rugosus*) is 2.1 mm long and contains about 1,920 hair cells (Köppl 1988). There is a smaller apical segment and a long basal segment containing on average 1,645 hair cells. Unlike most other lizard papillae, but typical for skinks, there is no significant unidirectional hair cell area (Fig. 4.9; see also Fig. 4.3). Whereas the basal region contains BDT-type hair cells and is simply bidirectionally oriented, the apical area is more complex. The innervational pattern and cytological structure of the apical area, is nonetheless clearly that of UDT hair cells. All hair cells are covered by unattached tectorial material. The apical area has a single, enormous tectorial structure (culmen). Over the basal hair cell area, there is a chain of about 80 sallets.

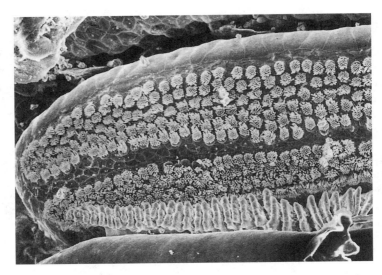

FIGURE 4.8. A scanning electron micrograph of the apical end of the basilar papilla (bidirectional-type hair cells) of the Australian gecko *Gehyra variegata*. Neural is to the bottom, apical to the left. The longitudinal hiatus between the preaxial hair cell strip (below) and the postaxial hair cell strip (above) can be clearly seen. It is also clear that the postaxial hair cells form clearer transverse rows than do preaxial hair cells. The tectorial material has been removed from all hair cell bundles. Remnants of tectorial structures can be seen at the bottom of the picture. The bulge at the bottom of the figure is the overhanging neural limbus typical of geckos. The papilla of *G. variegata* reaches its maximum width near the basal end at about 200 μm. (Original micrograph courtesy of C. Köppl.)

In the apical, low-frequency segment, the number of stereovilli changes *across* the papilla. In the basal, high-frequency papillar area, stereovillar numbers and the height of the bundles change *along* the papilla (Fig. 4.10).

2.1.6. A Comparison Across Species

All the above species in which the anatomy has been studied in some detail show relatively consistent patterns of stereovillar numbers and heights. In low-frequency areas, either the number of stereovilli changes, or both the number and the height of the stereovilli bundles changes *across* the papilla. *G. gecko* is the exception, in that in this species the bundle properties of the low-frequency area change *along* the papilla. Bundle properties also change along the length of the high-frequency area in all species, such that toward the highest frequencies, for example, the maximal height of the stereovillar bundles decreases (Fig. 4.10). The micromechanical significance of these differences is discussed below (Section 3.3).

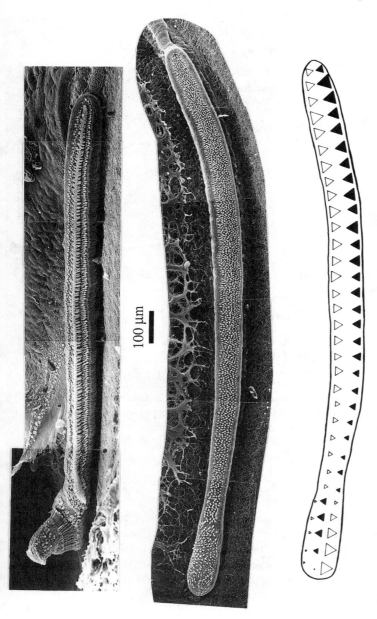

100 μm

FIGURE 4.9. A scanning electron micrograph of the lateral surface of the basilar papilla of *Tiliqua rugosa* with (top) and without (center) tectorial structures. This papilla is nearly 2 mm long and contains over 1,900 hair cells. To the left is the apical region, where a huge tectorial mass covers the hair cells. The much longer basal region is covered by a chain of about 80 tectorial sallets. Neither tectorial structure is connected to the neural limbus. A sketch of the hair cell orientation is shown at the bottom of the figure. The filled triangles represent neural stereovillar-bundle orientation, the open triangles represent abneural orientation. The size of each triangle indicates what percentage of the hair cell bundles in each area have which orientation, relative to the total number of hair cells. It is clear that the entire papilla is bidirectionally oriented, although the extreme apex (left) is almost exclusively abneurally oriented, as in the low-frequency regions of most lizard papillae (cf. Fig. 4.3). (After Köppl 1988 and Manley 1990.)

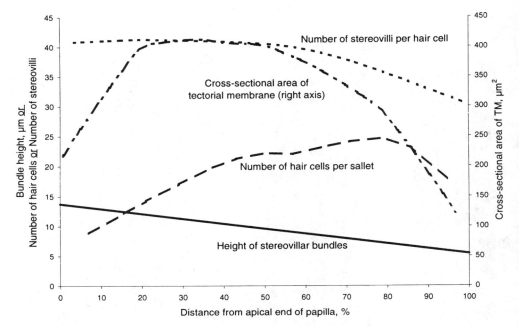

FIGURE 4.10. Diagram to illustrate the dimensional changes in selected anatomical parameters along the basal region of the basilar papilla of *Tiliqua rugosa*. The only data set referring to the right ordinate is the cross-sectional area of the tectorial structures (sallets). All other data sets refer to the left ordinate. These data were used to calculate the resonant frequencies of each location along the basal part of the papilla and, in spite of the nonlinear changes of some parameters along the papilla and the different behavior of the various parameters, the calculation resulted in a monotonic frequency map (see text and Fig. 4.13B). (Data from Köppl 1988.)

2.2. *Trends in Innervation Patterns*

Afferent nerve fibers in lizard species vary in diameter from 0.8 to 6.0 μm and are unimodally distributed in size. The average diameter of afferent fibers is between 2 and 4 μm (Miller 1985; Miller and Beck 1988). Although the number of *hair cells* in lizard papillae varies by a factor of more than 20, from below 100 to 2,000, the number of *afferent nerve fibers* varies by less than a factor of four. This implies that the innervational patterns of the hair cells in large and small papillae are quite different, a fact that could have important functional consequences.

The transmission electron microscopy studies of Miller and Beck (1990) showed that hair cell innervation patterns in lizards vary with the size of the papilla, with the numerical ratio of nerve fibers to hair cells, and thus with the taxonomic status of the species in question. The ratio of the number of nerve fibers to the number of hair cells in lizard families varies from about 1:1 in families with the largest papillae, to between 3.5:1 and 11:1

in families with smaller papillae (Miller 1985; Miller and Beck 1988; Table 4.3). In the distal auditory nerve of one papilla of *Gerrhonotus*, for example, Mulroy and Oblak (1985) found 891 nerve fibers, 29% innervated the apical region and 71% the basal region. With a total of 193 hair cells (151 basal, 42 apical), this worked out to a ratio of 6.1 fibers per hair cell apically and 4.2 basally, 4.6 per hair cell on average (Table 4.3).

According to Miller (1985), Teresi (1985), and Miller and Beck (1988, 1990), there is a tendency toward exclusive afferent innervation of hair cells in small papillae (one fiber only innervates one hair cell) and to nonexclusive innervation in large papillae. In the iguanid lizards studied, all hair cells were exclusively innervated, but the number of synapses varied with the cell's location on the papilla, being greater in UDT hair cells than in BDT hair cells. The greater number of synapses on UDT hair cells was associated with a larger mean afferent-fiber diameter in this low-frequency region (Miller 1985). In species of lacertids and teiids with papillae of medium length, Miller and Beck (1990) refer to a mixed innervation pattern, some hair cells being exclusively, others nonexclusively innervated. In *Varanus exanthematicus*, the hair cell-to-nerve fiber ratio is close to 1:1 (Miller and Beck 1988, 1990), but both UDT and BDT hair cells are nonexclusively innervated, each afferent nerve fiber innervating 3 or 4 hair cells. The number of afferent synapses on UDT is larger than the number on BDT hair cells. In a study in which single afferent fibers were stained in the skink *T. rugosa*, all fibers branched within the basilar papilla to innervate between 4 and 14 hair cells (Köppl and Manley 1990a; Fig. 4.11). Apical fibers tended to innervate hair cells with the same morphological polarity. Basal fibers, in contrast, innervated about equal numbers of hair cells of opposing polarity and were restricted in their longitudinal branching (Fig. 4.11, right) to hair cells beneath one or two sallets.

These facts on innervation patterns of different papillae are best understood in the context of the way in which localized frequency tuning is achieved in the different species. In the high-frequency regions of papillae in families such as iguanids, which completely lack a tectorial structure, each individual hair cell is micromechanically tuned to a particular frequency. Such hair cells are exclusively innervated, thus preserving the best possible frequency selectivity in the afferent fibers of the auditory nerve. In species in which the high-frequency hair cells are mechanically linked by sallets into micromechanically tuned groups (e.g., in geckos and skinks), the hair cells so coupled are jointly (i.e., nonexclusively) innervated. This innervation pattern does not degrade frequency selectivity in the auditory nerve fibers, since the hair cells are micromechanically coupled anyway. This mechanical coupling can actually increase frequency selectivity (see Section 3.4.2). Hair cells that are linked by a continuous tectorial membrane are, as might be expected, also nonexclusively innervated. Thus the fact that Miller and Beck (1990) found a correlation between the innervation patterns in lizards and papillar size and with the taxonomic status can be explained by realizing that papillar size and the species' taxonomic status are also

Apical

Basal

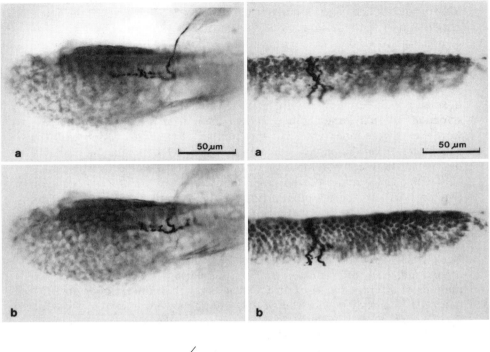

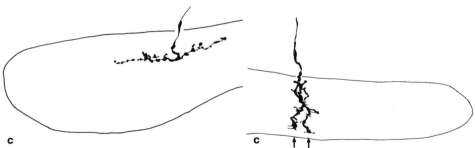

FIGURE 4.11. The innervation patterns of (left column) an apical afferent nerve fiber and (right column) a basal afferent nerve fiber in *Tiliqua rugosa*. The right photographs (**a** and **b**) are taken at two focus levels and show the whole of the apical papillar area. A stained afferent fiber enters from the top and innervates a number of hair cells along the length of the apical region. A drawing incorporating all focal levels is shown in **c** at bottom left. The left photographs **a** and **b** are taken at two focal depths and show an afferent innervating hair cells across a limited stretch near the base of the papilla, beneath sallets. The drawing below shows the entire course of innervation and suggests that the innervated hair cells were confined to one or two sallets. (After Köppl and Manley 1990a.)

correlated with tectorial hair cell coupling patterns and that the tectorial coupling determines to some extent the innervation pattern (see also Table 4.3).

Efferent nerve fibers, about whose physiology we still know nothing, are only found in contact with UDT and not with BDT hair cells in lizards (Miller and Beck 1988). In some species, such as *Calotes versicolor* (Bagger-Sjöbäck 1976) and, indeed probably in all other agamid lizards, there are no efferent fibers to the hearing organ at all (Miller and Beck 1990).

The afferent fibers leaving the auditory papilla and entering the brain stem branch and innervate the regions of the cochlear nuclei differentially. Although details of the termination pattern are only available for *Gerrhonotus* (Szpir et al. 1990), there are indications that the pattern may be general (see Section 5). These authors describe a medial and a lateral region of both cochlear nuclei, the nucleus magnocellularis (i.e. NMM and NML) and the nucleus angularis (NAM and NAL). In *Gerrhonotus*, there are differences between the low- and high-frequency auditory afferents. Low-frequency afferents project primarily to NML, NMM, and NAL. High-frequency fibers, in contrast, project primarily to NAM, although some send additional, thin branches to NML (see also Carr and Code, Chapter 5).

2.3. Correlation with Function: Frequency Range and Frequency Maps

Tonotopicity is one of the fundamental principles underlying the organization of tetrapod hearing organs (Manley 1986). Encoding a wide spectrum of frequencies is always achieved by breaking up the spectrum into a continuum of narrow ranges, each of which is encoded by a number of hair cells. This spectral analysis is achieved by the creation of developmental gradients that manifest themselves in systematically varying anatomical (e.g., cell morphology, height of stereovillar bundles) and/or electrophysiological (e.g., ion-channel characteristics) properties of the linear hair cell array (see Section 3).

All lizard species, whether they have a very well developed hearing organ with 2,000 hair cells on a long papilla or less than 100 hair cells on a very short papilla cover very much the same frequency range of hearing (Fig. 4.12), covering 5 to 6 octaves beginning at about 100 Hz. At the high-frequency end, sensitivity is apparently partly limited by the middle ear (Manley 1990). The neural audiograms of the different species are not all the same, however, and some detailed structure in the thresholds can be seen at frequencies where the frequency mapping on the papilla changes between UDT and BDT hair cell areas. This might explain, for example, the sensitivity dip in the audiogram of *Gerrhonotus* near 1 kHz.

Although the frequency range coded in large and small papillae is similar, there is a large and systematic difference in the capacity for the parallel

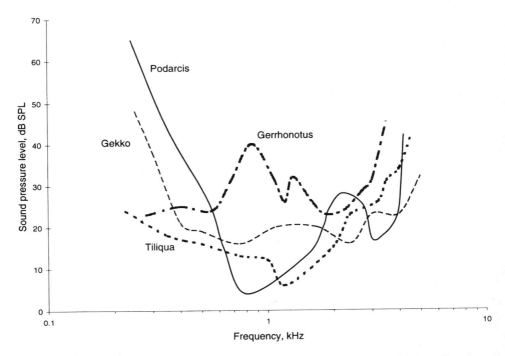

FIGURE 4.12. Neural audiograms derived from eighth-nerve data in four lizard species whose papillae exhibit up to a fivefold difference in length, to compare their sensitivities and frequency range. The lines join the CF thresholds of the most sensitive fibers encountered in the auditory nerve (see also Fig. 4.16E–H). The *Tiliqua rugosa* (scincid) data are from Manley, Köppl, and Johnstone 1990, the *Gekko gecko* (gekkonid) data from Eatock, Manley, and Pawson 1981, the *P. muralis* and *P. sicula* (lacertid) data from Manley 1990, and the *G. multicarinatus* (anguid) data from Holton and Weiss 1983. The discontinuity in the thresholds of some audiograms can be traced to specific features of the papillae. Thus the raised thresholds in *Gerrhonotus* near 1 kHz separates fibers originating from the two regions of the papilla. The raised thresholds in *Podarcis* near 2.5 kHz indicates the separation of the sub-papillae in this species.

coding of information: A papilla with 100 hair cells could, using micromechanical tuning, code maximally for 100 different characteristic frequencies (CFs); in fact, however, there are several hair-cells across each location of the papilla, so the maximum number of steps along the papilla is more likely to be about 25, i.e., only five steps per octave of hearing range. A much larger papilla could have correspondingly more, and of course finer, frequency steps. The number of CF steps would be smaller if neighboring hair cells are combined by nonexclusive afferent innervation or by mechanical coupling via tectorial connections. It is interesting to note that in very small papillae, the hair cells of the high-frequency, BDT area are generally not connected by a tectorial membrane. Their stereovillar bundles can thus move more independently of each other, and the frequency tuning of neighboring hair cells can differ more easily. Similarly, hair cells linked together via a tectorial sallet would vibrate as a unit, but with less coupling along the papilla than would be the case in the presence of a continuous tectorial membrane.

Frequency maps of various degrees of completeness are available for the following lizard species (using either recordings from single nerve fibers close to the papilla or staining and tracing single nerve afferent fibers; examples are shown in Fig. 4.13): *Gerrhonotus* (Weiss et al. 1976), *V. bengalensis* (Manley 1977), *S. orcutti* (Turner et al. 1981), *T. rugosa* (Köppl and Manley 1990a), *Podarcis* (Köppl and Manley 1992), and *G. gecko* (Manley et al. 1999). In all species, the papilla is clearly tonotopically organized, with characteristic frequencies (CFs) of the nerve fibers from the UDT area always being below 1 kHz and those from BDT areas being between about 1 kHz and about 4–6 kHz. In *S. orcutti* the BDT areas are functional mirror images (Turner 1987; Turner et al. 1981). In *Podarcis*, the two BDT hair cell areas, which are in different subpapillae, are not functional mirror images. The mid-frequency range from about 1 kHz up to about 2.7 kHz is localized in the basal part of the *basal* subpapilla, and CFs above about 2.7 kHz emanate from the *apical* subpapilla. The frequency-space mapping constants of the two subpapillae are also quite different. Whereas the basal subpapilla encompasses more than two octaves (<90 μm/octave), the apical subpapilla has less than one octave (>300 μm/octave). A similar differential distribution of frequencies was found in the divided papilla of *V. bengalensis* (Manley 1977, 1990). This reduces the redundancy in the frequency mapping on such papillae and increases the amount of space for analysis of the higher-frequency range and thus the number of hair cells devoted to each octave. This prominent improvement in high-frequency coding can be regarded as the evolutionary selective pressure that led to the development of limbic bridges in three families of lizards. This event occurred independently twice or perhaps even three times.

In general, UDT areas devote less space to a frequency octave than do BDT areas, but there are very substantial differences between species. In *S. orcutti*, less than 0.05 mm of the BDT areas are devoted to one octave— a row of fewer than five hair cells (Table 4.4). In *Gerrhonotus*, this value is 0.15 mm/octave. In the BDT (high-frequency) areas of some other lizards,

TABLE 4.4. Space constants of tonotopic mapping of lizard papillae

Species with neural maps	CFs < 1 kHz	CFs > 1 kHz
Sceloporus orcutti (granite spiny lizard)	30 μm/oct.	55 μm/oct.
Gerrhonotus multicarinatus (alligator lizard)	65 μm/oct.	120 μm/oct.
Podarcis sp. (European lizards)	90 μm/oct.	160 μm/oct.
Tiliqua rugosa (bobtail skink)	150 μm/oct.	750 μm/oct.
Gekko gecko (tokay gecko)	200 μm/oct.	450 μm/oct.
Varanus bengalensis (monitor lizard)	390 μm/oct.	1,000 μm/oct.

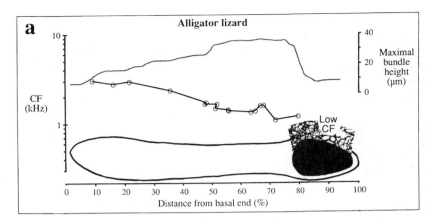

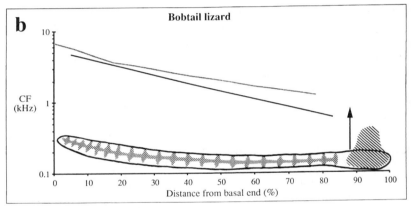

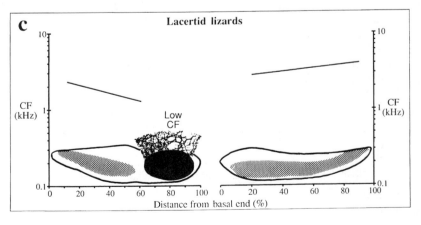

the values are closer to those known for birds and some mammals (Manley 1990). Thus in *T. rugosa*, one octave at higher frequencies occupies 750 μm (Köppl and Manley 1990a).

Low-frequency responses are localized to the small UDT area. In *T. rugosa*, the low frequencies are organized *across* the papilla, increasing from abneural to neural. At these low frequencies (below 800 Hz) one octave occupies only 150 μm (Köppl and Manley 1990a). Anatomical data in some other species, such as *Gerrhonotus*, also suggest that the low frequencies are arranged *across* the UDT region of the papilla. In *G. gecko*, this is not the case, but rather the low frequencies are arranged *along* the papilla, like the high frequencies (Manley et al. 1999).

3. How Is Frequency Tuning Achieved?

The structural and cellular basis of frequency tuning and frequency selectivity have been the most important question motivating auditory research over the past few decades. In these studies, the lizard ear has played a minor role, since most auditory scientists work on mammals. However, the papillar structure is so variable in lizards that these species present a unique opportunity to study structure-function relationships. The data collected have been used to model function in a quite successful manner.

3.1. Middle Ear and Basilar Membrane Mechanics

Of course, the inner ear can be no more sensitive than the middle ear, and it is reasonable to expect that the general pattern of sensitivity across

---

FIGURE 4.13. Diagrammatic presentation of the relationship between papillar morphology and the frequency-response organization of the basilar papillae of four lizard species: (**A**) *G. multicarinatus*; (**B**) *Tiliqua rugosa*; (**C**) a composite for *Podarcis muralis* and *P. sicula*. For each species, a schematic drawing of the outline of the papilla (apical is to the left, neural to the top) is placed along the abscissa of the graph. Tectorial structures attached to the neural limbus are drawn black, those not attached are stippled. There is no tectorial structure along the basal area in *Gerrhonotus*, and the basal tectorial structure in *T. rugosa* is a chain of sallets. The graphs in each case show, as a continuous line, the tonotopic organization of the papilla, in **A** measured by recording very close to the papilla, the line simply joins the data points (data from Holton and Weiss, 1983b). In **B** and **C** the data were gathered by tracing stained, single auditory afferents to the papilla; the black lines in **B** and **C** are linear fits to the data. In addition, the graph in **A** shows as a thin, stippled line the maximal height of the hair cell stereovillar bundles at each location and illustrates the relationship between bundle height and frequency along the basal region. In **B**, the gray line shows the tonotopic organization predicted by the model of Manley, Köppl, and Yates 1989 (see text). (From Köppl and Manley 1992.)

frequency will be strongly influenced by the middle ear response (Saunders et al., Chapter 2).

Peake and Ling (1980) measured the motion of the basilar membrane beneath the papilla in *Gerrhonotus* in response to different frequencies and found essentially the same frequency dependence at different positions along its length. The velocity of the basilar membrane was about equal to that of the middle ear extracolumella over the frequency range 0.5–6 kHz, demonstrating that the tuning of basilar membrane motion was dominated by the tuning of the middle ear. Had these authors used the columellar footplate velocity rather than that of the extracolumella, the match would probably have extended over an even wider frequency range (Manley 1990).

Similarly, measurements of the mechanical response of the middle ear and of the basilar membrane of *T. rugosa* revealed that, to a large extent, the transfer function of the middle ear determined the frequency response of the basilar membrane (Manley et al. 1988, 1989). In the same preparation, it was also shown that the frequency selectivity of the auditory nerve fibers is, around their most sensitive frequency, much greater than than that of the basilar membrane (Fig. 4.14). For the same papillar location in *T. rugosa* it was shown that the ratio of the tuning selectivity of the basilar membrane and auditory nerve fibers resembled a highpass filter with a distinct resonance (Fig. 4.15). At frequencies far from CF, the slopes of the neural curves strongly resembled the poor selectivity of the basilar membrane. The difference in selectivity was well modeled by assuming the presence of local, resonant hair cell and tectorial sallet groups with individual, sharp tuning characteristics that were superimposed on the broad tuning of the basilar membrane (Manley et al. 1988, 1989; see Section 3.4).

There was no systematic, place-dependent frequency selectivity of the basilar membrane mechanics in *T. rugosa*, which, as in *Gerrhonotus*, contrasts sharply with the clear and systematic tonotopic organization of the nerve fibers innervating the different regions along the papilla (Fig. 4.13).

3.2. Frequency-Tuning Characteristics Determined from the Auditory Nerve

In the alligator lizard *G. multicarinatus*, frequency selectivity measured in single auditory afferent fibers has been compared to that measured in hair cells. In each case, the sound pressure levels (SPL) necessary at different frequencies to elicit a standard receptor potential change in a hair cell or a standard increase in discharge rate in a nerve fiber, respectively, define a frequency tuning curve. Recordings of membrane potentials from hair cells during pure-tone stimulation showed that the frequency tuning curves were "V"-shaped, and quite similar to the tuning of afferent nerve fibers (Holton and Weiss 1983). To a first approximation, the auditory nerve fibers are

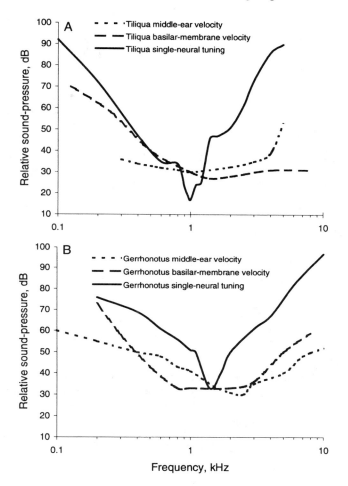

FIGURE 4.14. A comparison of frequency selectivity at three stages (middle ear, basilar membrane, and auditory afferent neuron) in two lizard species, (**A**) *Tiliqua rugosa* and (**B**) *G. multicarinatus*. While the middle ear and basilar membrane velocity frequency responses are broad and similar to each other, in both cases the neural afferent frequency selectivity shows two regions. Along the flanks at higher sound-pressure levels (SPL), the neural tuning curves have similar flank slopes to those of the basilar membrane. Near their center frequency, however, both neural tuning curves show a sharply tuned region of enhanced sensitivity (tuning curve "tip"). See text for description of the models of micromechanical tuning in these two species. For the neural tuning curves, the ordinate is in dB SPL. The middle ear and basilar membrane curves are arranged arbitrarily on the sensitivity axis. (*T. rugosa* data are from Manley et al. 1989, 1990a, and unpublished, *Gerrhonotus* data are from Weiss et al. 1976, Holton and Weiss 1983.)

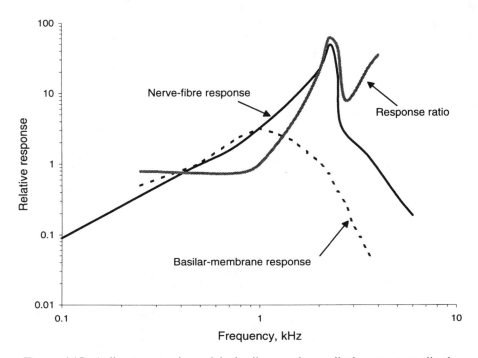

FIGURE 4.15. A direct comparison of the basilar membrane displacement-amplitude response (dashed line) in *T. rugosa* and the threshold of a single auditory afferent fiber (continuous black line), recorded from the same location in the papilla as the basilar-membrane measurement and immediately after it. The curves are arranged so that the low-frequency flanks coincide. The ratio between these curves (thick gray line) shows that the sharpening process underlying the neural tuning behaves like a highpass filter. (After Manley et al. 1989.)

simply reflecting the tuning selectivity of the hair cells with which they synapse, and this can be assumed to be the case in all lizards.

Primary afferent nerve fiber tuning has been studied in a variety of lizards: *G. multicarinatus*, *S. orcutti*, *Podarcis* sp., *V. bengalensis*, *G. gecko* and *T. rugosa*. In all cases, the resulting frequency tuning curves could be divided into two groups with different characteristics (Fig. 4.16A–D), that

FIGURE 4.16. Sensitivity and frequency selectivity at the level of the auditory nerve. (**A–D**) A selection of typical eighth-nerve, single auditory-afferent frequency tuning curves for four lizard species: (**A**) *Gerrhonotus multicarinatus*; (**B**) *Tiliqua rugosa*; (**C**) *Podarcis muralis* and (**D**) *Gekko gecko*. (**E–H**) Scattergrams showing the best thresholds (i.e., at characteristic frequency) for larger numbers of single auditory nerve fibers in the same species as in **A–D**. The data sources are given in the legend to Fig. 4.12.

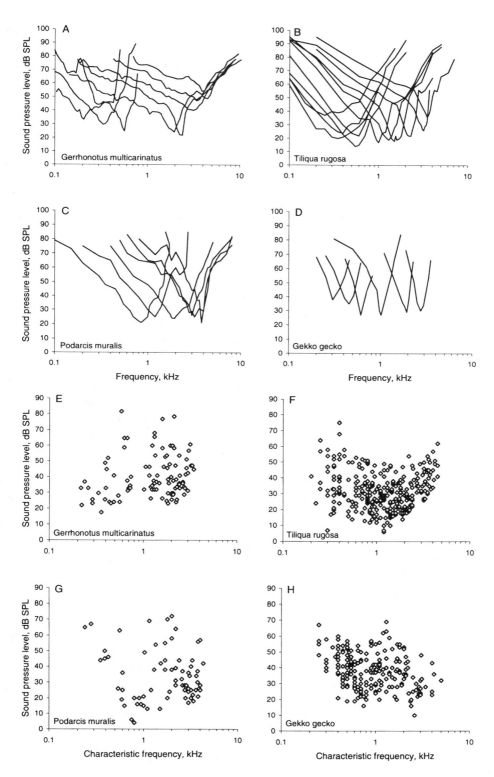

correspond to the two anatomical areas described above (Section 2). There is a low-frequency and a high-frequency group, which are separated near a characteristic frequency of about 1 kHz. The precise border frequency is species specific and will also depend on the animal's temperature. The high-frequency limit of the upper frequency group lies between about 4 and 6 kHz, again depending on the species and the temperature. All frequency tuning curves measured to date were "V"-shaped, but the symmetry of the flanks of the "V" can be different in different species (Manley 1990; Sams-Dodd and Capranica 1994).

In general, the best sensitivity of the hearing organ is about 10 dB SPL (Fig. 4.16E–H). In the two lacertid species *P. sicula* and *P. muralis*, and in *G. gecko*, the response thresholds are somewhat higher for fibers with intermediate CFs (Fig. 4.16G, H; Köppl and Manley, reported in Manley 1990; Köppl and Manley 1992). In *T. rugosa* it was shown that the tuning sharpness and the thresholds of both low- and high-CF fibers were sensitive to hypoxia, the loss of sensitivity during hypoxia being greatest at CF (Manley, Köppl, and Johnstone 1990a).

In most species, the sharpness coefficient of the frequency selectivity of tuning in single auditory nerve afferent fibers (measured as the Q_{10dB} value) increases with the CF of the nerve fiber, although the increase is not necessarily monotonic with CF (Köppl and Manley 1992). It often shows different behavior for CF below and above 1 kHz (Fig. 4.17), that is, the frequency at the division between the different papillar areas (Köppl and Manley 1992). The largest and clearest increases in Q_{10dB} value are seen in *T. rugosa* and *G. gecko*. In *Gerrhonotus*, in contrast, the Q_{10dB} range is essentially the same in the low- and the high-frequency population (Weiss et al. 1976) and in *S. orcutti* the average Q_{10dB} value is actually higher in the low-frequency population (Turner 1987).

In *Gerrhonotus*, the tuning curves of the low-CF units were steeper on the high-frequency flank than were the tuning curves of the high-CF units. This difference may be related to the absence of a tectorial membrane in the high-CF area, for the tectorial membrane is implicated in the strong suppression seen on the high-frequency flanks of the low-frequency tuning curves in the presence of second tones (see also two-tone rate suppression, Section 5.2). There is evidence from models of the micromechanics of lizard papillae that a tectorial membrane contributes significantly to tuning sharpness (Authier and Manley 1995; see Section 3.4.2). In *S. orcutti* (Turner 1987) the details of the shape of tuning curves, Q_{10dB} sharpness coefficients, etc. are remarkably similar to data for *Gerrhonotus*. The most sharply tuned tuning curves in lizards found to date were in *G. gecko* (Eatock et al. 1981). The average Q_{10dB} increased with CF from about 2 at low CF to near 10 at high CF (Fig. 4.17). In *G. gecko*, fibers below and above a CF of about 1 kHz each have an approximately constant but different 10 dB-bandwidth of 200 Hz and 400 Hz, respectively. This produces different rates of increase of Q_{10dB} with CF over these frequency ranges, correlating with the morphological differences between these two hair cell areas (Köppl and Authier

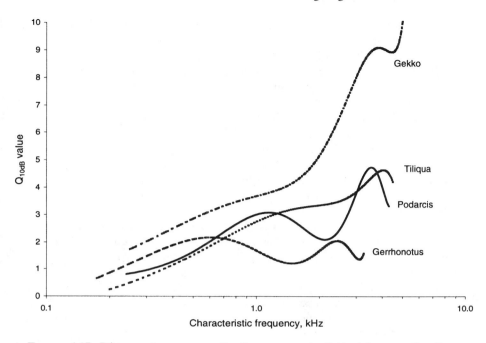

FIGURE 4.17. Diagram to compare the frequency selectivity (given as the Q_{10dB} value) for auditory-nerve fibers of four different lizard species. The lines are fifth-order polynomial fits to the data in each species. Whereas the selectivity for CFs below about 1 kHz are similar, there are large differences above this frequency. At higher frequencies, the tuning sharpness in *Gekko* is about five times higher than in *Gerrhonotus*. The data sources are given in the legend to Fig. 4.12.

1995). A new method of determining frequency selectivity of tuned elements in the hearing epithelium is by means of suppressing otoacoustic emissions (see Section 4.1.1.3).

3.3. Tuning and the Micromechanics of the Hair Cell Bundles

Frishkopf and DeRosier (1983) and Holton and Hudspeth (1983) studied the motion of the "free-standing" stereovillar bundles of basal hair cells (high-frequency area) in extirpated in vitro preparations of the papilla of *G. multicarinatus*. The basilar papilla moved under these conditions in a side-to-side rocking motion, pivoting at the neural limbus. At high frequencies, the largest relative motion between bundles and hair cell apices (i.e., the presumed stimulatory movement) was observed basally, where the hair-cell bundles are short. At lower frequencies, the bundles of the hair cells with the longest stereovilli (those at the apical end of the BDT area) had the largest relative movement. These data thus suggest that the tonotopic organization of the high-frequency part of this papilla is based on the micromechanically determined, frequency-selective motion of the stereovillar bundles, and this is determined by variations in the number and height of stereovilli in the bundles, which determine the bundle stiffness. This concept, and the rocking motion of the papilla, is central to all models of tuning in lizards.

Stimulation of auditory hair cells is achieved by a relative motion between the hair cell and the hair cell bundles with the fluid surrounding them or the tectorial material attached to them. Where a tectorial membrane is present, the resonant frequency is determined by the stiffness and mass of the hair cell bundles and the tectorial structure (Section 3.4). The relationship between bundle height and resonant frequency must thus be different in hair cells lacking a tectorial membrane (i.e., basal hair cells in *G. multicarinatus*) from that of apical hair cells in the same species, which do have a tectorial membrane and, of course, from high-frequency hair cells with a tectorial covering in other species. Although the apical hair cells in *G. multicarinatus* have the shortest bundles in this species, their CFs are *lower* than any in the basal area, because the tectorial membrane of the apical area, through its mass, plays a critical role in reducing the resonant frequencies. In fact, all other things being equal, a simple calculation shows that the tectorial membrane of this region reduces the resonant frequencies of hair cells to about one fifth of what they would be if the bundles were free-standing (see also Section 3.4.2). As the height of the stereovillar bundles of the apical area varies *across* the papilla in both *S. orcutti* and *G. multicarinatus* (e.g., Mulroy and Williams 1987) there is likely to be a tonotopic organization running at right angles to the length of the papilla, similar to that shown in the low-frequency region of *T. rugosa* (Köppl and Manley 1990a).

3.4. Models of Micromechanical Tuning

3.4.1. Models of Frequency Gradients

Modeling mechanical tuning is a complex process, for a number of structures can play a role in tuning—the basilar membrane, tectorial membrane and the hair cell stereovillar bundles being the most obvious candidates. The concept that there are micromechanical resonances in stereovillar bundles and their associated tectorial structures has been successfully used to model the tonotopic arrangement of high-frequency areas in *G. multicarinatus*, *S. orcutti*, *Podarcis*, *T. rugosa*, and *G. gecko*. Data and models correspond well, so that it can be taken as established that tuning in the high-frequency areas of lizard papillae is determined in this relatively simple fashion.

In all species where the tonotopic organization is known, there are obvious correlations between this frequency organization on the one hand

and the various morphological gradients along the basilar papilla on the other (Turner et al. 1981; Frishkopf and DeRosier 1983; Holton and Hudspeth 1983; Weiss and Leong 1985a, 1985b; Köppl 1988; Turner 1987; Manley et al. 1989; Köppl and Manley 1990a). Modeling free-standing stereovillar bundles (without a tectorial membrane) as mechanical spring-loaded pendula, Weiss and Leong (1985a) and Freeman and Weiss (1990) obtained good correspondences between the gradients of characteristic frequency for the high-frequency region of the alligator lizard papilla and the frequency gradient calculated from the model.

In *T. rugosa*, Manley et al. (1988) and Manley et al. (1989) modeled the frequency responses along a hair cell area covered with tectorial sallets. Morphologically defined units, each made up of a specific number of hair cells that had a certain number of stereovilli of a specific maximal height (features that determine the bundle stiffness) and connected by their own sallets (that varied in size from one end of the high-frequency area to the other, and thus determine the mass of the overlying tectorial structure), showed calculated resonances in the correct frequency range (Fig. 4.13B). Together, the inverse trends of mass and stiffness predict resonance frequencies along the basal segment (Manley et al. 1988, 1989) that closely resembled neurophysiological data (Fig. 4.13B, at 30°C; Köppl and Manley 1990a), suggesting that fine tuning in this area is only due to such micromechanical resonances. The gradient of resonance frequencies is mainly determined by the variation in stereovillar height.

When comparing the papillae and the response frequencies in *G. carinatus* or *S. orcutti* and in *T. rugosa*, it is obvious that the gradient in height of the stereovilli between different hair cells along an area without a tectorial membrane (e.g., from 12 to 37 μm) is steeper than the same gradient on hair cells covered by a tectorial membrane (e.g., from 5.5 to 14 μm). The mass provided by the tectorial membrane lowers the response frequencies of hair cells considerably, enabling the same low frequencies to be coded by hair cells with much shorter bundles.

3.4.2. Models of Frequency Selectivity and the Influence of the Tectorial Membrane

As noted above, neural tuning in *T. rugosa* and *G. carinatus* was not only place-specific, but also substantially sharper in frequency selectivity than was the tuning of the basilar membrane (Fig. 4.14A). In addition to the broad frequency response of the basilar membrane, which was manifest in the very similar slopes of the upper flanks of the tuning curves, each neural curve displayed a sensitive "tip" portion having its own private CF. These tip portions had a depth of up to 40 dB below the "predicted" basilar membrane response curve. The difference between these two functions in *T. rugosa* resembles the response characteristics of a relatively simple high-pass resonance filter (Fig. 4.15), whose physical equivalent was presumed to be the coupling of the individual hair cells to each other via the salletal tectorial membrane structures in the high-CF region. By adding a standard highpass filter function of differing center frequency to an average basilar membrane tuning function, it was possible to model auditory nerve tuning curves (Manley et al. 1988). The models we proposed for tuning in the papilla of *T. rugosa* (Manley et al. 1988; Manley et al. 1989) and *G. gecko* (Authier and Manley 1995) not only suggested a role for the tectorial membrane in determining the resonance frequency of localized salletal-hair-cell groups, they also implicated the tectorial membrane in sharpening tuning curves through the coupling of hair cells. In both *Gerrhonotus* and *S. orcutti*, the high-frequency population of nerve fibers (innervating "free-standing" hair cells) is not very sharply tuned (Weiss et al. 1976) when compared to fibers of similar CF in other lizard species that have similarly sized papillae with tectorial membranes, for example, *Podarcis* (Köppl and Manley 1992) and also when compared to data from lizards with larger papillae (see Fig. 4.17). The absence of a tectorial membrane over high-CF hair-cells in *Gerrhonotus* and *S. orcutti* results in shallower high-frequency flanks of the tuning curves.

In their model of the responses of free-standing hair cells in *Gerrhonotus*, Freeman and Weiss (1990) showed that both fluid inertia and viscosity play a key role in bundle mechanics. To achieve the measured tuning via a resonance, however, the modeled stereovillar bundles of *Gerrhonotus* would have to have significantly more mass than they actually possess (Weiss and Leong 1985a). Alternatively, the frictional coefficient would have to be at least one order of magnitude smaller than the value they used (Authier and Manley 1995). It is important to note in this context that hair cells themselves generate mechanical activity (see Section 4.1), and this has not yet been incorporated into any model of lizard papillae. Such mechanical activity due to an active process is perhaps the most important factor working against the damping effects of the cochlear fluid, making the assumption of a strongly reduced frictional coefficient (equivalent to adding a force working against damping) a realistic one, for internal energy is available to work against friction.

To examine the influence of the sallets of *G. gecko* on the resonance frequencies, we carried out a calculation for "free" hair cell bundles in *G. gecko* as if in this species there were no sallets (Authier and Manley 1995). Free-standing hair bundles of the height found in *G. gecko* would resonate over the range from 3 kHz at the basal end of the apical area to 20 kHz at the apical end, that is, at much higher frequencies and over a broader range than is the case when the model includes sallets. In order to cover the same frequency range as that of the sallet-bundle system as described above, we calculated that free hair cell bundles in *G. gecko* would have to have heights between 31 and 17 μm, values that match very well those that actually exist in the free-standing regions of *Gerrhonotus* and *S. orcutti* (Mulroy and Williams 1987; Turner 1987) over roughly the same frequency range as in *G. gecko*. The damping coefficients of the hypothetical "free" bundles in

G. gecko would also be about three times larger (Authier and Manley 1995), and the amplitude maxima would be significantly less sharp than those for the sallet-bundle system. Thus the sallets, in addition to reducing the resonance frequency, appear to substantially sharpen the resonances (Authier and Manley 1995). Auditory afferent nerve fibers of *G. gecko* are in fact more sharply tuned than seen in other lizard species (see Fig. 4.17; Eatock et al. 1981; Manley 1990; Köppl and Manley 1992), especially those seen in "free-standing" hair cell bundle regions of other species (e.g., Weiss et al. 1976).

3.5. Electrical Tuning

Electrical tuning is a primordial property of hair cells and of nerve cell membranes that contain appropriate combinations of ion channels. A general description of electrical tuning can be found in Chapter 3, Section 2.1.1, by Gleich and Manley on hearing in birds, and in Fettiplace (1987) and Wu et al. (1995). The morphological characteristics of the apical segment of the basilar papilla in lizards are quite different from those for the basal segment, as gradients are much less obvious. The lack of prominent morphological gradients in UDT hair cell areas of lizards suggests that a major part of the peripheral tuning mechanism resides in the properties of individual cell membranes of the sensory cells, that is, in electrical tuning. In contrast, electrical tuning would not be expected to play a significant role in high-frequency hair cells. In vitro studies of isolated hair cells from the high-frequency region of the *G. multicarinatus* papilla suggest that electrical tuning does not play a role in such cells (Eatock et al. 1993). Only a rigorous analysis of the properties of the hair cells and the systematic organization of their membrane characteristics can confirm that hair cells are electrically tuned. To date, such an analysis of a hearing organ has only been carried out in turtles (for review see, e.g., Fettiplace 1987) and not in lizards. The presence of preferred intervals in spontaneous activity and of a temperature sensitivity in the tuning of primary auditory fibers, however, provides some evidence for the presence of electrical tuning in many non-mammalian hair cells, at least at low CF (Klinke 1979; Manley 1979, 1981, 1986; Eatock and Manley 1981; Manley and Gleich 1984).

In *T. rugosa* and perhaps other species, the rate of increase of tuning sharpness with CF is different at CFs below and above 1 kHz (Köppl and Manley 1992). This could be due to a changeover from electrical to micromechanical tuning mechanisms, because there is an intrinsic upper frequency limitation in electrical tuning mechanisms (Wu et al. 1995). This necessitates switching to micromechanical mechanisms for the analysis of higher CFs. As suggested by Rose and Weiss (1988) for *G. multicarinatus*, phase-locked timing information is present—and is indeed better—in low-CF fibers (see Section 5.3). The usage of this information about the frequency of the stimulus (and not only of place-dependent tuning information) in low-CF fibers might compensate for the small amount of space devoted to low-frequency octaves. Thus, evidence for electrical tuning in lizard hair cells is only circumstantial and suggestive and based on the idea that the low-frequency region of lizard papillae is probably homologous to the whole of the turtle's papilla. A more rigorous analysis will be necessary in the future.

4. Active Feedback by Hair Cells: Evidence from Otoacoustic Emissions

The discovery that sensory hair cells of vertebrates are not just passive sound receivers but have a metabolically driven active process capable of amplifying weak stimuli opened new ways to study the ear and to understand some old problems concerning the function of the auditory papilla (Manley and Köppl 1998). There is earlier evidence that hair cell bundles are not purely passive structures, since Crawford and Fettiplace (1985) showed that bundle resonances in turtle basilar papilla hair cells depend on an active contribution from the cell body. Thus all frequency-selectivity mechanisms that depend on bundle properties cannot be regarded as resulting from purely passive mechanical characteristics. Active adaptational mechanisms that exert force to move bundles along the axis of sensitivity have been described in saccular hair cells (Hudspeth and Gillespie 1994).

4.1. Spontaneous Otoacoustic Emissions (SOAE) in Lizards

Otoacoustic emissions are sounds in the normal audio range, and of very low amplitude, that are measurable in the external ear canal either spontaneously or when elicited by externally applied tones. They reflect mechanical activity of hair cells of the basilar papilla or cochlea that is associated with normal frequency-tuning mechanisms. Emissions of one or more of the four types (spontaneous, delayed-evoked, simultaneous-evoked, and distortion-product) have been detected in all classes of terrestrial vertebrates (for reviews see Zurek 1985; Probst et al. 1991; Köppl 1995; Whitehead et al. 1996), strongly suggesting that they are a primitive property. It is, however, unclear what the generation mechanism or mechanisms are (Hudspeth 1997).

Spontaneous otoacoustic emissions (SOAE) result from self-sustained oscillations in hearing organs, for which a source of energy—thought to be in the hair cell metabolism—is required (e.g., Talmadge et al. 1991). Evidence from amplitude studies in frogs and mammals—and more recently in lizards and a bird (van Dijk et al. 1996)—indicates that such oscillators produce SOAE in all species, including lizards. The characteristics of these

SOAE are remarkably similar to those of mammals and humans (Köppl 1995; van Dijk et al. 1996). Such mechanical activity would of course strongly influence the passive mechanical resonances that form the basis of models of lizard papilla frequency tuning (see Section 3.4). In mammals, the demonstration of fast in vitro motility of the outer hair cells (Brownell et al. 1985) provides an attractive potential mechanism for such emissions. However, the highly specialized structure of mammalian outer hair cells and their unique anatomical relationship to other cochlear structures suggests that this kind of motility may only be found in mammals. Apart from mammals, SOAE having very similar characteristics have been reported from all tetrapod groups, each having a quite different papillar morphology; from frogs, lizards, and birds (Köppl 1995). While the mechanisms underlying the generation of otoacoustic emissions may not be exactly the same in all species, it is reasonable from an evolutionary point of view to initially postulate a common basis. Nonetheless, it is entirely possible that two different mechanisms driving cochlear amplifiers exist, perhaps working together in some animal groups (Hudspeth 1997; Manley and Köppl 1998).

4.1.1. General Features of SOAE in Lizards

Otoacoustic emissions have been described in many lizard species, including *T. rugosa* (Köppl and Manley 1993b, 1994), *G. gecko* and *Eublepharis macularius* (Manley et al. 1996), *Callopistes maculatus* and *V. exanthematicus* (van Dijk et al. 1996; Manley et al. in preparation), *Tupinambis teguixin* (Manley 1997), and *Anolis sagrei* and *G. leiocephalus* (Manley 1997; Manley and Gallo 1997). Peaks in the averaged spectra of the lizard ear canal sound field (Fig. 4.18) can be identified as SOAE by their consistent presence, their temperature- and hypoxia-sensitivity and their suppressibility by external tones. Spontaneous otoacoustic emissions are found in a very high percentage of lizard ears (often 100%), with between 2 and 15 SOAE per ear and peak levels between −10 and 27 dB SPL. In some species, for example the bobtail skink *T. rugosa* and the geckos, the peaks are sometimes superimposed on a broad baseline emission (Fig. 4.18B, D, E). In general, the instantaneous frequency of any emission varies rapidly within its bandwidth limits. Because frequencies in the center of the band occur most commonly, but not with higher levels than frequencies on the periphery of the band, these rapid variations lead to the broad peaks observed after averaging. The 3-dB bandwidths of SOAE peaks in *T. rugosa* lie between 44 and 225 Hz.

4.1.1.1. The Origin of SOAE in the High-Frequency Segment of the Papilla

The spectral peaks of SOAE in lizards have center frequencies between 1 and 7.5 kHz, which indicates that they are being generated by hair cells of

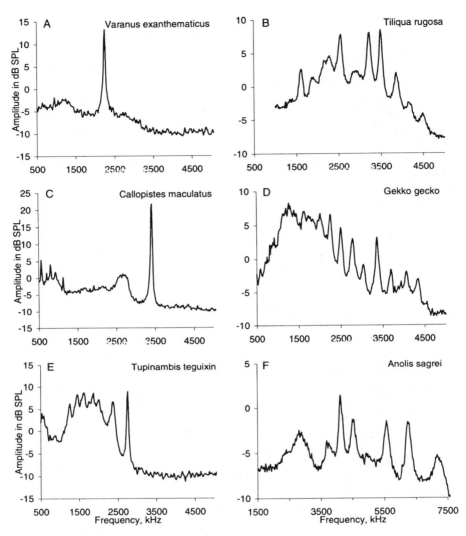

FIGURE 4.18. Typical ear canal sound spectra in quiet for six lizard species, to illustrate the species-typical spectral patterns of spontaneous otoacoustic emissions (SOAE). (**A**) *Varanus exanthematicus*; (**B**) *Tiliqua rugosa*; (**C**) the teeid *Calopistes multicarinatus*; (**D**) *Gekko gecko*; (**E**) the gold tegu, *Tupinambis teguixin*, also a teiid; (**F**) the Bahamian brown anole, *Anolis sagrei*. The three species on the left show fewer, larger SOAE than the three on the right.

the high-frequency hair cell area. In *T. rugosa* the frequencies of SOAE peaks were found to correspond exactly to the frequency range of the basal segment of the papilla as known from single-unit neural data. Also, at equivalent body temperatures, the frequency of the highest SOAE and the highest nerve fiber CF (4.5 kHz; Manley et al. 1990a) corresponded almost exactly to the highest frequencies suppressible by added tones. In contrast, no SOAE were found in the frequency range processed by the apical, low-frequency segment of the papilla. This dichotomy between the two papillar segments is very interesting, given the probability that the high-frequency segment of the basilar papilla operates only with micromechanical tuning, whereas the low-frequency segment may also, or primarily, rely on electrical tuning (see Sections 3.4 and 3.5). The low- and high-frequency segments of the papilla of lizards differ anatomically in three respects that are relevant to the generation of SOAE: (1) the hair cells of the high-frequency hair cell area are bidirectionally oriented and often in roughly equally sized populations on the two sides of the papilla; (2) the hair cells of the high-frequency segment receive no efferent innervation (Miller 1980, 1992; Miller and Beck 1988, 1990), precluding an involvement of the efferent system in the generation of SOAE; (3) the tectorial structures often differ in their morphology.

4.1.1.2. The Temperature Dependence of SOAE in Lizards

Spontaneous otoacoustic emissions are temperature-dependent in all species, including lizards. The SOAE peaks shift to higher frequencies when the animals are warmed and to lower frequencies when cooled, the rate of change often differing between species (Manley 1997). The degree of frequency shift is generally dependent not only on the center frequency of the SOAE (lower-frequency SOAE show a smaller shift) but also on the temperature range in question (shifts are greater at lower temperatures; Fig. 4.19). The smallest shifts were observed in the temperature range around the species' ecological optima. The temperature dependence is also related to the papillar structure (Section 4.1.1.1). Suppression tuning curves of SOAE (Section 4.1.1.3) also shifted with temperature, the largest effects being near the center frequency in the tuning curve's tip region. These findings are very similar to the effect of temperature on SOAE of frogs and mammals (Köppl 1995). In contrast to these effects on frequency, there is no consistent effect of temperature on SOAE amplitudes. The amplitude seems to vary unsystematically with temperature, except that at temperature extremes, the SOAE gradually disappear into the noise.

4.1.1.3. Suppression Effects on SOAE in Lizards

Three basic types of effects were observed when lizard SOAE were exposed to external tones: suppression (reduction of the emission's amplitude), facilitation (an increase in the emission's amplitude), and frequency shifting:

1. The suppressive effect was always highly frequency selective, and the frequency with the lowest suppression threshold was close to the emission's own frequency. From the input-output functions of suppression by a range of frequencies of external tones, iso-suppression tuning curves (STC) can be derived. These curves are V-shaped and strongly resemble primary

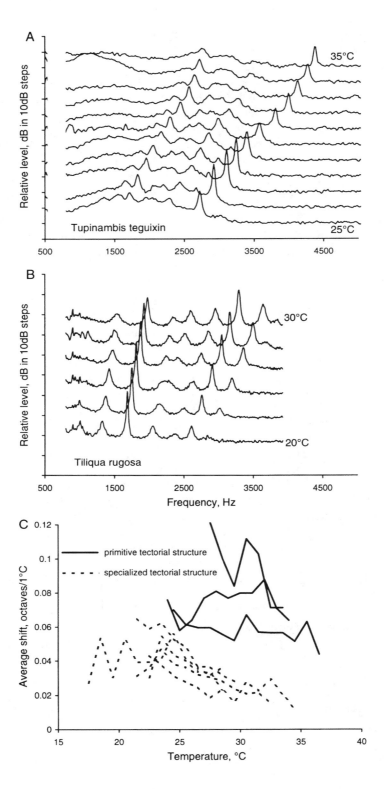

neural tuning curves in threshold, in their tuning coefficient Q_{10dB}, and in the slopes of the flanks (an example for *T. rugosa* is shown in Fig. 4.20). This fact, combined with the fact of the hypoxia-sensitivity of SOAE, suggests that the mechanism(s) generating SOAE are intimately related to or part of those responsible for the frequency tuning and sensitivity of the high-frequency part of these hearing organs.

2. In restricted frequency ranges, facilitation of between 2 and 20 dB also occurs at sound pressure levels below those that suppress at those particular frequencies. In *T. rugosa* the most commonly observed facilitation frequency range lies between 0.2 and 0.6 octaves above the emission's own frequency and this coincides with a characteristic notch in the iso-suppression tuning curves and in the tuning curves of auditory afferents (Manley, Köppl, and Johnstone 1990a). In the same narrow frequency range, the input/output functions of amplitude suppression always showed a pronounced increase in slope.

3. External tones often cause shifts in the frequency of SOAE; frequency "pushing," where emissions moved their own frequency away from that of an external tone, are more common than "pulling." The maximal frequency shifts are large (greater than 300 Hz) and more pronounced in the downward direction.

4.1.1.4. The Relationship Between SOAE Characteristics and Papillar Structure

In comparing the various characteristics of SOAE from eight species of lizards, it became obvious that it was possible to classify the lizards into two broadly defined groups (Manley 1997). One group consists of *A. sagrei* (Manley and Gallo 1997), *G. leiocephalus* (Manley 1997), *G. gecko* and *E. macularius* (Manley et al. 1996), and *T. rugosa* (Köppl and Manley 1993b, 1994; Manley and Köppl 1994). These animals are characterized by a specialized tectorial structure over the high-frequency hair cell area. In

---

FIGURE 4.19. Diagrams to illustrate the differences in the temperature sensitivity of the center frequency of spontaneous otoacoustic emissions (SOAE) in different lizard species. Those with a relatively primitive tectorial membrane structure (continuous lines in **C**; see also Fig. 4.18A, C, E) show a greater frequency shift as a result of temperature change. This is illustrated for one species of this group in **A** by a waterfall display of successively recorded spectra taken at intervals of 1°C change in mouth temperature. Those with either no tectorial structure over the high-frequency area or a specialized structure (broken lines in **C**; see also Fig. 4.18B, D, F), such as sallets, show a lesser sensitivity. This is illustrated for one species from this group by the waterfall display in **B**; in this case the spectra were taken at 2°C intervals in mouth temperature. Both **A** and **B** were measured over a 10°C range. The degree of difference in temperature shift in these two lizard groups is shown in **C**. (Some data from Manley 1997.)

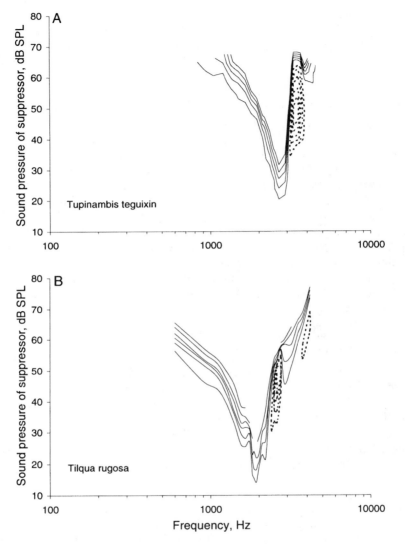

FIGURE 4.20. Two examples of suppression and facilitation effects on SOAE in lizards. Both diagrams show several iso-suppression contours (continuous lines, for 1–6 dB suppression in the bobtail skink in B, or 1–5 dB suppression for the gold tegu in **A**. The dashed lines are iso-facilitation contours for similar degrees of facilitation. Such contours are generally closed. In the case of *Tiliqua rugosa*, a second region of facilitation is seen at a higher frequency. Whatever causes this facilitation is presumably also responsible for the steepening of the high-frequency flank of these iso-suppression curves and, as is known in *T. rugosa* at least, the same steep flanks seen in single auditory-neural tuning curves as shown in Fig. 4.16B.

A. sagrei and *G. leiocephalus*, the tectorial membrane is missing, whereas in *G. gecko*, *E. macularius*, and *T. rugosa* it is partly or completely in the form of sallets. In this group, there are on average 8.6 SOAE per ear, with a maximal amplitude of 6.5 dB SPL. The degree of frequency shift with temperature change is small, on average 0.03 octave/°C.

The second group is made up of two teiid species (*C. maculatus* and *T. eguixin*), and of *V. exanthematicus* (Manley et al. in preparation). These animals are characterized by a continuous tectorial structure over their high-frequency hair cell area, the primitive condition in lizards. They have fewer (on average 4.8) SOAE in each ear, that are of higher amplitude (average species maximum 22 dB SPL). The temperature dependence of the SOAE frequency is almost three times higher, or about 0.08 octave/°C. The larger SOAE and the greater temperature dependence in the monitor lizard and the teiids may be due to stronger coupling of hair cells through the tectorial membrane (Manley 1997).

4.2. Distortion Product Otoacoustic Emissions (DPOAE) in Lizards

Distortion product otoacoustic emissions (DPOAE) have only been reported in *G. multicarinatus* (Rosowski et al. 1984) and *T. rugosa* (Manley et al. 1990b, 1993; Manley and Köppl 1992; Köppl and Manley 1993a), although detailed low-level measurements were only carried out in *T. rugosa* (Fig. 4.21). In *T. rugosa*, DPOAE were measurable throughout the whole hearing range, including the low frequencies not represented in SOAE. DPOAE in different frequency ranges—arising in the two segments of the basilar papilla—differed in a number of respects.

In spite of substantial differences in hearing organ structure and function, DPOAE patterns in *T. rugosa* are very similar to those of mammals. This suggests that the mechanisms generating the DPOAE patterns are based on basic hair cell properties and not on special structural constellations of hair cells or their accessory structures (Manley et al. 1993).

The frequency selectivity of the DPOAE $2f_1-f_2$ and $2f_2-f_1$ in *T. rugosa* was studied in two ways: firstly, by measuring thresholds of DPOAE detection while varying the frequency ratio of the primary tones, and secondly, by constructing iso-suppression tuning curves. Both the DPOAE $2f_1-f_2$ and $2f_2-f_1$ of *T. rugosa* are suppressed by tones in a frequency-selective way (Fig. 4.21B, C, D). The characteristics of tuning in high-frequency DPOAE showed some remarkable similarities to the tuning curves of single nerve fibers (Fig. 4.21B, C, D), suggesting that the characteristics of acoustic DPOAE produced by low-level primary tones were determined by the tuning characteristics of the local hair-cell area. Iso-suppression tuning

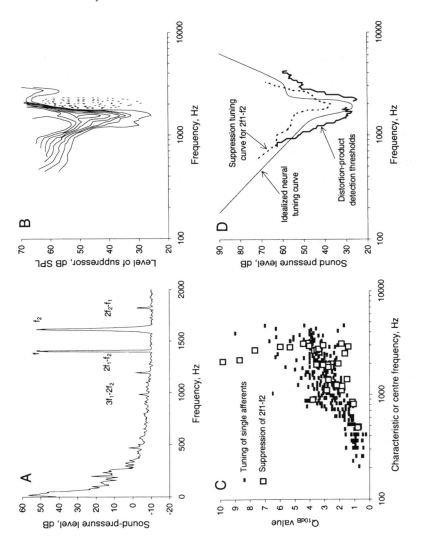

curves for DPOAE generated by low-level primary tones resemble excitatory iso-rate tuning curves for single auditory-nerve fibers of *T. rugosa*. Iso-suppression tuning curves of low- and high-frequency DPOAE mirror the characteristic differences between low- and high-frequency neural tuning curves in this lizard (Manley et al. 1990a; Köppl and Manley 1993a), strongly supporting the assumption of an origin of the DPOAE in the corresponding segments of the basilar papilla (Köppl and Manley 1992). When generated by equal-level primary tones, the best suppressive frequencies for the DPOAE $2f_1-f_2$ at low primary-tone levels were near the first primary tone (f_1), those for $2f_2-f_1$ lay near or above f_2. Facilitation of DPOAE resembled that seen in SOAE (Fig. 4.21B). Low-level DPOAE were sensitive to hypoxia.

5. Time and Intensity Information Coding by Afferent Nerve Fibers

Our deliberations so far have concerned the mechanisms by which the sensory cells achieve their sensitivity and frequency selectivity. Of course, this information has to be coded into trains of action potentials that pass along the afferent nerve fibers to the first auditory brain centers in the medulla oblongata. In this section, we will consider the discharge activity patterns of these afferent fibers and see how the information coding takes place.

FIGURE 4.21. Diagrams to illustrate the characteristics of distortion-product otoacoustic emissions (DPOAE) in *Tiliqua rugosa*. In **A** is shown an ear canal spectrum measured during the presentation of two tonal stimuli (f_1 and f_2) of moderate sound-pressure level. The three distortion peaks shown ($2f_1-f_2$, $2f_2-f_1$, $3f_2-2f_1$) only occur during tonal stimulation and originate in the hearing organ. In **B** is shown an example of a series of iso-suppression contours (continuous lines) for suppression of the distortion product $2f_1-f_2$ by 1–8 dB. The distortion product was facilitated in its amplitude by some combinations of frequency and sound pressure (dashed lines in **B**; cf. similar contours for SOAE in Fig. 4.20). The frequency selectivity of iso-suppression curves can be compared to rate-threshold curves for auditory nerve afferents by measuring the Q_{10dB} value in both cases. A comparison of such data is shown in **C**, where the short bars represent single nerve fibers and the open boxes are suppression tuning curves for DPOAE. A further comparison of tuning selectivity is shown in **D**, where an idealized auditory nerve fiber tuning curve (thin continuous line) with a CF near 2 kHz is compared to one suppression tuning curve for a DPOAE (dashed line) and the detection thresholds for DPOAE (thick continuous lines) both for an f_1 at 1.9 kHz. In the latter case, the data represent the minimal level of the primary tones at which a distortion product ($2f_1-f_2$) can be detected. (Data from Köppl and Manley 1993a; Manley et al. 1993.)

5.1. Spontaneous Activity of Auditory Nerve Afferents

Spontaneous activity in auditory afferents of lizards—action potential production in the absence of auditory stimuli—has not been studied in great detail. Most of the available information concerns only the spontaneous rates, whereby it is important to remember that the temperatures of the animals during the recordings differed, as do the optimal temperatures of the various species. In all species, the spontaneous activity is irregular in nature, resembling a Poisson distribution that is truncated at very short time intervals due to the duration of the action potential and its the refractory period. In *Gerrhonotus*, primary afferent nerve fibers have spontaneous discharge rates up to 80 spikes/s (Weiss et al. 1976). Rates of *Podarcis* were from 0 to 71 spikes/s (Manley 1990), of *G. gecko* between 1.6 and 40 spikes/s (Eatock and Manley 1981; Eatock et al. 1981) or show means in *G. gecko* between 8 and 58 spikes/s (Dodd and Capranica 1992), in *Varanus* between 0.65 and 52.1 spikes/s, and in *T. rugosa* between 0.1 and 123.7 spikes/s (Köppl and Manley 1990b).

In *G. gecko*, fibers with lower CF were found to have a more complex discharge pattern than the typical truncated Poisson distribution. Instead, the time-interval histograms (TIH) showed preferred intervals (i.e., intervals that occurred more often than expected in a Poisson distribution), separated by intervals that seldom occurred. These preferred intervals appeared as more-or-less prominent peaks in the TIH. Their duration was close to the reciprocal of the CF-frequency, that is, the period of the CF. This pattern that could be due to uncontrolled noise, possibly indicating that the activity recorded was not truly spontaneous. There are a number of good reasons for believing that this activity is in fact spontaneous (Eatock and Manley 1981), including the very high thresholds of some units showing this pattern and the fact that in some cells the period of the preferred intervals did not in fact correspond to the period of the cell's CF.

As demonstrated in turtle basilar papilla auditory afferent fibers (Fettiplace 1987), such preferred intervals may be the result of spontaneous oscillations in the resting potentials of the hair cells. These oscillations can be due to random ion-channel openings, with this membrane noise being filtered by the oscillation characteristics of the cell's ion channels (Manley 1979; Eatock and Manley 1981). The fact that preferred intervals are confined to low frequencies may be due to the fact that only UDT hair cells are likely to be electrically tuned, and in *G. gecko* only UDT hair cells are exclusively innervated. In *T. rugosa* preferred intervals were not normally seen, perhaps due to the lack of exclusive innervation of hair cells in this species (Köppl and Manley 1990b).

5.2. Responses to Tonal Stimuli: Rate-Intensity Functions; PSTH and TTRS

Based upon data from *G. multicarinatus, T. rugosa, G. gecko,* and *V. bengalensis*, it was suggested that the rate-level functions for tonal stimulation of nerve fibers innervating the low- and high-frequency segments in lizards are different (Manley 1977; Eatock et al. 1981; Eatock and Weiss 1986; Köppl and Manley 1990b).

In *G. multicarinatus*, apical (low-frequency) fibers showed a steeper growth of rate-intensity functions (narrower dynamic range) at CF than high-CF fibers. The steeper rate of growth may be related to the fact that apical fibers make about three times as many afferent synapses than do basal (free-standing, high-frequency) fibers (Mulroy 1986). In high-frequency fibers, the saturation rates and the slope of the rate-level functions are independent of frequency, whereas in low-frequency fibers, the discharge rate saturates at lower rates for frequencies above CF. The dynamic ranges of different fibers ranged from 14 to 40 dB (Eatock and Weiss 1986). Similar data were reported for *G. gecko* by Sams-Dodd and Capranica (1994). In total, however, there are insufficient data in lizard rate-intensity functions for a systematic analysis, and this analysis would be difficult due to different methods of measuring the dynamic range.

In *T. rugosa*, primary fibers mostly responded to sound with a discharge rate increase, and driven activity rates sometimes exceeded 500 spikes/s. The response patterns as indicated by the shape of peri-stimulus-time histograms (PSTH) varied (Fig. 4.22). Whereas almost all low-CF cells (CF <0.65 kHz) showed more tonic discharge patterns at their CF (Fig. 4.22A), higher-CF cells showed strongly phasic or intermediate types (Fig. 4.22B; Köppl and Manley 1990b). For low-CF cells, the form of the PSTH and the maximal discharge rate varied with stimulus frequency; the highest rates being found at the CF. Below the CF, the cells tended to phase lock to the stimulus, producing lower rates. Above CF, primary suppression played an important role in shaping both the nonmonotonic increase in discharge rate and the very phasic response to tones at the upper frequency range of such fibers (Fig. 4.22A). The average discharge rate at these frequencies was sometimes lower than the spontaneous rate. In contrast, the PSTH of higher-CF cells (CF 0.55–4 kHz) were similar at all stimulus frequencies and either phasic-tonic ("primary-like") or highly phasic (Fig. 4.22B). As in *G. gecko* and *V. bengalensis*, the highly phasic cells very often showed sharp peaks (1 to 10) at the beginning of the histogram (showing a "chopper" pattern seen when viewed with high time resolution), with a mean separation of about 2 ms. Here, the cells are apparently discharging at their highest possible rate, and the interpeak interval likely represents the fiber's refractory period. As in mammalian cochlear-nucleus stellate cells, this pattern may originate from temporal summation in the

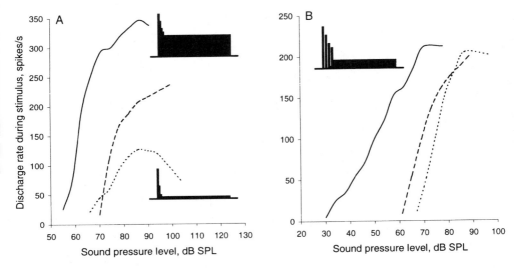

FIGURE 4.22. Diagrams to illustrate the differences in discharge patterns shown by (**A**) low-frequency auditory nerve afferents and (**B**) high-frequency (above about 1 kHz) afferents. The lines represent rate-level functions for the characteristic frequency (continuous line in both cases), for a frequency below characteristic frequency (CF) (dashed lines) and a frequency above CF (dotted line). The slope of the lines in **A** and **B** depend differently on the frequency. Three idealized examples of peri-stimulus-time histograms (PSTH) are also shown: At the top in **A**, a high-level CF stimulus, illustrating a phasic-tonic discharge. At the bottom in **A**, a high-level stimulus at a frequency above CF, illustrating the strongly phasic nature of the response in the nonmonotonic rate-level functions. In **B** is shown a PSTH for a high-level CF stimulus, illustrating the rapid chopper-type response at stimulus onset. (After Köppl and Manley 1992.)

nerve fiber (Köppl and Manley 1990b). A similar distribution of PSTH types was also found in *V. bengalensis* and *G. gecko* (Manley 1990).

In some species, primary auditory nerve fibers have also been tested for the presence of two-tone rate suppression (TTRS) by simultaneous presentation of a CF tone burst and a second tone. In *G. multicarinatus*, TTRS was found for second-tone frequencies both above and below CF, but only in fibers emanating from the apical, low-frequency area (Holton and Weiss 1978) and not for any basal, high-frequency fibers (Holton 1980). These differences between the responses of apical and basal fibers may be due to the absence of a tectorial membrane in the basal area. The TTRS patterns in *G. multicarinatus* are strongly reminiscent of those from *T. rugosa* (Köppl and Manley 1990b). In *T. rugosa* also, only the low-CF group of fibers showed TTRS, even though a tectorial membrane in the form of a chain of sallets is present over the high-CF area. TTRS in *T. rugosa* was only found when using suppressor frequencies higher than CF, but not lower than CF.

In *G. gecko* TTRS was observed in fibers from both UDT and BDT areas (Eatock et al. 1981; Sams-Dodd and Capranica 1994), usually both at frequencies above and below the CF of the excitatory tuning curve. Lizard TTRS data are thus varied and difficult to interpret.

5.3. Phase Locking in Primary Auditory Afferents

Single auditory nerve fibers of *G. multicarinatus* of both the low- and high-CF populations showed phase locking only to frequencies below 1 kHz (Rose and Weiss 1988; Weiss and Rose 1988). Low-CF fibers showed better synchronization and a higher corner frequency than basal fibers. This is related to the fact that apical (low-CF) fibers make more synapses per fiber on each hair cell than basal (high-CF) fibers (Miller and Beck 1990). Mulroy (1986) suggested for *G. multicarinatus* that a greater number of synapses per afferent fiber may be correlated with a better phase-locking behavior of the fibers. Evidence in support of this possibility was described (see below) by Rose and Weiss (1988), and similar phase data were found in *T. rugosa* (Manley et al. 1990) and *G. gecko* (Sams-Dodd and Capranica 1994). It has already been noted above that low-frequency fibers have steeper rate-level functions, possibly for the same reason. Rose and Weiss (1988) suggested that low-CF fibers subserve the function of transmitting information requiring the measurement of temporal synchronization at low sound levels ("timing pathway"), such as the localization of low-frequency sounds.

In *T. rugosa* all auditory nerve fibers showed phase locking at low frequencies, optimal frequencies being near 300–400 Hz (Fig. 4.23A, B). Phase locking was very weak above 1 kHz at 30°C body temperature. Low-CF cells phase locked better than high-CF cells (Fig. 4.23B). The corner frequency of the low-CF group was 0.73 kHz, that of the high-CF group was 0.51 kHz (at 30°C; Manley et al. 1990). It was suggested that the membrane time constants of high-CF fibers are longer than those of low-CF fibers, a feature that is related to the PSTH patterns of these fibers, that show a series of rapid peaks at their onset (Fig. 4.22B). These data thus bear a close resemblance to those described above from *G. multicarinatus* and those reported for *G. gecko* (Sams-Dodd and Capranica 1994).

In the *T. rugosa* data, about a quarter of the cells responded to two phases of the signal, producing a phase histogram with two prominent peaks that lay 180° apart, even when measured at levels below the audio-visual rate threshold for the cell. The presence of two peaks separated by 180° is probably due to the innervation of hair cells of opposing polarities by these fibers (Köppl and Manley 1990a). Each peak is likely to be the result of the activity of one group of hair cells of the same orientation.

After correcting for the phase-delay characteristics of the sound system, the response phase of a nerve fiber shifts with reference to the sound source as the frequency is changed. The amount of this shift depends on the CF of the fiber in a systematic way; the lower-CF fibers show a more rapid change

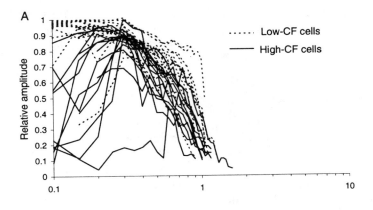

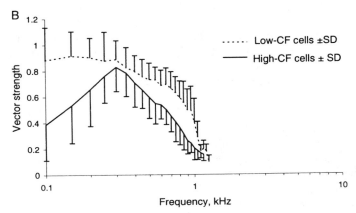

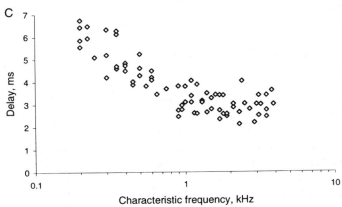

with frequency than do the higher-CF fibers. In mammals, such systematic changes of delay with CF have been attributed to the travel time of traveling waves in the basilar membrane organ of Corti complex. In *T. rugosa*, in which there is no such traveling wave, these delay times (near 3 ms for high-CF cells to 6 ms for low-CF cells; Fig. 4.23C) strongly resemble those previously reported for other vertebrate auditory nerve fibers, including those of mammals (Manley et al. 1990). It seems likely that the CF-dependent delays are due to mechanisms intrinsic to the hair cell(s), perhaps also to mechanisms involved in the active response of hair cells to stimulation.

6. Physiological Ecology of Hearing

6.1. A Behavioral Audiogram

A great variety of methods has been developed to derive behavioral audiograms from vertebrates (Fay 1988). Lizards have, however, remained rather intractable with respect to behavioral training, and so far no published behavioral audiogram is available for any lizard. We are thus in the somewhat peculiar situation of having a great deal of anatomical and physiological data available on the function of the lizard hearing organ, but are unable to say much about what roles this sensory capacity plays in the normal physiological ecology of the different species.

To my knowledge, there have been a number of (unsuccessful and unpublished) attempts to use various kinds of conditioning techniques on lizards. One older, successful report was, however, largely ignored. In 1924, Berger reported some interesting observations on sound perception in reptiles. Her lacertid lizards would bask in the sun with their eyes closed, but open their

FIGURE 4.23. Diagrams to illustrate the phase-locking behavior of low- and high-frequency auditory-nerve fibers in *Tiliqua rugosa*. In **A** are shown curves for a number of single auditory afferents that illustrate the relative strength of their phase locking over a range of frequencies. Low-frequency cell responses are plotted as dotted lines, high-frequency cells as continuous lines. In **B**, the data from these two cell populations have been averaged to show the population response as a function of stimulus frequency (means ± SD). It can be seen that the low-frequency population shows stronger phase locking at almost all frequencies. This is presumably due to the fiber time constants that are partially revealed in the PSTH patterns (see Fig. 4.22). In **C** are shown the response-time delays calculated from the phase-versus-frequency characteristics of many single auditory nerve fibers, after correction for acoustic and middle ear delays, as a function of the cells' characteristic frequencies (CFs). These data can be approximated by a sloping straight line through the data up to 1 kHz, and a flat straight line for the data above 1 kHz. Thus the delays essentially only change for CFs below 1 kHz. (Data from Manley et al. 1990.)

eyes in response to sounds she produced from tuning forks and musical instruments, even when she went to great lengths to prevent the lizards having any other cues available. Using different instruments, she studied the range of tones over which a response could be elicited (Berger 1924). Not being able to calibrate the sound-pressure levels or the spectral content of her stimuli, however, she was only able to give a rough indication of the frequency-response range. The upper frequency limit for *Lacerta agilis* (the sand lizard) was found to be 8 kHz, a figure in agreement with recent physiological data from other lizards of this family, when relatively high-level stimuli are used (data of C. Köppl on the lacertid genus *Podarcis*, described in Manley 1990).

We have repeated Berger's experiment using one juvenile individual of *T. rugosa*, while calibrating the sound signals and observing both eyes remotely with a video camera. Pure-tone stimuli from a loudspeaker mounted 30 cm to the side of the animal were delivered to the lizard in a wire cage suspended in a noise-attenuating chamber. The testing procedure was begun when the animal had positioned itself on a warmed brick and had closed its eyes. A response was scored if the animal opened one or both eyes within 5 s of stimulus onset. To test the rate at which the animal opened its eyes spontaneously, 1,050 trials and 700 blind tests (trials with the loudspeaker turned off) were carried out. Statistical tests compared experimental data to those of the blind runs. The audiogram of this animal derived from these data is shown in Figure 4.24, and this is compared to the sensitivity of primary nerve fiber responses in *T. rugosa* measured using a closed acoustic system. The behavioral data points lie within the scatter of neural thresholds, even on the high-frequency side, where the thresholds are rapidly increasing. Thus the method used seems to give a reasonable estimate of the hearing ability of this species, but the match of the data is better in the upper frequency range. The thresholds of the most sensitive nerve fibers are much better than the behavioral thresholds at low frequencies, there being a difference of over 20 dB at 0.55 kHz. One explanation for this difference between low- and high-frequency ranges is that in a closed sound system such as that used for the neural studies, the effective sound pressure is higher than in an open sound system (see, e.g., Manley and Johnstone 1974).

The sensitivity differences in the low- and high-frequency ranges may, however, also be the result of evolutionary specializations of function. The physiological data outlined above strongly indicate that the mechanisms of sound analysis in the two papillar areas are different. This difference may well have resulted in a differentiation of function of the two areas with respect to the animal's behavior. As higher-frequency signals are more easily localized, the responses of high-frequency afferents may be "used" to alert the animal to the presence of a moving object.

According to this hypothesis, the low-frequency area responses would not alert the animal as readily and thus not induce eye opening. In addition, there are often significantly higher levels of low-frequency sound in the

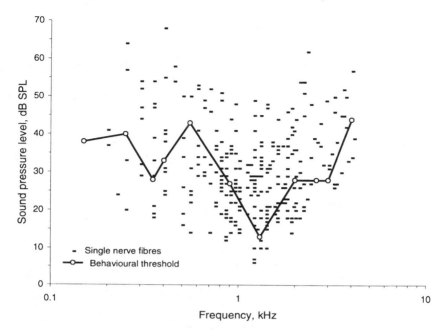

FIGURE 4.24. A comparison between the sensitivity at characteristic frequency (CF) for a large number of auditory nerve afferents in *Tiliqua rugosa* (short bars) and the behavioral audiogram derived from one individual animal (continuous line). Some of the difference at low frequencies may be explained by the use of a closed sound system for the neural recordings and an open sound system for the behavioral measurements.

natural environment (e.g., wind noise), which would interfere with the detection of low-frequency signals. It is interesting to note that there is also evidence that low and high frequencies are probably processed separately in the brain. Miller (1975) noted that the size of one region of the cochlear nuclei (the medial region of nucleus magnocellularis, or NM) is related to the size of the unidirectionally oriented hair cell area. In the cochlear nucleus of *V. bengalensis*, Manley (1976, 1981) noted that the CF range of cells of the NM only covered the lower frequencies, whereas cells of the nucleus angularis responded over the whole frequency range. In *G. multicarinatus*, Szpir, Sento, and Ryugo (1990) found that the afferent supply to cells of the NM is derived from hair cells of the low-frequency region of this species' papilla. These anatomical and physiological findings may form part of the basis of a future explanation of the behavioral responses to sounds of different frequency. While no conclusions can be drawn from the audiogram of one individual animal, the present data are presented because they indicate that Berger's (1924) method is appropriate for studies of audiograms in lizard species that show basking behavior. Emulation is strongly encouraged!

6.2. Seasonal Effects and Their Potential Pitfalls

In 1969, Johnstone and Johnstone first reported seasonal variation in the hearing of *T. rugosa*, in which the size of the summating potential in scala tympani showed a 10-fold increase in October, when the animals are naturally most active (Johnstone and Johnstone 1969a). This was accompanied by a sharp rise in the number of active, recordable auditory nerve fibers (Johnstone and Johnstone 1969b). Holmes and Johnstone (1984) also reported an enhanced auditory sensitivity of *T. rugosa* from September to November. In an unrelated study on the discharge properties of primary auditory nerve fibers of *T. rugosa* (Köppl and Manley 1990a, 1990b; Manley et al. 1990a), experiments were performed between September and January and during March and April. In contrast to the earlier findings, no differences were found in the response characteristics from the different time periods (Köppl et al. 1990). There was, however, a strong seasonal effect on the amount of anesthetic necessary to produce surgical anesthesia, and it is apparent that this effect secondarily affected the data collected in the earlier studies. Anesthetic doses that are too large can, in extreme cases, lead to a (reversible) total loss of function in the inner ear. This effect is not so surprising in ectothermic animals that undergo cyclic changes in activity, changes associated with variation of steroid and plasma melatonin levels (for references see Köppl et al. 1990).

6.3. Temperature Effect on Tuning

Lizards are almost exclusively ectothermic, and many species undergo a substantial circadian change in body temperature, with profound consequences for metabolic rates. Such changes in temperature also affect the hearing system. Temperature influences the sensitivity of cochlear microphonics in lizards, maximal amplitudes being produced at an optimal temperature that is related to the species' ecological optimum (e.g., Werner 1972, 1976). In a study of primary auditory afferents in *G. gecko*, Eatock and Manley (1981) found that the whole tuning curve shifted reversibly up in frequency with warming and down with cooling, at a rate of 0.05 to 0.06 octaves/°C. For a 1 kHz cell at 20°C, this changed the CF to about 1.55 kHz at 30°C. Prominent and partly species-specific effects of temperature on SOAE have also been reported, and these have already been discussed in Section 4.1.1.2.

7. Conclusions

We now have a good understanding of the structure and the function of the basilar papilla of several species of lizards. Our understanding of the mechanisms underlying tonotopicity and frequency selectivity is as far advanced as it is for the mechanisms of mammals. The functional consequences of different morphological types are beginning to become clear, and the future holds promise of using different lizard species to clarify some of the basic questions in hearing. For example, what underlies the active mechanism(s) in SOAE? And what causes the shift in SOAE and neural tuning during a temperature change? The different lizard species with their extraordinary variety of hearing organ structure offer promising "natural experiments" to study such questions. There is a great similarity between neural and otoacoustic-emission data of lizards and those of other tetrapod vertebrates, including mammals. This makes it possible to use the data described in this chapter to examine the generality of functional hypotheses and models of auditory function, at least in amniotes. This approach can be strongly recommended to all those interested in achieving a more global understanding of the basic function of these remarkable sense organs.

Acknowledgments. Most of the work from my own group as reported here was supported by grants from the Deutsche Forschungsgemeinschaft (SFB 204). I thank Christine Köppl for comments on an earlier version of the manuscript and Georg Klump for pointing out the paper from Berger (1924).

References

Authier S, Manley GA (1995) A model of frequency tuning in the basilar papilla of the tokay gecko, *Gekko gecko*. Hear Res 82:1–13.

Bagger-Sjöbäck D (1976) The cellular organization and nervous supply of the basilar papilla in the lizard, *Calotes versicolor*. Cell Tiss Res 165:141–156.

Bagger-Sjöbäck D, Wersäll J (1973) The sensory hairs and tectorial membrane of the basilar papilla in the lizard *Calotes versicolor*. J Neurocytol 2:329–350.

Baird IL (1970) The anatomy of the reptilian ear. In: Gans C, Parsons TS (eds) Biology of the Reptilia Vol 2. New York, London: Academic Press, pp. 193–275.

Berger K (1924) Experimentelle Studien über Schallperzeption bei Reptilien. Z vergl Physiol 1:517–540.

Brownell WE, Bader CR, Bertrand D, de Ribaupierre Y (1985) Evoked mechanical responses of isolated cochlear outer hair cells. Science 227:194–196.

Crawford AC, Fettiplace R (1985) The mechanical properties of ciliary bundles of turtle cochlear hair cells. J Physiol 364:359–379.

Dodd F, Capranica RR (1992) A comparison of anesthetic agents and their effects on the response properties of the peripheral auditory system. Hear Res 62:173–180.

Eatock RA, Manley GA (1981) Auditory nerve fiber activity in the tokay gecko: II, temperature effect on tuning. J Comp Physiol A 142:219–226.

Eatock RA, Weiss TF (1986) Relation of discharge rate to sound-pressure level for cochlear nerve fibers in the alligator lizard. Abstracts 9th Mtg Assoc Res Otolaryngol pp. 63–64.

Eatock RA, Manley GA, Pawson L (1981) Auditory nerve fiber activity in the tokay gecko: I, implications for cochlear processing. J Comp Physiol A 142:203–218.

Eatock RA, Saeki M, Hutzler MJ (1993) Electrical resonance of isolated hair cells does not account for acoustic tuning in the free-standing region of the alligator lizard's cochlea. J Neurosci 13:1767–1783.

Eatock RA, Weiss T, Otto KL (1991) Dependence of discharge rate on sound pressure level in cochlear nerve fibers of the alligator lizard: Implications for cochlear mechanisms. J Neurophysiol 65:1580–1597.

Estes R, de Queiroz K, Gauthier J (1988) Phylogenetic relationships within Squamata. In: Estes R, Pregill, G (eds) Phylogenetic Relationships of the Lizard Families. Stanford: Stanford University Press, pp. 119–281.

Fay RR (1988) Hearing in Vertebrates. Chicago: Hill-Fay Associates.

Fettiplace R (1987) Electrical tuning of hair cells in the inner ear. Trends Neurosci 10:421–425.

Freeman DM, Weiss TF (1990) Hydrodynamic analysis of a two-dimensional model for micromechanical resonance of free-standing hair bundles. Hear Res 48:37–68.

Frishkopf LS, DeRosier DJ (1983) Mechanical tuning of free-standing stereociliary bundles and frequency analysis in the alligator lizard cochlea. Hear Res 12:393–404.

Holmes RM, Johnstone BM (1984) Gross potentials recorded from the cochlea of the skink *Tiliqua rugosa*. II. Increases in metabolic rate and hearing responsiveness during austral spring. J Comp Physiol A 154:729–738.

Holton T (1980) Relations between frequency selectivity and two-tone rate suppression in lizard cochlear-nerve fibers. Hear Res 2:21–38.

Holton T, Hudspeth AJ (1983) A micromechanical contribution to cochlear tuning and tonotopic organization. Science 222:508–510.

Holton T, Weiss TF (1978) Two-tone rate suppression in lizard cochlear nerve fibers, relation to receptor organ morphology. Brain Res 159:219–222.

Holton T, Weiss TF (1983) Frequency selectivity of hair cells and nerve fibers in the alligator lizard cochlea. J Physiol 345:241–260.

Hudspeth AJ (1997) Mechanical amplification of stimuli by hair cells. Curr Opin Neurobiol 7:480–486.

Hudspeth AJ, Gillespie PG (1994) Pulling strings to tune transduction: adaptation by hair cells. Neuron 12:1–9.

Johnstone JR, Johnstone BM (1969a) Electrophysiology of the lizard cochlea. Exp Neurol 24:99–109.

Johnstone JR, Johnstone BM (1969b) Unit responses from the lizard auditory nerve. Exp Neurol 24:528–537.

Klinke R (1979) Comparative physiology of primary auditory neurones. In: Hoke M, de Boer E (eds) Models of the Auditory System and Related Signal Processing Techniques. Scand Audiol Suppl 9:49–61.

Köppl C (1988) Morphology of the basilar papilla of the bobtail skink *Tiliqua rugosa*. Hear Res 35:209–228.

Köppl C. (1995) Otoacoustic emissions as an indicator for active cochlear mechanics: a primitive property of vertebrate auditory organs, In: Manley GA, Klump GM, Köppl C, Fastl H, Oeckinghaus H (eds) Advances in Hearing Research. Singapore: World Scientific, pp. 207–216.

Köppl C, Authier S (1995) Quantitative anatomical basis for a model of micromechanical frequency tuning in the tokay gecko, *Gekko gecko*. Hear Res 82:14–25.

Köppl C, Manley GA (1990a) Peripheral auditory processing in the bobtail lizard *Tiliqua rugosa*: II. Tonotopic organization and innervation pattern of the basilar papilla. J Comp Physiol A 167:101–112.

Köppl C, Manley GA (1990b) Peripheral auditory processing in the bobtail lizard *Tiliqua rugosa*: III. Patterns of spontaneous and tone-evoked nerve-fiber activity. J Comp Physiol A 167:113–127.

Köppl C, Manley GA (1992) Functional consequences of morphological trends in the evolution of lizard hearing organs. In: Webster DB, Fay RR, Popper AN (eds) The Evolutionary Biology of Hearing. New York: Springer-Verlag, pp. 489–509.

Köppl C, Manley GA (1993a) Distortion-product otoacoustic emissions in the bobtail lizard. 2: Suppression tuning characteristics. J Acoust Soc Amer 93:2834–2944.

Köppl C, Manley GA (1993b) Spontaneous otoacoustic emissions in the bobtail lizard. I. General characteristics. Hear Res 71:157–169.

Köppl C, Manley GA (1994) Spontaneous otoacoustic emissions in the bobtail lizard. II: Interactions with external tones. Hear Res 72:159–170.

Köppl C, Manley GA, Johnstone BM (1990) Peripheral auditory processing in the bobtail lizard *Tiliqua rugosa*: V. Seasonal effects of anaesthesia. J Comp Physiol A 167:139–144.

Lee MSY (1997) The phylogeny of varanoid lizards and the affinities of snakes. Phil Trans R Soc Lond B 352:53–91.

Manley GA (1976) Auditory responses from the medulla of the monitor lizard *Varanus bengalensis*. Brain Res 102:329–324.

Manley GA (1977) Response patterns and peripheral origin of auditory nerve fibers in the monitor lizard, *Varanus bengalensis*. J Comp Physiol A 118:249–260.

Manley GA (1979) Preferred intervals in the spontaneous activity of primary auditory neurones. Naturwiss 66:582.

Manley GA (1981) A review of the auditory physiology of the reptiles. Progr Sens Physiol 2:49–134.

Manley GA (1986) The evolution of the mechanisms of frequency selectivity in vertebrates. In: Moore BCJ, Patterson RD (eds) Auditory Frequency Selectivity. New York, London: Plenum Press, pp. 63–72.

Manley GA (1990) Peripheral Hearing Mechanisms in Reptiles and Birds. Heidelberg: Springer-Verlag.

Manley GA (1997) Diversity in hearing-organ structure and the characteristics of spontaneous otoacoustic emissions in lizards. In: Lewis ER, Long GR, Lyon RF, Narins PM, Steele CR, Hecht-Poinar E (eds) Diversity in Auditory Mechanics. Singapore: World Scientific, pp. 32–38.

Manley GA, Gallo L (1997) Otoacoustic emissions, hair cells, and myosin motors. J Acoust Soc Am 102:1049–1055.

Manley GA, Gleich O (1984) Avian primary auditory neurones: the relationship between characteristic frequency and preferred intervals. Naturwiss 71:592–594.

Manley GA, Johnstone BM (1974) Middle-ear function in the guinea pig. J Acoust Soc Am 56:571–576.

Manley GA, Köppl C (1992) A comparison of peripheral tuning measures: primary afferent tuning curves versus suppression tuning curves of spontaneous and distortion-product otoacoustic emissions. In: Cazals Y, Demany L, Horner K (eds) Auditory Physiology and Perception. Oxford, New York: Pergamon Press, pp. 151–157.

Manley GA, Köppl C (1994) Spontaneous otoacoustic emissions in the bobtail lizard. III: Temperature effects. Hear Res 72:171–180.

Manley GA, Köppl C (1998) Phylogenetic development of the cochlea and its innervation. Curr Opin Neurobiol 8:468–474.

Manley GA, Yates GK, Köppl C (1988) Auditory peripheral tuning: evidence for a simple resonance phenomenon in the lizard *Tiliqua*. Hear Res 33:181–190.

Manley GA, Köppl C, Yates GK (1989) Micromechanical basis of high-frequency tuning in the bobtail lizard. In: Wilson JP, Kemp D (eds), Cochlear Mechanisms—Structure, Function and Models. New York: Plenum Press, pp. 143–150.

Manley GA, Köppl C, Johnstone BM (1990a) Peripheral auditory processing in the bobtail lizard *Tiliqua rugosa*: I. Frequency tuning of auditory-nerve fibers. J Comp Physiol A 167:89–99.

Manley GA, Köppl C, Johnstone BM (1990b) Components of the $2f_1–f_2$ distortion product in the ear canal of the bobtail lizard. In: Dallos P, Geisler CD, Matthews JW, Ruggero M, Steele CR (eds) Mechanics and Biophysics of Hearing. New York: Springer-Verlag, pp. 210–217.

Manley GA, Yates GK, Köppl C and Johnstone BM (1990) Peripheral auditory processing in the bobtail lizard *Tiliqua rugosa*: IV. Phase locking of auditory-nerve fibers. J Comp Physiol A 167:129–138.

Manley GA, Köppl C, Johnstone BM (1993) Distortion-product otoacoustic emissions in the bobtail lizard. 1: General characteristics. J Acoust Soc Am 93:2820–2933.

Manley GA, Gallo L, Köppl C (1996) Spontaneous otoacoustic emissions in two gecko species, *Gekko gecko* and *Eublepharis macularius*. J Acoust Soc Am 99:1588–1603.

Manley GA, Köppl C, Sneary M (1999) Reversed tonotopic map of the basilar papilla in *Gekko gecko*. Hear Res 131:107–116.

Marcellini D (1977) Acoustic and visual display behavior of gekkonid lizards. Am Zool 17:251–260.

Miller MR (1973a) Scanning electron microscope studies of some lizard basilar papillae. Am J Anat 138:301–330.

Miller MR (1973b) A scanning electron microscope study of the papilla basilaris of *Gekko gecko*. Z Zellforsch 136:307–328.

Miller MR (1975) The cochlear nuclei of lizards. J Comp Neurol 159:375–406.

Miller MR (1978) Further scanning electron microscope studies of lizard auditory papillae. J Morphol 156:381–418.

Miller MR (1980) The reptilian cochlear duct. In: Popper AN, Fay RR (eds) Comparative Studies of Hearing in Vertebrates. New York: Springer-Verlag, pp. 169–204.

Miller MR (1985) Quantitative studies of auditory hair cells and nerves in lizards. J Comp Neurol 232:1–24.

Miller MR (1992) The evolutionary implications of the structural variations in the auditory papilla of lizards. In: Webster DB, Fay RR, Popper AN (eds) The Evolutionary Biology of Hearing. New York: Springer-Verlag, pp. 463–487.

Miller MR, Beck J (1988) Auditory hair cell innervational patterns in lizards. J Comp Neurol 271:604–628.

Miller MR, Beck J (1990) Further serial transmission electron microscopy studies of auditory hair cell innervation in lizards and a snake. Amer J Anat 188:175–184.

Mulroy MJ (1974) Cochlear anatomy of the alligator lizard. Brain Behav Evol 10:69–87.

Mulroy MJ (1986) Patterns of afferent synaptic contacts in the alligator lizard's cochlea. J Comp Neurol 248:263–271.

Mulroy MJ, Oblak TG (1985) Cochlear nerve of the alligator lizard. J Comp Neurol 233:463–472.

Mulroy MJ, Williams RS (1987) Auditory stereocilia in the alligator lizard. Hear Res 25:11–21.

Peake WT, Ling A (1980) Basilar-membrane motion in the alligator lizard: its relation to tonotopic organization and frequency selectivity. J Acoust Soc Am 67:1736–1745.

Probst R, Lonsbury-Martin BL, Martin GK (1991) A review of otoacoustic emissions. J Acoust Soc Am 89:2027–2067.

Ricci AJ, Fettiplace R (1998) Calcium permeation of the turtle hair cell mechanotransducer channel and its relation to the composition of endolymph. J Physiol 506:159–173.

Rose C, Weiss TF (1988) Frequency dependence of synchronization of cochlear nerve fibers in the alligator lizard: evidence for a cochlear origin of timing and non-timing neural pathways. Hear Res 33:151–166.

Rosowski JJ, Peake WT, White JR (1984) Cochlear nonlinearities inferred from two-tone distortion products in the ear canal of the alligator lizard. Hear Res 13:141–158.

Sams-Dodd F, Capranica RR (1994) Representation of acoustic signals in the eighth nerve of the tokay gecko: I. Pure tones. Hear Res 76:16–30.

Szpir MR, Sento S, Ryugo DK (1990) Central projections of cochlear nerve fibers in the alligator lizard. J Comp Physiol 295:530–547.

Talmadge CL, Tubis A, Wit HP, Long GR (1991) Are spontaneous otoacoustic emissions generated by self-sustained cochlear oscillations? J Acoust Soc Am 89:2391–2399.

Teresi PV (1985) Hair cell innervation patterns in the papilla basilaris of the fence lizard *Sceloporus occidentalis*. Dissertation, University of California, San Francisco.

Turner RG (1987) Neural tuning in the granite spiny lizard. Hear Res 26:287–299.

Turner RG, Muraski AA, Nielsen DW (1981) Cilium length: influence on neural tonotopic organization. Science 213:1519–1521.

van Dijk P, Manley GA, Gallo L, Pavusa A (1996) Statistical properties of spontaneous otoacoustic emissions in one bird and three lizard species. J Acoust Soc Am 99:1588–1603.

van Dijk P, Manley GA, Gallo L (1998) Correlated amplitude fluctuations of spontaneous otoacoustic emissions in six lizard species. J Acoust Soc Am 104:1559–1564.

van Dijk P, Manley GA, Gallo L, Pavusa A, Taschenberger G (1996) Statistical properties of spontaneous otoacoustic emissions in one bird and three lizard species. J Acoust Soc Am 100:2220–2227.

Weiss TF, Leong R (1985a) A model for signal transmission in an ear having hair cells with free-standing stereocilia. III. Micromechanical stage. Hear Res 20:157–174.

Weiss TF, Leong R (1985b) A model for signal transmission in an ear having hair cells with free-standing stereocilia. IV. Mechanoelectric transduction stage. Hear Res 20:175–195.

Weiss TF, Rose C (1988) Stages of degradation of timing information in the cochlea: a comparison of hair-cell and nerve-fiber responses in the alligator lizard. Hear Res 33:167–174.

Weiss TF, Mulroy MJ, Turner RG, Pike CL (1976) Tuning of single fibers in the cochlear nerve of the alligator lizard: relation to receptor morphology. Brain Res 115:71–90.

Werner YL (1972) Temperature effects on inner-ear sensitivity in six species of iguanid lizards. J Herpetol 6:147–177.

Werner YL (1976) Optimal temperatures for inner-ear performance in gekkonid lizards. J Exp Zool 195:319–352.

Werner YL, Frankenberg E, Adar O (1978) Further observations on the distinctive vocal repertoire of *Ptyodactylus hasselquistii* CF. hasselquistii (Reptilia: Gekkonidae). Israel J Zool 27:176–188.

Wever EG (1978) The Reptile Ear. Princeton, N.J.: Princeton University Press.

Whitehead ML, Lonsbury-Martin B, Martin GK, McCoy MJ (1996) Otoacoustic emissions: animal models and clinical observations. In: Van De Water T, Popper A, Fay R (eds) Clinical Aspects of Hearing. New York: Springer-Verlag, pp. 199–257.

Wu Y-C, Art JJ, Goodman MB, Fettiplace R (1995) A kinetic description of the calcium-activated potassium channel and its application to electrical tuning of hair cells. Prog Biophys Molec Biol 63:131–158.

Zurek PM (1985) Acoustic emissions from the ear: a summary of results from humans and animals. J Acoust Soc Am 78:340–344.

5
The Central Auditory System of Reptiles and Birds

CATHERINE E. CARR and REBECCA A. CODE

1. Introduction

The central auditory systems of both birds and reptiles (jointly known as sauropsids) are organized along a common plan. The similarities among the sauropsids are presumably due to the conserved nature of the auditory sense and to the close phylogenetic relationships within the group. The common organization of the auditory system has allowed us to organize this chapter by auditory nucleus from hindbrain to forebrain. The embryology, anatomy, and physiology of the auditory nuclei of the turtles, snakes and lizards, crocodiles, and birds will be described, with attention paid to both conserved and derived features. A more extensive list of the older literature may be found in Carr (1992).

The morphotype or conserved pattern of organization (Northcutt 1995; Fig. 5.1A) of the sauropsid auditory system may be constructed as follows: The auditory nerve projects tonotopically to two cochlear nuclei, the nucleus magnocellularis and the nucleus angularis. Nucleus magnocellularis projects to the second order nucleus laminaris that in turn projects to the superior olive, the lemniscal nuclei, and the central nucleus of the auditory midbrain (torus semicircularis, nucleus mesencephalicus lateralis dorsalis, inferior colliculus). Nucleus angularis projects to the superior olive, the lemniscal nuclei, and the central nucleus of the auditory midbrain. The parallel ascending projections of nucleus angularis and nucleus laminaris may or may not overlap with one another, and probably do overlap in the primitive condition. Hindbrain auditory connections are generally bilateral, although contralateral projections are stronger. The lemniscal nuclei project to midbrain, thalamic, and forebrain targets. The central nucleus of the auditory midbrain projects bilaterally to its thalamic target (nucleus medialis or reuniens in reptiles, nucleus ovoidalis in birds; see Section 8). The auditory thalamus projects to the auditory region of the forebrain (medial dorsal ventricular ridge in reptiles, field L in birds; see Section 8). Field L projects to other forebrain nuclei that may be involved in the control of song and other vocalizations. Descending projections from the forebrain archi-

MORPHOTYPE

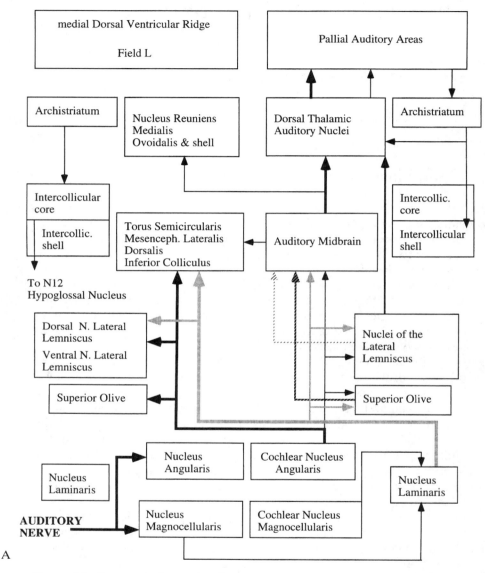

A

FIGURE 5.1. Summary of central auditory connections and origins. (**A**) Morphotype with conserved connections. (**B**) A schematic saggittal view of *Gallus domesticus* hindbrain nuclei shows the rhombomeric origins of each nucleus. Ventrodorsal position correlates roughly with the normal topography. (From Marin and Puelles 1995. Copyright European Neuroscience Association.)

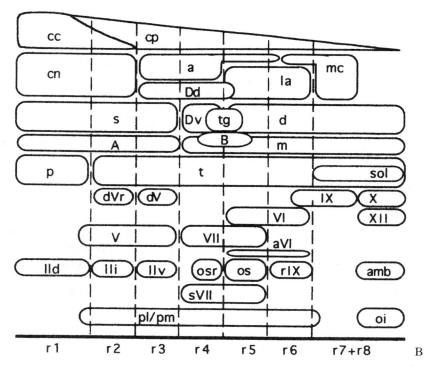

FIGURE 5.1. *Continued*

striatum to the intercollicular area and the hypoglossal nucleus in the hindbrain appear to mediate vocalization. There have been many independent developments superimposed upon this conserved pattern, and some nuclei may more correctly be described as field homologues among the different groups, in that they contain both homologous and independently derived cell types.

Major differences among central auditory structures are few. Those differences that have been discovered are mostly found in the hindbrain or in the auditory forebrain. Selective forces driving these changes have been ascribed to the development of new end organs in the auditory periphery and/or the selection resulting from the increased use of sound for communication (Wilczynski 1984). The formation of a new division of the cochlear nucleus angularis in lizards coincident with the development of a new population of high-frequency hair cells (Szpir et al. 1990) is an example of a central change that appears to result from changes in the periphery. Examples of specializations for communication may be found in song birds and geckos, while specializations for sound localization characterize the owls (see Klump, Chapter 6, and Dooling et al., Chapter 7).

Not enough is known about the comparative physiology of the central auditory nuclei to determine if major differences have arisen in auditory processing in amniotes. The transition from water to land was clearly marked by such changes, but the sauropsid common ancestors are assumed to be adapted to process airborne sound (see Manley, Chapter 4) and probably had a tonotopically organized basilar papilla. The ancestral condition appears to be sensitivity to low-frequency sound. The main features of the auditory stimulus (frequency, intensity, and phase or timing of a signal) appear to be encoded in the central nervous system in a similar way in all vertebrates, although the physiological investigations of some groups have not been extensive. Frequency is encoded in the auditory periphery, and tonotopic order is maintained in auditory nuclei from the level of the cochlear nuclei to the forebrain. The intensity of the sound is generally encoded by changes in spike rate, while phase-locked spikes encode the timing of the stimulus. It appears that labeled-line inputs convey computed information (Konishi 1986). When sound becomes important for communication, it appears that the ability to perceive and differentiate among complex sounds becomes more important (see Dooling et al., Chapter 7).

2. Cochlear Nuclei

The auditory nerve projects to two cochlear nuclei, located in a rostro-caudal column in the dorsal hindbrain with angularis rostral to magnocellularis. In specialized birds like the barn owl, this separation into two cochlear nuclei reflects the parallel processing of temporal and loudness information in the auditory stimulus. The projection to nucleus magnocellularis preserves the phase-locked response to the stimulus from the auditory nerve via specialized end bulb synapses, while loudness information is preserved in the projection to nucleus angularis. Since the two cochlear nuclei exist in the morphotype, it is likely that the separation into two anatomical channels represents the primitive condition. The separation into two physiological channels processing time and sound level information appears to have occurred to varying degrees in the birds and reptiles.

The variations in the auditory system may eventually be understood in terms of its embryological origins. A segmental analysis of the hindbrain nuclei has been carried out by homotopic grafting of quail rhombomeres onto the hindbrain of chicken embryos (Marin and Puelles 1995; Fig. 1B). Nucleus angularis is derived from rhombomeres 3 through 6, nucleus laminaris from 5 through 6, nucleus magnocellularis from 6 through 7/8, and superior olive (SO) from 5. The nuclei of the lateral lemniscus originate from rhombomeres 1 through 3. Future embryological and gene expression studies may be used to determine the fate and origins of the elements of the auditory system in other sauropsids and in mammals. Embryological studies of the rhombomeric origins of individual auditory nuclei in other amniotes might resolve the issues of homology versus homoplasy in comparative studies of the auditory hindbrain.

2.1. Turtles and Tortoises (Chelonia)

The turtles and tortoises have been regarded as the most primitive of the modern reptiles, principally because of the lack of temporal openings in their skulls. Recent controversial analyses have, however, placed turtles as the sister group of an alligator and chicken (archosaur) clade rather than a sister group between mammals and birds, and have rejected the placement of turtles as the most basal living amniotes (Zardoya and Meyer 1998). Other recent analyses also support the diapsid affinities of turtles (Rieppel and deBraga 1996). Regardless of their position among the sauropsids, turtles are sensitive to fairly low sound frequencies, and this is probably the primitive condition. Cochlear microphonics, hair cell, and auditory nerve recordings show best sensitivities between 100–1,000 Hz (Wever 1978; Crawford and Fettiplace 1980; Manley et al. 1990; see Manley, Chapter 4). The auditory endorgan is tonotopically organized (see Chapter 4). Thus the auditory nerve conveys information about sound frequency, phase and loudness to the two cochlear nuclei, and possibly to nucleus laminaris. The nucleus magnocellularis is large and located caudal to a smaller nucleus angularis (Miller and Kasahara 1979). Nucleus magnocellularis contains large, round principal cells with few dendrites, and multipolar cells (Browner and Marbey 1988). The principal cells receive end bulb of Held terminals from the auditory axons and project to nucleus laminaris in the conserved pattern found in most reptiles and in all birds (Browner and Marbey 1988). Neither nucleus angularis nor nucleus magnocellularis are as well developed in turtles as in the other reptiles, and nucleus angularis is poorly demarcated. The cochlear nucleus complex, including nucleus laminaris, forms a predominantly contralateral projection to the superior olive, the nuclei of the lateral lemniscus, and the torus semicircularis (Kunzle 1986). It is parsimonious to assume that nucleus angularis contributes to this ascending projection, and that nucleus magnocellularis projects to nucleus laminaris but these connections require further study.

2.2. Lizards and Snakes (Lepidosauria)

The hindbrain auditory nuclei of lizards and snakes differ from those of all other sauropsids because in many cases the basilar papilla projects to a complex of four, rather than two, cochlear nuclei (Szpir et al. 1990). These differences in the central auditory projections appear to result from changes in the auditory periphery (Miller 1975). The effect of peripheral changes has been discussed by Wilczynski (1984), who made the argument that since new peripheral structures are specializations of pre-existing structures, their central systems must also be specializations of pre-existing areas. The newly specialized areas would be related to the pre-existing area either as field homologs or true homologs. Thus, a major (genetic) change in the periphery need not require a similar genetic change because the "new" inputs in the periphery should cause epigenetic rearrangements at other levels of a functional system. These rearrangements might then be followed by later genetic changes. Wilczynski's model has been used here to explain the differences observed between the cochlear nuclei of lizards and other sauropsids.

Lizards are unique in that they have two populations of hair cells in the basilar papilla, a unidirectional, low center frequency (CF) population and a bidirectional, high CF population (Manley et al. 1990). Each population is cytologically distinct (Weiss et al. 1976; Miller and Beck 1988) and there are significant physiological differences between the responses from the two populations (Köppl and Manley 1990a, 1990b, 1992; Manley et al. 1990). Auditory nerve fibers that receive input from the unidirectional population have low center frequencies (100–800 Hz), low spontaneous discharge rates, and sharp, asymmetric tuning curves, that is, responses like those in other sauropsids and mammals. These unidirectional type hair cells are contacted by auditory nerve fibers that appear to correspond to the mammalian type 1 and avian cochlear nerve fibers (Szpir et al. 1990). This arrangement appears to have been present in the stem reptiles from which modern reptiles, birds, and mammals evolved. The bidirectional population of hair cells is physiologically and morphologically distinct, and a unique feature of the lizard auditory system (Wever 1978; Miller 1980). In the alligator lizard, fibers from the bidirectional hair population have high center frequencies (900–4,000 Hz), high spontaneous rates, and broad symmetric tuning curves (for review see Szpir et al. 1990). In the bobtail lizard, *Tiliqua rugosa*, these fibers have high CFs (900–4,500 Hz), high spontaneous rates and V-shaped tuning curves with obvious sharp tips around CF (Köppl and Manley 1990a, 1996b; Manley et al. 1990; see Gleich and Manley, Chapter 3).

The considerable interspecific variation in the cochlear nuclei of the lizards appears to be related to the changes in the auditory periphery. Auditory nerve fibers whose peripheral processes contact unidirectional hair cells project to three of the four divisions of the cochlear nuclei, the medial and lateral nucleus magnocellularis and the lateral nucleus angularis, forming end bulbs of Held and terminal boutons in nucleus magnocellularis, and small boutons in nucleus angularis (Szpir et al. 1990). Neurons that contact the new bidirectional population of hair cells project primarily to the medial nucleus angularis (Szpir et al. 1990). It may be concluded that the unique organization and morphological characteristics of the first-order auditory nuclei in lizards reflect the variation in the bidirectional hair cell population (Köppl and Manley 1990a, 1990b). Thus, elaboration of the bidirectional hair cell population in some lizard species is correlated with substantial changes in the central auditory system.

Recordings from the cochlear nuclei in the monitor lizard reflect the different responses from the uni- and bidirectional hair cell populations. Low best-frequency responses were obtained from the lateral region of nucleus angularis and from nucleus magnocellularis, while high-frequency responses were only found in nucleus angularis (Manley 1977). Tuning curves from the cochlear nuclei also fall into two groups. Units with best frequencies below 1.1 kHz had sharp V-shaped tuning curves, while units with best frequencies from 1.1–2.8 kHz had more complex tuning curves (Manley 1977). Despite these differences, both populations yield primary-like responses to auditory stimuli (Manley 1974, 1977). Recording from the auditory nerve in *Gambelia*, the leopard lizard, and *Gekko gecko*, the Tokay gecko, produced indications of tonotopic order, with primary-like and onset responses (Manley 1970b; Sams-Dodd and Capranica 1994).

2.3. Crocodilians

Crocodilians share an archosaur common ancestor with the birds, and the central auditory pathways of the two groups are very similar, although the hearing range of crocodilians is lower (20–2,800 Hz for *Caiman crocodilus*). The inner ear is large, with a long basilar membrane and unidirectional population of hair cells covered by a tectorial membrane. Auditory nerve units are relatively sharply tuned, and phase-lock to frequencies up to 1.5 kHz (Manley 1970a; Smolders and Klinke 1986). The auditory nerve projects to nucleus angularis and nucleus magnocellularis (Leake 1974) and has recently been reported to project to nucleus laminaris as well (Soares et al. 1999). Nucleus angularis lies anterior to the root of the auditory nerve and is composed of large and small ovoid cells. Nucleus magnocellularis contains characteristic large round principal cells and is much larger than nucleus angularis (Leake 1974). Nucleus magnocellularis has lateral and medial divisions, with the caudal part of the medial division capped by a small-celled component. There is a similar small-celled region of nucleus magnocellularis of the bird that receives low best-frequency auditory nerve fibers (Köppl 1994; Köppl and Carr 1996). The cochlear nuclei of *C. crocodilus* are tonotopically organized in a similar fashion to that of birds (Konishi 1970; Manley 1970a). Recordings from the cochlear nucleus in the caiman produced primary-like responses that were very similar to those in the leopard lizard and many birds (Manley 1970b; 1974).

2.4. Birds

Birds use sound for communication and generally hear higher frequencies than reptiles (see Klump, Chapter 6, and Dooling et al., Chapter 7). Most birds hear up to 5–6 kHz, while *Tyto alba*, the barn owl, has exceptional high frequency hearing, with characteristic frequencies of 9–10 kHz in the auditory nerve (Konishi 1973; Köppl 1997). Some landbirds (pigeons, chickens, and guinea fowl) are also sensitive to infrasound, below 20 Hz (see Dooling et al., Chapter 7). Infrasound signals may travel over great distances, and Kreithen and Quine (1979) have suggested that pigeons, *Columba livia*, may use infrasound for orientation. Responses to infrasound stimuli have been found throughout the central auditory system (Theurich et al. 1984; Warchol and Dallos 1989; Schermuly and Klinke 1990).

2.4.1. Auditory Nerve Projections

The auditory nerve projects to nucleus magnocellularis and nucleus angularis (Boord and Rasmussen 1963; see Rubel and Parks, 1988). Each auditory nerve fiber divides to form varicose terminals in the nucleus angularis, and end bulbs of Held on the principal cells of nucleus magnocellularis (Jhaveri and Morest 1982; Köppl 1994). The projections to the cochlear nuclei are tonotopically organized (Rubel and Parks 1975). In the owl, nucleus magnocellularis is the origin of a neural pathway that encodes timing information, while a parallel pathway for encoding sound level originates with nucleus angularis (Sullivan and Konishi 1984; Takahashi et al. 1984). Recordings in the chicken cochlear nuclei have found a similar but less clear segregation of function, where nucleus magnocellularis units phase-lock better than nucleus angularis units at frequencies above 1 kHz (Fig. 5.2; Warchol and Dallos 1990). The similarities between the owl and the chicken suggest that some degree of separation between time and level coding may be a common feature of the avian auditory system.

2.4.2. Temporal Coding in Nucleus Magnocellularis

The auditory system uses phase-locked spikes to encode the timing of the auditory stimulus (Fig. 5.2). Phase-locking underlies accurate detection of temporal information, including interaural time differences (Klump, Chapter 6), and is conspicuous in the owl. Owl auditory nerve fibers show phase-locked responses to acoustic stimuli up to 9 kHz (Sullivan and Konishi 1984; Köppl 1996), as opposed to about 2 kHz in the pigeon, *C. livia* (Hill et al. 1989), about 4 kHz in the starling, *Sturnus vulgaris* (Gleich and Narins 1988) and 5–6 kHz in the blackbird, *Agelaius phoeniceus*, and cat, *Felis cattus* (Woolf and Sachs 1977; Sachs and Sinnott 1978; Johnson 1980). Volman and Konishi (1990) provided a hypothesis to explain the development of high-frequency phase-locking in the owl. *T. alba* uses a bicoordinate system of sound localization, that is, it uses interaural level differences to localize sound in elevation, and interaural time differences to localize sound in azimuth (Konishi et al. 1985; see Klump, Chapter 6). Thus, *T. alba*

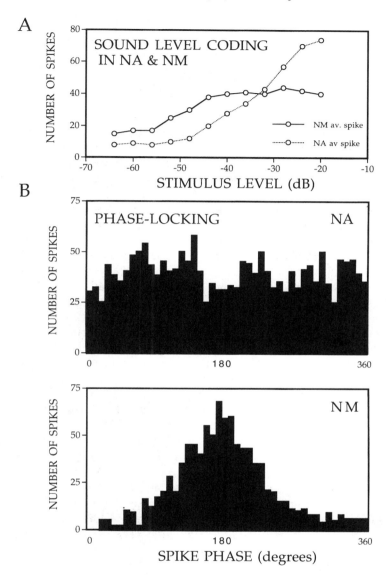

FIGURE 5.2. Parallel coding of phase and sound level in the cochlear nuclei of *Tyto alba* (adapted from Sullivan and Konishi 1984). Rate intensity plots show that nucleus magnocellularis and nucleus angularis units have similar dynamic range, although some angularis units may have a wider range (see example rate intensity plot for NA). Period histograms from the cochlear nucleus neurons show that nucleus magnocellularis cells phase lock to the auditory stimulus (frequency = 3.3 kHz), while nucleus angularis units do so poorly above 1 kHz, and do not show phase-locking above 3.5 kHz (frequency = 2.2 kHz).

(and perhaps other birds) processes time and level information in parallel to determine the location of a sound source. Since interaural level differences become more prominent at higher frequencies (Coles and Guppy 1988; Moiseff 1989a, 1989b), the pairing of time and level may have provided the selective pressure for the evolution of phase-locking to high frequencies in the owl.

The end bulb synapse, with its multiple sites of synaptic contact, preserves the temporal relationship between phase-locked spikes in the auditory nerve and the postsynaptic nucleus magnocellularis cell. End bulbs do not, however, appear to be essential for transmittal of very low best-frequency (BF) stimuli. Auditory nerve fibers with CFs below 1 kHz form bouton terminals on the cell bodies and dendrites of nucleus magnocellularis neurons in the caudolateral region of nucleus magnocellularis (Köppl 1994). This low BF region contains a mixture of multipolar and small principal cells (Köppl and Carr 1997). The multipolar cells resemble the multipolar cells of the nucleus magnocellularis of turtles (Browner and Marbey 1988), lizards (Szpir et al. 1995), and crocodilians (Leake 1974), and it is possible that both multipolar and principal cells were found in the primitive nucleus magnocellularis. The principal cells are better suited to encoding higher frequency temporal information, and may be associated with the development of high-frequency hearing in a sauropsid ancestor.

Other specializations are associated with phase-locking in birds (Oertel 1997; Trussel 1997). Nucleus magnocellularis principal cells are large, which reduces jitter (Carr et al. 1996), and they tend to have few dendrites (Smith and Rubel 1979; Jhaveri and Morest 1982; Carr and Boudreau 1993). The glutamate receptors that mediate synaptic transmission in the auditory hindbrain also appear to be adapted to accurate transmission of phase-locked spikes. AMPA subtype glutamate receptors with rapid desensitization (0.5 msec) are expressed throughout the auditory hindbrain (Raman and Trussell 1992; Levin et al. 1997). AMPA receptors that mediate the temporally accurate responses to the auditory stimulus in the auditory hindbrain may reflect a specialization for phase-locking and other aspects of temporal coding (Raman et al. 1994; Zhang and Trussell 1994). Nucleus magnocellularis cells are also characterized by large K^+ conductances and short time constants (Reyes et al. 1994; Zhang and Trussell 1994).

2.4.3. Level Coding and Nucleus Angularis

Nucleus angularis is tonotopically organized with high best frequencies mapped dorso-lateral and low best frequencies mapped ventro-medial (Konishi 1970; Hotta 1971). There are five major cell types in the nucleus angularis of *C. livia*: large, medium, and small multipolar cells, bipolar cells, and stubby cells with many short dendrites (Hausler et al. 1999). The medium multipolar cells were most common and resemble the multipolar cells of the mammalian ventral cochlear nucleus (VCN). The similarity in cell types could either support homology between the multipolar cells of the mammalian VCN and nucleus angularis or convergent evolution of cells specialized for encoding changes in sound level. Whether the similarities between nucleus angularis and VCN owe more to convergent evolution or homology remains to be seen. In *T. alba*, the majority of nucleus angularis neurons show "chopper" firing patterns and sensitivity to changing sound levels (Sullivan 1985). The multipolar morphology is compatible with the loss of phase-locking observed in nucleus angularis (Fig. 5.2), while the summation of a large number of auditory nerve inputs onto dendritic trees might result in the observed wider dynamic range (Sullivan 1985). Sachs and Sinnott (1978) recorded from nucleus magnocellularis and nucleus angularis in *A. phoeniceus* and found very similar results to those obtained later for *T. alba* (Sullivan and Konishi 1984) and for chick, *Gallus domesticus* (Warschol and Dallos 1990). Nucleus magnocellularis contained primary-like units, while nucleus angularis neurons had a variety of response types that differed in their degree of inhibition (Sachs and Sinnott 1978).

3. The Nucleus Laminaris

Nucleus laminaris is a nucleus in the dorsal hindbrain responsible for the detection of interaural time differences (Carr and Konishi 1990; Overholt et al. 1992; see Klump, Chapter 6). Nucleus laminaris receives bilateral inputs from nucleus magnocellularis in birds, crocodilians, and some turtles and lizards. In birds, its organization conforms to a model proposed by Jeffress to explain sound localization (Jeffress 1948). Nucleus magnocellularis afferents to nucleus laminaris function as delay lines and nucleus laminaris neurons act as coincidence detectors to create a map of interaural time difference (Klump, Chapter 6). Nucleus laminaris is also the most variable of the auditory hindbrain nuclei and its neurons originate from rhombomeres 5 and 6 in *G. domesticus* (Puelles et al. 1994).

3.1. Turtles

The turtle nucleus laminaris does not resemble the large, well-developed nucleus of crocodilians and birds, but is a mass of fusiform cells just ventral and lateral to nucleus magnocellularis (Miller and Kasahara 1979). We assume that this ventrolateral complex characterizes the sauropsid common ancestor. The central connections of the cochlear nucleus complex (including nucleus laminaris) in the turtle have been described by Kunzle (1986). The complex forms a predominantly contralateral projection to the superior olive, the nuclei of the lateral lemniscus, and the torus semicircularis.

3.2. Snakes and Lizards

Nucleus laminaris is the most variable structure among medullary auditory nuclei, and may only be present in those lizards with a primitive basilar papilla and few bidirectional hair cells (Miller 1975). In lizards with more derived basilar papilla morphology, for example the alligator lizard, a distinct nucleus laminaris cannot be identified and may have been lost or greatly reduced in size (Miller 1975). Because of the interspecific variation, the auditory nuclei that project to the auditory midbrain should be homologized to nucleus angularis and nucleus laminaris, and those cochlear nucleus neurons that project locally should be homologized to nucleus magnocellularis. Injections of tracer into the central nucleus of the torus in the iguana, *Iguana iguana*, produced retrogradely labeled cells in two hindbrain nuclei (Foster and Hall 1978). These cells were mostly in nucleus angularis (only one division was recognized), but some were found in the "nucleus magnocellularis lateralis." Injections into the midbrain torus of *G. gecko* also produced labeled neurons in both nucleus angularis and the nucleus magnocellularis lateralis (Dodd and Capranica 1989, unpublished results). Parsimony would suggest that this lateral portion of nucleus magnocellularis, which may not receive first-order auditory input, is homologous to the nucleus laminaris described in all other sauropsids.

3.3. Crocodiles

The crocodilians are the only living reptiles to have a well-developed nucleus laminaris. Nucleus laminaris forms a layer sheet of bipolar spindle-shaped cells that is very similar to that seen in the basal land birds (Carr 1992). Nucleus angularis and nucleus laminaris project bilaterally to the torus (Brauth, personal communication), as is also the case in other sauropsids.

3.4. Birds

Nucleus laminaris is the first site of binaural interactions in the auditory system, and is where sensitivity to interaural time differences first appears (Sullivan and Konishi 1986; Carr and Konishi 1990; Klump, Chapter 6). The organization of nucleus laminaris in the chicken appears to reflect the primitive pattern found in most basal land birds and in the crocodilians, where nucleus laminaris is composed of a layer of bipolar neurons (Ramon y Cajal 1908; Rubel and Parks 1988). Nucleus laminaris cells receive tonotopically organized input from nucleus magnocellularis, with input from the ipsilateral nucleus magnocellularis onto their dorsal dendrites and input from the contralateral nucleus magnocellularis onto their ventral dendrites. Axons from the contralateral nucleus magnocellularis form an elongated band of endings along the mediolateral length of the nucleus, and act as delay lines to form a map of interaural time differences in nucleus laminaris (Parks and Rubel 1975; Young and Rubel 1983, 1986). Delays are detected by coincidence detection in nucleus laminaris neurons (Overholt et al. 1992). In *T. alba*, nucleus laminaris is vastly expanded from the monolayer structure typical of the chicken. Nucleus laminaris is no longer a flat sheet but a 1-mm-thick neuropil with the neurons sparsely distributed throughout a plexus of myelinated fibers. It contains four times more neurons than are found in other birds of similar size (*T. alba* 10,020 versus crow, *Corvus corvus* 2,540; Winter and Schwartzkopff 1961). Interaural time difference is mapped in the dorsoventral direction in the *T. alba* nucleus laminaris. There is an obvious adaptive significance in the development of many parallel maps of interaural time difference in nucleus laminaris, because the iteration of the same calculation should improve the sensitivity to interaural time differences. It seems likely that independent growth of nucleus laminaris has occurred many times (Carr et al. 1996).

Nucleus laminaris cells have low threshold K^+ channels, short time constants, and the ability to phase-lock to the auditory stimulus (Reyes et al. 1994; Zhang and Trussel 1994). An excitatory amino acid like glutamate mediates synaptic transmission from nucleus magnocellularis to nucleus laminaris (Zhang and Trussell 1994; Levin et al. 1997). Since nucleus laminaris neurons phase-lock to the auditory stimulus and act as coincidence detectors to encode interaural time differences with microsecond resolution, the high levels of AMPA receptors in nucleus laminaris may be required to mediate the temporally precise responses in nucleus laminaris neurons. Nucleus magnocellularis and nucleus laminaris are also characterized by high levels of calcium binding proteins, specifically calretinin in the owl and the chicken (Parks et al. 1997).

4. Superior Olivary Nucleus

There is a second noncochlear auditory nucleus in the ventrolateral medulla and pons, the superior olive, located below nucleus laminaris. In the chicken superior olive, neurons have their embryonic origins in rhombomere 5 and therefore are derived from the same segment as the rostral half of nucleus laminaris (Fig. 5.1B). Because little is known about the superior olive in reptiles, it will not be discussed, beyond stating that there are projections to a region in the ventral medulla, termed the superior olive, in both turtles and lizards.

4.1. Birds

The superior olive is a round nucleus distinguishable in the ventrolateral medulla and pons (Fig. 5.3). It is anatomically and physiologically hetero-

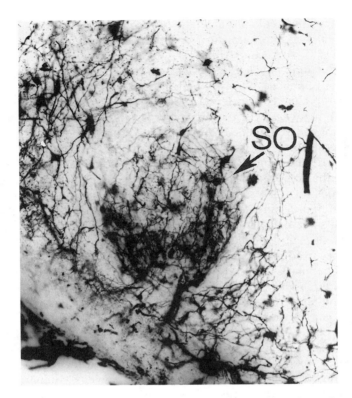

FIGURE 5.3. Morphology of the Superior Olive (SO). The SO (*arrow*) is a round nucleus distinguishable in the ventrolateral medulla and pons in Golgi material. One large multipolar cell type projects back to the cochlear nuclei and nucleus laminaris. *Gallus domesticus*, Rapid Golgi. Bar = 100 µm.

geneous (Moiseff and Konishi 1983; Carr et al. 1989) and stains for cytochrome oxidase (Dezso et al. 1993; Lachica et al. 1994). The superior olive is immunoreactive for calcium-binding proteins (Takahashi et al. 1987; Adolphs 1993), GABA (Muller 1987; Carr et al. 1989; Lachica et al. 1994), and dynorphin (Code 1996). There is no convincing evidence that the avian superior olive is homologous to either the mammalian medial superior olive or the lateral superior olive. Based on its connections, however, the avian superior olive does not appear homologous to the lateral superior olive (Westerberg and Schwarz 1995).

The superior olive receives bilateral ascending inputs from nucleus laminaris and nucleus angularis with that from the nucleus angularis being heavier (Conlee and Parks 1986; Takahashi and Konishi 1988a; Westerberg and Schwarz 1995; Fig. 5.4). The superior olive sends ascending projections to the contralateral dorsal nucleus of the lateral lemniscus (LLDp) and to the shell region of the central nucleus of the inferior colliculus (Westerberg and Schwarz 1995). If these projections to the auditory midbrain are not exclusively contralateral (Westerberg and Schwarz 1995), the ipsilateral projections are sparse (Conlee and Parks 1986). Cells in the superior olive also provide descending projections to the cochlear nuclei and nucleus laminaris (Lachica et al. 1994; Westerberg and Schwarz 1995) and appear to be a major source of GABAergic inputs to the nucleus magnocellularis and nucleus laminaris (Code et al. 1989; von Bartheld et al. 1989; Carr et al. 1989; Lachica et al. 1994). These GABAergic projections from the superior olive to nucleus laminaris have been proposed to function as gain control elements, or a negative feedback to protect nucleus laminaris neurons from losing their sensitivity to interaural time differences at high sound intensities (Peña et al. 1996).

5. Cochlear Efferents

The efferent auditory system refers to a set of neurons in the brain stem that send descending axons to the cochlea. Although their precise function remains to be elucidated, they seem to be involved in the modulation of afferent auditory input. An efferent auditory system can be found in all classes of vertebrates and in some invertebrates. An excellent overview of the efferent innervation of the vertebrate inner ear has been provided by Roberts and Meredith (1992). Although some cochlear efferent neurons in birds are located near the superior olivary nucleus, they have an extensive rostrocaudal distribution beyond the superior olive. Thus, we have not applied the term *olivocochlear* to these neurons because of the implied spatial limitation. Instead, we refer to these cells simply as cochlear efferent neurons (Code and Carr 1994).

5.1. Birds

The cochlear efferent system in the CNS of birds has been studied to a greater extent than in reptiles, although data are still sparse. The cochlear efferent system has been characterized best in *G. domesticus, C. livia, M. undulatus*, and *S. vulgaris*. For a more extensive review of the cochlear efferent innervation of the avian basilar papilla, see Code (1997).

5.1.1. Distribution of Cochlear Efferent Neurons in the Hindbrain

Neurons that project to the avian basilar papilla (BP) are located bilaterally in the hindbrain in a loosely organized column of cells that extends

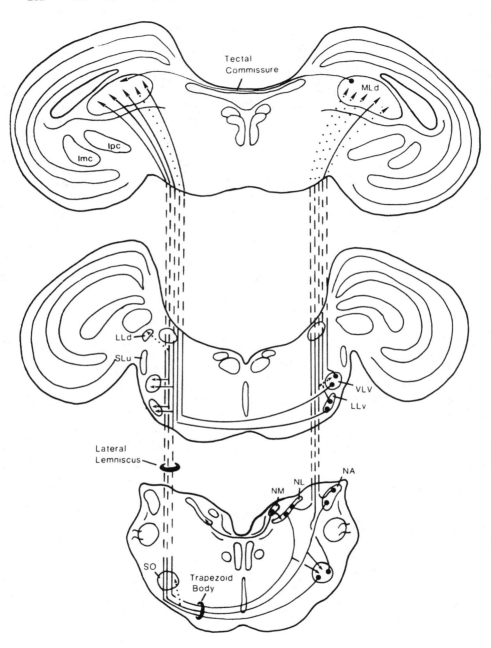

from the superior olive (SO) caudally to the nucleus subceruleus ventralis rostrally (Whitehead and Morest 1981; Strutz and Schmidt 1982; Cole and Gummer 1990; Schwarz et al. 1992; Code and Carr 1994; Code 1997). The cochlear efferent neurons can be divided into two spatially distinct cell groups: the ventrolateral and the dorsomedial cell groups (Fig. 5.5). Cochlear efferent neurons in the ventrolateral group are located medial to the superior olive (SO) and the ventral facial motor nucleus (VIIv) and lateral to the nucleus pontis lateralis (PL). The dorsomedial group is found in the pontine reticular formation (RP) ventromedial to the dorsal facial motor nucleus (VIId) and lateral to the abducens nerve root (N. VI). Neurons in the ventrolateral group appear to project only to the basilar papilla (Code 1995), while neurons in the dorsomedial group project to the BP and to the vestibular end organs, the semicircular canals, and the macula lagena (Schwarz et al. 1978, 1981; Whitehead and Morest 1981; Code 1995). It is not known, however, whether an individual cochlear efferent neuron in the dorsomedial group projects exclusively to the lagena or the BP, or whether it sends axon collaterals to both sensory epithelia. About 50–65% of cochlear efferent neurons project to the contralateral ear (Whitehead and Morest 1981; Strutz and Schmidt 1982; Cole and Gummer 1990; Schwarz et al. 1992; Code and Carr 1994), while few or no efferents project to both ears (Schwarz et al. 1992; Cole and Gummer 1990).

Most cochlear efferent neurons have a multipolar morphology but some are more elongated or polygonal in shape (Whitehead and Morest 1981; Code and Carr 1994). Cells located in the ventrolateral group appear to be larger than those in the dorsomedial group (Strutz and Schmidt 1982).

FIGURE 5.4. Ascending connections of the hindbrain auditory nuclei in *G. domesticus*. Nucleus angularis sends an extensive projection to the contralateral auditory midbrain (MLd) and a smaller projection to the rostral pole of the ipsilateral MLd. Nucleus angularis also projects bilaterally to the superior olive and dorsal nucleus of the lateral lemniscus (VLV in figure) and to the contralateral ventral nucleus of the lateral lemniscus. Projections from nucleus laminaris were demonstrated to the ipsilateral superior olive, to the contralateral lemniscal nuclei, and a small medial region in the auditory midbrain bilaterally; the contralateral projection is much denser than the ipsilateral one. Other nuclei having ascending connections with the midbrain include the contralateral superior olive, the ipsilateral dorsal nucleus of the lateral lemniscus, the contralateral ventral nucleus of the lateral lemniscus, and the contralateral midbrain. The ipsilateral superior olive and ventral nucleus of the lateral lemniscus also project to the midbrain but much more sparsely than in their contralateral projection. The bilateral connections of the cochlear nuclei with the auditory midbrain demonstrates an important avian similarity with the hindbrain auditory pathways of other terrestrial vertebrates. (From Conlee and Parks 1986, used by permission.)

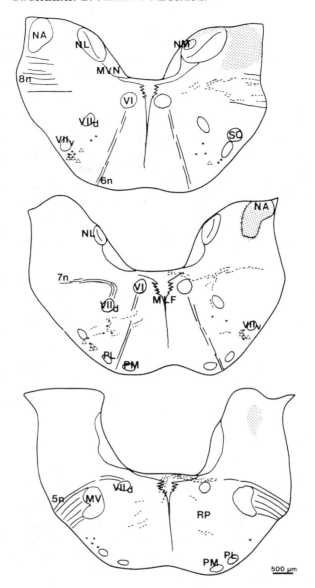

FIGURE 5.5. Efferent neurons (*triangles*) that project to the basilar papilla (BP) in *Gallus domesticus* are located in two loosely organized cell groups mainly within the pons: the ventrolateral and the dorsomedial cell groups. The ventrolateral group includes cells that are ventromedial to the superior olivary nucleus (SO) and the ventral facial nucleus (VIIv) and lateral to the nucleus pontis lateralis (PL). This group is actually a column of cells that extends from the SO caudally to the nucleus subceruleus rostrally. Cells in the dorsomedial group are found in the pontine reticular formation (RP), ventromedial to the dorsal facial nucleus (VIId) and lateral to the abducens nerve root (N VI). Neurons in the dorsomedial group also project to the vestibular end organs, the semicircular canals, and the macula lagena. Most cochlear efferent neurons are ChAT-immunoreactive. (From Code and Carr 1994; used by permission.)

These differences in somal size may be related to the two diameters of cochlear efferent fibers found within the vestibular nerve root and in the basilar papilla (Whitehead and Morest 1981; Fischer et al. 1994: Keppler et al. 1994). An absolute correlation has not yet been made, however, between the thick efferent fibers and the larger neurons in the ventrolateral group, and between the thin efferent fibers and the smaller neurons in the dorsomedial group. Additionally, two to three morphologically distinct types of efferent nerve terminals have been described in the avian basilar papilla (Takasaka and Smith 1971; Tanaka and Smith 1978; Whitehead and Morest 1978; Firbas and Muller 1983; Zidanic and Fuchs 1996). The cells of origin, however, of each type of cochlear efferent terminal have not been correlated with specific morphological cell types in either the ventrolateral or dorsomedial groups.

5.1.2. Physiology of Cochlear Efferent Neurons

Given that a certain population of hair cells in the avian BP has no afferent auditory innervation whatsoever, but only efferent endings (Fischer 1992, 1994), it has become compelling to determine the function of their efferent innervation. Activation of the cochlear efferent system inhibits responses of the auditory nerve and raises auditory thresholds (Desmedt and Delwaide 1963). A comprehensive study of the physiology of individual cochlear efferent neurons has been performed in the chick (Kaiser and Manley 1994). Cochlear efferent cells are more sensitive to sound stimulation of the contralateral, rather than the ipsilateral, ear and have latencies greater than 5 msec. They have poorer frequency selectivity than nearby ascending auditory neurons, with Q_{10dB} values of their tuning curves being less than 2.5. Their average spontaneous activity is fewer than 30 spikes/sec, much lower than that of ascending auditory neurons, while 28% of efferent neurons show no spontaneous activity. Chick cochlear efferent neurons exhibit chopper and primary-like excitatory responses (similar to mammalian olivocochlear neurons) that may be correlated with the thin and thick efferent fibers. Additionally, they display a unique suppression type that may be associated specifically with the functioning of the basilar papilla (Kaiser and Manley 1994).

5.2. Comparison of Avian and Reptilian Cochlear Efferent Systems

Only a few studies of the reptilian cochlear efferent system have been carried out, specifically in *C. crocodilus*, *T. ornata*, and *V. exanthematicus* (Strutz 1981, 1982; Barbas-Henry and Lohman 1988). Briefly, these studies show that, as in most vertebrates studied, reptilian efferent nerve cell bodies are located on both sides of the brain stem in a column of cells extending approximately from the abducens nucleus to the rostral pole of the facial motor nucleus. Different groups of reptiles, however, exhibit variation as to whether the column of cells is divided into two spatially distinct groups or not. For example, in the turtle, a more evolutionarily primitive reptile, cochlear efferent neurons are located in one cell column in the medial reticular formation (Strutz 1982) similar to those in teleosts (Strutz et al. 1980) and amphibians (Fritzsch 1981). In *V. exanthematicus* and the more highly advanced *C. crocodilus*, however, cochlear efferents are found in separate dorsal and ventral cell groups resembling the avian organization (Strutz 1981; Barbas-Henry and Lohman 1988).

Additionally, the segregation of efferent neurons into cell groups projecting to the BP or to the vestibular end organs is found in lizards, crocodilians, and birds but not in turtles. For example, in *G. domesticus*, the dorsomedial group of cochlear efferent neurons sends axons to the macula lagena while those in the ventrolateral group project to the basilar papilla (Code 1995) similar to efferent projections in *C. crocodilus* and *V. exanthematicus*. In the turtle, however, acoustic and vestibular efferent neurons appear to be dispersed throughout the single efferent cell column (Strutz 1982).

Finally, differences in the side of origin of cochlear efferent neurons between birds and lizards and turtles can be noted. The majority of avian cochlear efferent neurons project to the contralateral BP as do those of the caiman (Strutz 1981). In contrast, most efferent neurons in the turtle and monitor lizard are located ipsilaterally (Strutz 1981; Barbas-Henry and Lohman 1988). Thus, the anatomical distribution of avian cochlear efferent neurons appears to have more features in common with the *C. crocodilus* than with the turtle.

5.2.1. Comparison of Avian and Mammalian Cochlear Efferent Systems: Anatomy

A comprehensive text on the neuroanatomy, physiology, and neuropharmacolgy of the mammalian efferent auditory system has recently been published (Sahley et al. 1997). In mammals, two separate systems of olivocochlear neurons can be distinguished based on their locations in or near the superior olivary complex and on their projections to the inner ear (for review, see Warr 1992). Strict comparisons between these and avian cochlear efferent systems are difficult to make because of the limited amount of data in birds. One might first be inclined to homologize the avian dorsomedial efferent cell group with the mammalian medial olivocochlear neurons based on the location of their cells bodies within the hindbrain. This should not be done, because avian dorsomedial cells differ from mammalian olivocochlear cells in their projections and cell size.

The avian cochlear efferent and the mammalian olivocochlear systems differ in the number of cell types upon which they terminate in the inner ear. In birds, thin efferent fibers terminate on tall, intermediate, and short hair cells (Keppler et al. 1994; Zidanic and Fuchs 1996). Mammalian lateral olivocochlear axons, however, predominantly innervate just one type of hair cell, the inner hair cell, which is considered by some to be homologous to avian tall hair cells. Thick fibers of avian efferent neurons synapse on two cell types, the short hair cells and the hyaline cells (a type of support cell; Keppler et al. 1994; Zidanic and Fuchs 1996), while the thick medial olivocochlear axons synapse beneath outer hair cells only. Finally, the mammalian olivocochlear and the avian cochlear efferent systems differ in the innervation patterns of their axon terminals. In general, mammalian inner hair cells do not receive efferent synapses directly on their cell bodies, rather olivocochlear terminals synapse on auditory afferent endings beneath inner hair cells (Hashimoto et al. 1990, Nadol 1990). This type of axodendritic synapse rarely occurs in the avian BP (Firbas and Muller 1983; Chandler 1984; Smith 1985; Fischer 1992). In contrast, both the mammalian outer hair cell and the avian short hair cell receive direct efferent innervation via large cup-like terminals onto their cell bodies.

5.2.2. Comparison of Avian and Mammalian Cochlear Efferent Systems: Immunohistochemistry

Most, if not all, avian cochlear efferent nerve terminals contain acetylcholine (ACh; Zidanic and Fuchs 1995), as do mammalian olivocochlear neurons. In addition, most cochlear efferent neurons in the chick hindbrain are immunoreactive for choline acetyltransferase (ChAT), the biosynthetic enzyme for ACh (Code and Carr 1994). There are, however, several immunohistochemical differences between the avian and mammalian cochlear efferent systems. In mammals, calcitonin gene-related peptide-immunoreactivity (CGRP-I) is found primarily in the lateral olivocochlear, but not medial olivocochlear system (see Altschuler and Fex 1986; Roberts and Meredith 1992; Eybalin 1993). In the chick, however, no cochlear efferent neuron is immunoreactive for CGRP (Code et al. 1996; Zidanic and Fuchs 1996). Interestingly, CGRP-I has been demonstrated in efferent fibers of the lateral line system of the frog (Adams et al. 1987) and eel (Roberts et al. 1994). Additionally, the cell bodies of most mammalian lateral olivocochlear neurons that are immunoreactive for ChAT are also enkephalin-immunoreactive (Enk-I). In *G. domesticus*, however, cochlear efferent somata do not appear to be Enk-I but instead are located in regions that have a high density of Enk-I terminals (Code and Carr 1995). The source of these enkephalinergic terminals may be a population of Enk-I neurons in the nuclei of the lateral lemniscus. Finally, in contrast to mammalian lateral olivocochlear neurons, chick cochlear efferent neurons are not immunoreactive for the opioid peptide dynorphin (Code 1996). Also, they are not likely to be GABAergic since no

GABA- or GAD-immunolabeled efferent terminals were found in the BP (Zidanic and Fuchs 1996). Thus, there are many striking differences in the immunohistochemical organization of the avian and mammalian cochlear efferent systems that may be related to different functional roles in the inner ear.

Although further investigations are needed to fill in the gaps of our knowledge about birds, there appear to be obvious dissimilarities in the organization of avian cochlear efferent and the mammalian olivocochlear systems. These data suggest that there may be significant differences in how these two systems modulate incoming auditory information. These differences may reflect and/or result from the independent but parallel evolution of the basilar papilla and the organ of Corti (Manley et al. 1989).

5.3. Cochlear Efferents and Hair Cell Regeneration

It is a well-documented fact that hair cells in the avian cochlea can reproduce themselves after damage (for reviews, see Cotanche et al. 1994, 1997). Much research has focused on understanding the factors that contribute to the regeneration of the sensory epithelium and the subsequent restoration of auditory function. In that regard, many studies have examined in detail the regeneration of new hair cells in the BP and their reinnervation by afferent auditory fibers. Only recently, however, has much attention been paid to the possible role that efferent innervation may play in the process of hair cell regeneration and functional recovery. Because the details of hair cell regeneration are too numerous to be described here, we will briefly summarize recent studies on the efferent innervation to the cochlea as it relates to hair cell regeneration.

Cochlear efferent nerve terminals in the avian BP can be damaged by either exposure to intense sound or to aminoglycoside antibiotics, although different types of damage are produced by each treatment (Ofsie et al. 1997). Additionally, efferent terminals reinnervate the BP much more slowly after gentamycin-induced damage than after acoustic trauma (Hennig and Cotanche 1998). Nevertheless, studies of cochlear damage in birds show that regenerating hair cells appear in the BP before efferent nerve terminals (Duckert and Rubel 1990). Functional recovery of hearing, however, is delayed (Tucci and Rubel 1990; Saunders et al. 1992). The reinnervation of the BP by efferent terminals is also delayed, suggesting a correlation between efferent synaptogenesis and normal auditory function. These data imply that, although cochlear efferents may not be necessary for hair cell regeneration to proceed (Wang and Raphael 1996), they may be needed for recovery of normal hearing. Reinnervation of the BP by cochlear efferents, however, is just one of several factors (including changes in hair cell morphology, stereocilia bundle orientation, and others) contributing to the complex process of functional recovery (Duckert and Rubel 1993).

6. Nuclei of the Lateral Lemniscus

The lemniscal nuclei occur in a column in the ventral hindbrain, rostral to the superior olive and ventral to the auditory midbrain. There are two identified lemniscal nuclei in reptiles, dorsal and ventral, and three in birds, dorsal, intermediate, and ventral. These names are the same as those of the lemniscal nuclei in mammals, but at present the nuclei should not be considered homologous, because the outgroup comparison with the turtle does not reveal three nuclei. In embryological studies in the chick *G. domesticus*, the lemniscal nuclei were shown to orginate from the hindbrain, with contributions from rhombomeres 1 through 3 (Fig. 5.1). Each of the first three hindbrain rhombomeres contributes to a lemniscal nucleus with dorsal (r1), intermediate (r2), and ventral (r3). Since the superior olive originates from r5, these ventral auditory structures do not form a continuous column, but one interrupted at r4. The number and origin of lemniscal nuclei among other sauropsids is uncertain.

6.1. Reptiles

In lizards, the medullary auditory nuclei project to both the torus semicircularis and the lemniscal nuclei. The nuclei of the lateral lemniscus are divided into dorsal and ventral divisions and are made up of very small to medium-sized cells that project to the central nucleus of the torus (ten Donkelaar et al. 1987).

6.2. Birds

A new nomenclature was introduced for the lemniscal nuclei by Arends (1981; Arends and Ziegler 1986; Wild 1987). In this terminology, there is a ventral nucleus of the lateral lemniscus (LLV = LLv of Karten and Hodos 1967), an intermediate nucleus of the lateral lemniscus (LLI), and a slightly more rostrally situated dorsal nucleus of the lateral lemniscus (LLD = VLVa of Leibler 1975 and VLVp of Karten and Hodos 1967).

6.2.1. Connections of the Lemniscal Nuclei

Nucleus angularis projects bilaterally to LLDp, LLI, and LLV (Fig. 5.1A; Conlee and Parks 1986; Takahashi and Konishi 1988b). Nucleus laminaris projects contralaterally to LLV and LLDa. Nucleus laminaris terminal fields in LLDa are delineated by calretinin immunoreactivity (Takahashi et al. 1987). In *G. gallus*, Conlee and Parks's original study did not differentiate between subdivisions of LLD, and so it is not clear if part of LLD is innervated purely by nucleus laminaris, and another part by nucleus angularis, as it is in *T. alba*. The separation of time and intensity information may be less distinct in the lemniscal projections of the chicken nucleus angularis and nucleus laminaris, although some studies support functional segregation (Leibler 1975; Heil and Scheich 1986).

The lemniscal nuclei have diverse ascending connections to midbrain, thalamus, and forebrain. In *T. alba*, LLDa relays interaural time difference information to the contralateral core of the auditory midbrain, while LLDp encodes interaural intensity differences (see below) and projects bilaterally to the lateral shell of the auditory midbrain. LLI projects to the basal forebrain, to or around the nucleus basalis (Arends and Zeigler 1986; Wild et al. in preparation). This projection accounts for the short latency auditory potentials that have been recorded in the nucleus basalis of many birds (Kirsch et al. 1980; Maekawa 1987; Kubke et al. 1998). LLD and LLV also project to the auditory thalamus, primarily to the nucleus semilunaris parovoidalis (SPO; Wild 1987; Durand et al. 1992; Proctor 1997) and both receive descending projections from the archistriatum (Aivm; Wild et al. 1993; see below).

6.2.2. Physiology of Lemniscal Nuclei

LLDp is the only lemniscal nucleus to be studied extensively, although LLDa neurons have been shown to encode interaural time difference (Moiseff and Konishi 1983). LLDp mediates detection of interaural level differences in the barn owl (Manley et al. 1988; Mogdans and Knudsen 1994a, 1994b). Interaural level differences are produced by the shadowing effect of the head when a sound source originates from off the midline (see Klump, Chapter 6). Some owls experience larger than predicted differences because their external ears are also oriented asymmetrically in the vertical plane. Because of this asymmetry, interaural level differences vary more with the elevation of the sound source than with azimuth. This asymmetry allows owls to use interaural level differences to localize sounds in elevation, while they use interaural time differences to determine the azimuthal location of a sound (Norberg 1978; Moiseff 1989b; Olsen et al. 1989).

In *T. alba*, the level pathway begins with nucleus angularis, which responds to changing sound level over about a 30-dB range (Sullivan 1985; Fig. 5.2A). Each nucleus angularis projects to contralateral LLDp. The cells of LLDp are excited by stimulation of the contralateral ear and inhibited by stimulation of the ipsilateral ear (Moiseff and Konishi 1983; Manley et al. 1988; Mogdans and Knudsen 1994a, 1994b). The source of the inhibition is the contralateral LLDp (Takahashi et al. 1988). Mapping of interaural level differences begins in LLDp, with neurons organized according to their preferred interaural level difference (Manley et al. 1988). The dorsal region of the nucleus contains neurons that are strongly inhibited by ipsilateral stimuli, and thus prefer a louder sound in the contralateral ear (Fig. 5.6). This preference changes gradually with depth in the nucleus because the strength of the inhibition decreases (Manley et al. 1988). Thus a map of preferred interaural level difference is created within LLDp. LLDp neurons do not encode elevation unambiguously, however, because they are not immune to changes in sound level. The encoding of elevation improves in the auditory midbrain (see below).

Neurons in LLDp show regular discharge patterns, time-locked to the onset of a stimulus, similar to the chopper pattern observed in the mammalian lateral superior olive and cochlear nuclei (Mogdans and Knudsen 1994). The functional similarities between LLDp and the lateral superior olive suggest that both *T. alba* and mammals employ similar physiological mechanisms for detecting and encoding interaural level difference (Mogdans and Knudsen 1994).

7. Auditory Midbrain

The auditory midbrain receives ascending input and projects to the thalamus. It is termed the torus semicircularis in reptiles and in birds either the nucleus mesencephalicus lateralis, pars dorsalis (MLd; Karten 1967) or the inferior colliculus (IC; Knudsen 1983), or the torus (Puelles et al. 1994). It is surrounded rostrally and laterally by an intercollicular area that receives descending input from the archistriatum, and medially by a less defined caudal preisthmic superfical area (Puelles et al. 1994; Fig. 5.7). The auditory midbrain has different numbers of subdivisions among the different sauropsids. It mediates auditory processing in all cases, while the intercollicular area appears to mediate vocalization and other auditory-motor behaviors. The function of the superficial area is less clear, and may not be auditory. During embryogenesis, the auditory midbrain is formed when a caudodorsal portion of the mesencephalic vesicle bends inward and rostral and thickens to form a large bulge into the tectal ventricle (Puelles et al. 1994).

7.1. Turtles

The torus is divided into central and laminar nuclei. Both nuclei have a number of distinct cell types (Browner et al. 1981). The major portion of the ascending auditory input is confined to the central nucleus. In both pond turtles and the tortoises, the cells of the central nucleus of the torus project bilaterally to the nucleus reuniens of the dorsal thalamus (Belekhova et al. 1985).

7.2. Lizards and Snakes

The medullary auditory nuclei project to the torus semicircularis and to the lemniscal nuclei. The torus and its connections have been described in several lizards (Foster and Hall 1978; Kennedy and Browner 1981; ten Donkelaar et al. 1987). There are varying reports of these connections. In the iguanid, the cochlear nuclear complex forms a mostly contralateral pro-

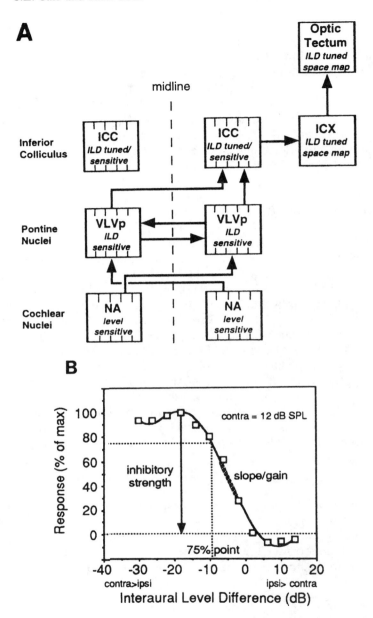

jection to the central nucleus of the torus. The central nucleus of the torus also receives inputs from the contralateral superior olive and from the ipsilateral nucleus of the lateral lemniscus. In the monitor lizard, *Varanus exanthematicus*, the projections from nucleus angularis and nucleus laminaris have been described as predominantly ipsilateral, while there are extensive ascending bilateral projections from superior olive (ten Donkelaar et al. 1987).

The torus is divided into central, laminar, and superficial nuclei. The central nucleus extends from the lateral part of the laminar nucleus in a mediocaudal direction and is the primary source of the commissural projection to the contralateral central nucleus (ten Donkelaar et al. 1987). The central nucleus contains a variety of complex cell types (Kennedy and Browner 1981) and gives rise to ascending projections to nucleus medialis of the dorsal thalamus (Foster and Hall 1978). The laminar nucleus forms part of the compact periventricular cell layer that extends throughout the mesencephalon. As is also the case in the red-eared turtle, *Chrysemys scripta elegans*, this region does not receive ascending auditory projections, but spinal afferents in *Iguana iguana* (Foster and Hall 1978). The laminar nucleus has been homologized to the intercollicular area of birds because of its connections and role in vocalization (Kennedy and Browner 1981). It forms part of the compact periventricular cell layer that extends throughout the mesencephalon. As is also the case in the red-eared turtle, the *I. iguana* laminar nucleus does not receive ascending auditory projections, but spinal afferents (Foster and Hall 1978).

Auditory responses to tones of 50–1,000 Hz have been reported in the midbrain tectum of three families of snake (Hartline and Campbell 1969). Auditory responses have been also recorded from the torus in *G. gecko*, probably from the central nucleus (Kennedy 1974; see Manley 1981). Although the inputs to the laminar nucleus in *G. gecko* are unknown, stimulation of this region evokes species-specific vocalizations (Kennedy 1975).

FIGURE 5.6. Interaural level difference computation in LLDp (VLVp) in *Tyto alba*. (**A**) Ascending emergence of interaural level difference (ILD) sensitivity. Boxes with tick marks represent nuclei that are organized tonotopically. Nucleus angularis neurons are level sensitive, while LLDp (VLVp in figure) neurons are interaural level difference sensitive and compare the relative levels of sound at the two ears within narrow frequency channels. ICC neurons become interaural level difference tuned as they project to the external nucleus (ICX). (**B**) LLDp units are sensitive to changing interaural level difference cues. The contralateral ear sound level was held constant at 20 dB above threshold, and the sound level at the ipsilateral ear was varied randomly to produce this plot. The relative strengths of the excitatory and inhibitory inputs onto the neuron determines how sensitive it is to changing interaural level difference. (Redrawn from Mogdans and Knudsen, 1994a, 1994b; used by permission).

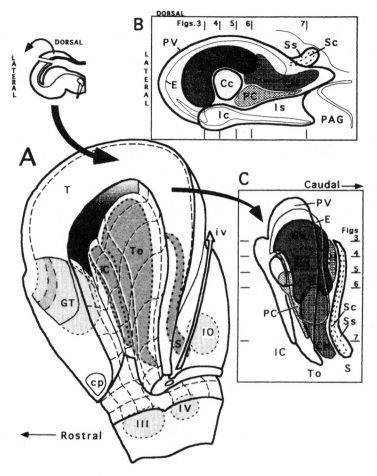

FIGURE 5.7. Organization of the auditory midbrain (torus). (**A**) Schema of the right torus bulging into the tectal ventricle, illustrating its topographic relationship with tectum (T), tectal gray (GT), and isthmic region (IO). The relative positions of the three main subdivisions (IC, To, S) are indicated in *shaded, dotted contours*. The *black arrow* leads to the nuclear subdivisions shown in C. (**B**) Schema of nuclear subdivisions of the right torus viewed from in front in transverse plane show the external nucleus (E), the periventricular lamina or superficial nucleus (PV), lateral shell of the central nucleus (Cs), the central core (Cc), the paracentral toral nucleus (PC), the caudomedial or medial shell (CM), the core and shell of the intercollicular area (Ic and Is), the periaqueductal gray (PAG), and the core and shell of the preisthmic superficial area (Ss and Sc). (**C**) Schema of toral subdivisions seen in the same perspective as A, to visualize the location of the central core. (From Puelles et al. 1994; © 1994 John Wiley & Sons, Inc. Reprinted by permission of John Wiley & Sons, Inc.)

Geckos are the most vocal lizards, and are reported to be as sensitive to sound as some birds and mammals (Marcellini 1978).

7.3. Crocodilians

The cytology and connections of the midbrain and thalamic components of the auditory system of *C. crocodilus* have been described by Pritz (1974a, 1974b). The torus is composed of two distinct regions in *C. crocodilus*, an external nucleus continuous with the deep layers of the optic tectum, and a central nucleus. The central nucleus is like that of lizards and birds, while it is not yet clear how the external nucleus is related to the laminar and superficial nuclei of the lizard torus. The external nucleus resembles the superficial nucleus of the avian midbrain, which is also continuous with the deep layers of the optic tectum. The central nucleus of the torus contains a number of different cell types and is tonotopically organized (Pritz 1974b; Manley 1971). Again, like birds, the characteristic frequencies varied with recording depth in the torus with low frequencies (70 Hz) represented superficially and high frequencies (1,850 Hz) represented deep. Most units recorded in the torus were of the onset type (Manley 1971).

7.4. Birds

7.4.1. Anatomy and Connections

In birds the auditory midbrain is divided into, from lateral to medial, an external nucleus and a central nucleus, which is further divided into a lateral shell, central core, and medial shell (Fig. 5.7A, B). Both nucleus angularis and nucleus laminaris project to the central nucleus (Boord 1968; Leibler 1975; Conlee and Parks 1986; Takahashi and Konishi 1988a; Fig. 5.3). These projections are segregated within the central nucleus, as are the projections from the lemniscal nuclei (Boord 1968; Leibler 1975; Conlee and Parks 1986; Fig. 5.3), although this segregation appears to be greater in *T. alba* than in *G. gallus* (Conlee and Parks 1986). In *T. alba*, the medial shell receives input from the contralateral nucleus angularis and from the contralateral LLDp (Adolphs 1993) and the central core receives interaural time difference input from nucleus laminaris and LLD (Takahashi et al. 1987; Takahashi and Konishi 1988a; Adolphs 1993). The lateral shell receives converging inputs from the contralateral nucleus angularis, interaural time difference input from the contralateral core, and bilateral interaural level difference input from LLDp (Adolphs 1993). These interaural time difference and interaural level difference signals are combined in various ways in the lateral shell and the combinations conveyed to the external nucleus, which contains a map of auditory space (see below; Knudsen and Konishi 1978; Mazer 1998).

The central nucleus projects to both the external nucleus and the nucleus ovoidalis of the thalamus (Ov). The external nucleus projects to the optic tectum and also receives a descending input from the paleostriatum of the forebrain (Bonke et al. 1979; cf. Puelles et al. 1994) while the archistriatum (Aivm) projects to the intercollicular area. Recordings from *T. alba* archistriatum and paleostriatum augmentatum show that these areas are senstive to spatial cues, although they do not contain an ordered space map (Cohen and Knudsen 1994, 1995).

7.4.2. Physiology

Studies of the avian auditory midbrain have shown that most neurons are binaural, excited by inputs from the contralateral ear, and inhibited by the ipsilateral ear, although bilateral excitation and contralateral excitation are also present. In the barn owl, many neurons are sensitive to changes in interaural level and time difference. The tonotopic organization is consistent with the tonotopy observed in lizards and crocodiles, with low best frequencies superficial (Coles and Aitkin 1979; Heil and Scheich 1986).

Interaural time differences are first computed in the ICC core (see Klump, Chapter 6). Nucleus laminaris neurons do not code unambiguously for time differences because they encode interaural phase differences (IPDs) at one stimulus frequency, and their responses therefore show phase ambiguity. This ambiguity is resolved in the projections of nucleus laminaris to the core where all neurons in a penetration perpendicular to isofrequency laminae encode one interaural time difference. Thus an interaural time difference is conserved in an array of neurons, not in any single neuron (Wagner et al. 1987). Each array projects to interaural time difference specific neurons in the lateral shell, which project to "space-specific" neurons in the external nucleus. The response of the population produces the space-specific neuron with a preferred interaural time difference (Takahashi et al. 1988; Mazer 1997).

Interaural level difference is computed in the lateral shell that receives a contralateral projection from LLDp (Figs. 5.6 and 5.7). LLDp output does not unambiuously encode interaural level difference, because LLDp neurons encode a preference for sound from the contralateral ear, but also change their firing rate with changes in average binaural sound level. Some of their targets in the lateral shell also show these ambiguous responses, while others, perhaps at a higher level of processing, begin to show interaural level difference specific responses. Intermediate responses exist in the lateral shell where neurons with a peak and shoulder are common (Fig. 5.6; Majer 1995).

Space-specific responses appear to be created through the gradual emergence of relevant stimulus responses in the progression across the auditory midbrain. Less is known about responses to other stimulus features. Information about both interaural time and level differences are merged in the lateral shell to form "complex field" responses as a prelude to construction of the space map in the external nucleus (Mazer 1998). These complex field units project to the external nucleus, and each space-specific neuron receives inputs from a population of neurons tuned to different frequencies (Takahashi and Konishi 1986). The non-linear interactions of these different frequency channels act to remove phase ambiguity in the response to interaural time differences (Takahashi and Konishi 1986). The representation of auditory space is ordered, with most of the external nucleus devoted to the contralateral hemifield (Knudsen and Konishi 1978). The external nucleus projects topographically to the optic tectum, which contains maps of visual and auditory space that are in register (Knudsen and Knudsen 1983). Activity in the tectum directs the rapid head movements made by the owl in response to auditory and visual stimuli (Du Lac and Knudsen 1990). The sensory map of space in the tectum is transformed into an orthogonal coordinate system representing the direction of head movement (Masino and Knudsen 1990).

8. Thalamus and Forebrain

Conserved projections from the auditory midbrain to the dorsal thalamus to the forebrain characterize the sauropsid morphotype. The dorsal thalamic region is made up of a central nucleus surrounded by a shell. The core and shell receive ascending input from the central nucleus of the auditory midbrain and from the lemniscal nuclei, respectively (Fig. 5.9). In both birds and *C. crocodilus*, the core and shell thalamic regions project to separate, nonoverlapping regions of forebrain, suggesting a basic division of auditory input to the forebrain into two separate streams (Wild et al. 1993). The primary targets of these ascending thalamic inputs are the DVR in reptiles and field L in birds.

Controversy exists as to whether the primary ascending thalamic targets in mammals and sauropsids are homologous (Karten 1991; Smeets and Gonzalez 1994; Striedter 1997). Comparative studies of the embryological origins of the forebrain nuclei in birds and mammals are therefore an area of extensive investigation (Striedter et al. 1998). Striedter (1997) has proposed that thalamic sensory projections to derivatives of the lateral and dorsal pallia in sauropsids and mammals evolved independently, while Karten (1991) has proposed that, in mammals, the components of the DVR are incorporated into the overlying pallium to form the neocortex.

8.1. Turtles

The cells of the central nucleus of the torus project bilaterally to the nucleus reuniens of the dorsal thalamus (Belekhova et al. 1985). The neurons of the nucleus reuniens project to the ipsilateral medial part of the DVR as well as to the striatum. Although the major projection of the nucleus reuniens is to the DVR, it also projects to other thalamic nuclei. Belekhova et al. (1985) have also shown a partial overlap between the auditory and somatosensory pathways in the mesencephalic and thalamic auditory nuclei. The significance of this observation is not clear, although turtles are sensitive to sound transmitted through their shells (Wever 1978).

8.2. Lizards

The central nucleus of the torus projects ipsilaterally to the nucleus medialis of the dorsal thalamus. Nucleus medialis in turn projects ipsilaterally to the medial anterior portion of the DVR in a number of genera (Foster and Hall 1978; Bruce and Butler 1984). There is also a sparse projection to the striatum from the nucleus medialis. The projections of the lizard auditory system are comparable to those found in turtles (Hall and Ebner 1970; Parent 1976; Balaban and Ulinski 1981; Belekhova et al. 1985), *C. crocodilus* (Pritz 1974a, 1974b), and *C. livia* (Karten 1967, 1968).

8.3. Crocodiles

The central nucleus of the torus projects bilaterally to the central part of the nucleus reuniens (Pritz 1974b) and the central part of the nucleus reuniens projects to the caudomedial region of the ipsilateral DVR (Pritz 1974a). The medial region of the DVR contains neurons that are sensitive to sound stimuli (Pritz 1974a). Neurons in the area surrounding the nucleus reuniens project to the nearby caudocentral portion of the DVR (Pritz and Stritzel 1992), which suggests the presence of a second auditory region in the DVR. Pritz and Stritzel use the evidence for a second auditory area in the area surrounding the nucleus reuniens in caudocentral DVR to argue for homology between the crocodilian and avian ascending auditory pathways.

8.4. Birds

The central nucleus of the inferior colliculus (torus, MLd) projects ipsilaterally to the nucleus ovoidalis of the dorsal thalamus. Nucleus ovoidalis in turn projects ipsilaterally to field L. The connections of the auditory forebrain have been reevaluated to describe the major telencephalic components of the auditory system, and to show that they are distributed in three separate regions, the medial and lateral caudal neostriatum and the intermediate archistriatum (Fig. 5.9; Wild et al. 1993; see Puelles et al. 1994). Other recent anatomical studies include cytochrome oxidase studies of the auditory midbrain, thalamus, and forebrain of *G. gallus* (Dezso et al. 1993),

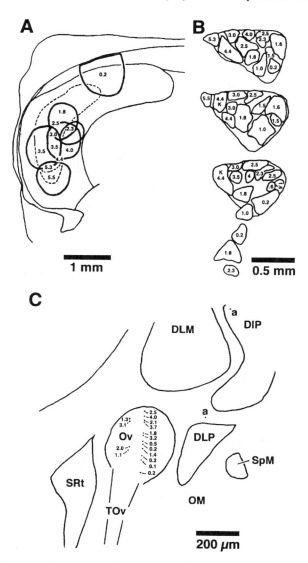

FIGURE 5.8. Tonotopic organization and tuning properties in starling nucleus ovoidalis. **A, B**: Retrograde label in nucleus ovoidalis after WGA-HRP injection into different regions of field L. (**A**) Injection sites with the best frequencies for the center of injections and overlapping zones in field L. (**B**) Corresponding zones of retrograde label in nucleus ovoidalis (Ov) linked to the corresponding best frequencies in field L. (**C**) Best frequencies observed during electrophysiological recordings of single units in nucleus ovoidalis. (**D**) Tuning curve of a type III neuron in Ov. Hatched areas show inhibition. (**E**) Rate-intensity functions at best frequency for three different Ov units a, b Type III units, c type IV unit. (From Hausler 1989; used by permission.)

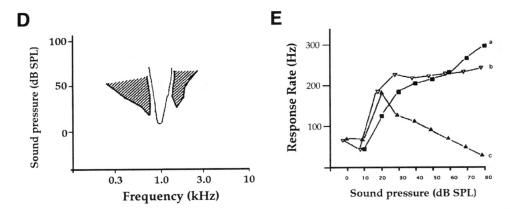

FIGURE 5.8. *Continued*

while several studies have described the physiology of the nucleus ovoidalis and field L (see below).

8.4.1. Nucleus Ovoidalis

The ovoidalis complex has been homologized to the mammalian medial geniculate nucleus (MGv; Karten 1967; Wild et al. 1993). Nucleus ovoidalis receives an ascending, ipsilateral projection from the central nucleus of the auditory midbrain. There is a parallel projection from the midbrain caudomedial shell region (cf. medial intercollicularis) to the nucleus semilunaris (SPO), a small nucleus that surrounds the ventrolateral edge of nucleus ovoidalis (Durand et al. 1992). The nucleus ovoidalis shell region receives ascending, short latency auditory input from LLD and LLV (Wild 1987). Nucleus ovoidalis projects to field L2a, while the surrounding nucleus semilunaris and shell regions project to the other layers of field L (Figs. 5.1, 5.8, and 5.9; Karten 1968; Bonke et al. 1979; Brauth et al. 1987; Vates et al. 1996). The shell region receives descending connections from the archistriatum (Fig. 5.9, Wild et al. 1993).

Nucleus ovoidalis is a large cell group encapsulated by fibers, lying lateral to the third ventricle and medial to the dorsolateral thalamic nucleus (Fig. 5.8C). It is tonotopically organized, with high best frequencies located dorsally, and low best frequencies ventrally (Bigalke-Kunz Rubsamen and Dorrscheidt 1987; Hausler 1983, 1984; Durand et al. 1992; Proctor and Konishi 1997). In the barn owl, ovoidalis contains 2 divisions based on tonotopic organization (Proctor and Konishi 1997). The medial two-thirds of the nucleus has a tonotopic organization inverted from the map of frequency in the auditory midbrain. There are also "out of order" neurons, or cells with best frequencies that differed greatly from their neighbors (Diekamp and Margoliash 1991; Proctor and Konishi 1997). In *T. alba*, all three divisions

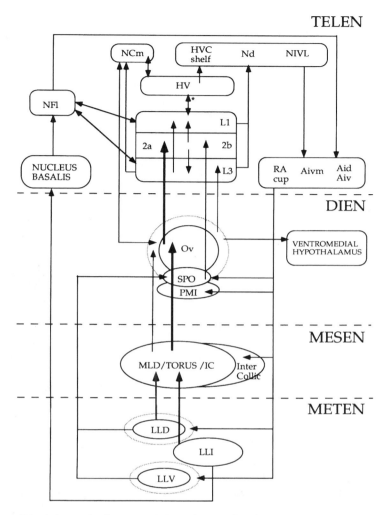

FIGURE 5.9. Schematic diagram of the avian auditory forebrain pathways, adapted from figure 17 of Wild, Karten, and Frost (1993), with additions taken from Durand et al. (1993), Vates et al. (1996), and Farabaugh and Wild (1997). Note the general pattern whereby nuclei of the caudomedial neostriatum (field L and NCM or caudomedial neostriatum) project to dorsal neostriatal nuclei (HVC or higher vocal center in oscines, Nd or dorsal neostriatum in *Columba livia* and NIVL or ventrolateral neostriatum in budgerigars) which in turn project to the nuclei of the intermediate archistriatum (Ai). The archistriatal targets include RA, the robust nucleus of the archistriatum in oscines, Aivm or ventromedial Ai in *C. livia*, and dorsal and ventral Ai in budgerigars. Note that the projection from HV (ventral hyperstriatum) to field L is reciprocal in oscines, but not in *C. livia* (Wild et al. 1993, Vates et al. 1996). The descending archistriatal projections do not project to the core of the hindbrain and midbrain auditory nuclei, but rather terminate in the surrounding shell.

of the central nucleus (lateral shell, core, and medial shell) project to ovoidalis, and the physiological responses in ovoidalis reflect this diverse array of inputs (Proctor 1998). Most neurons had responses to interaural time difference, interaural level difference, and stimulus frequency similar to those found in the lateral shell and core of the midbrain. In contrast to the mapping found in the torus, however, no systematic representation of sound localization cues was found in ovoidalis (Proctor and Konishi 1998). Nevertheless, sound localization and gaze control are mediated in parallel in the midbrain and forebrain of *T. alba* (Knudsen et al. 1993; Wagner 1993).

Studies of the starling nucleus ovoidalis emphasized spectrotemporal properties (Hausler 1983, 1988). Most neurons exhibited responses with inhibitory sidebands, and monotonic rate-intensity functions with tonic or phasic-tonic responses (Fig. 5.8D). Fewer units were observed with more complex tuning properties (Fig. 5.8; Hausler 1983, 1988). *S. vulgaris* ovoidalis neurons do not prefer song, and responses to complex auditory stimuli such as noise or species-specific song were predictable on the basis of their tuning properties. The distribution of selectivity for species-specific calls was not statistically different from the distribution observed for auditory nerve fibers (Hausler 1983). Auditory responses in *S. vulgaris* were not confined to ovoidalis, but extended into a shell region, and beyond the caudal borders of ovoidalis. Scattered auditory responses were also recorded from the dorsal thalamus both in *C. livia* and *S. vulgaris* (Hausler 1983; Durand et al. 1992).

8.4.2. Field L

Field L is the principal target of ascending input from the auditory dorsal thalamus. It is divided into three parallel layers, L1, L2, and L3, with L2 further divided into L2a and L2b (Figs. 5.8, and 5.9). There are parallel ascending projections from nucleus ovoidalis to field L2a, and from SPO and the nucleus ovoidalis shell to L2b, L1, and L3 (Durand et al. 1993; Wild et al. 1993). Field L is tonotopically organized (Fig. 5.8; Leppelsack and Schwartzkopff 1972; Zaretsky and Konishi 1976; Rubsamen and Dorrscheidt 1986; Wild et al. 1993). The highest best frequencies were recorded in ventral L2a. More broadband responses were recorded in L2b (Wild et al. 1993). Auditory units in L2 generally have narrow tuning curves with inhibitory sidebands, as might be expected from their direct input from dorsal thalamus, while the cells of L1 and L3 exhibit more complex responses (Fig. 5.8; Knudsen et al. 1977; Scheich et al. 1979).

The general avian pattern is that field L projects to the adjacent hyperstriatum and to other nuclei of the caudal neostriatum (Vates et al. 1996). Auditory neostriatal targets of field L (direct and indirect) include dorsal neostriatum in *C. livia*, HVC in songbirds, and ventrolateral neostriatum in budgerigars. These neostriatal nuclei project to the auditory areas of the archistriatum (AIVM, RA), which project back down to the auditory thalamus and midbrain (Fig. 5.9; see Wild et al. 1993). The songbird RA also projects to medullary motor nuclei.

8.4.3. Song System

Birds produce species-specific calls that are used for both species and individual recognition (Konishi 1985). The nature of birdsong suggests that auditory discrimination of songbirds may be particularly acute in the temporal and spectral domains. These observations have been supported by numerous psychophysical studies (Fay 1988; Dooling et al. 1990; Dooling et al., Chapter 7). The interest in structures specialized for the perception and production of song has driven studies of the auditory nuclei of the thalamus and forebrain. Song has evolved independently in several groups of birds, and comparisons of the vocal and auditory systems among taxa have shown that different groups may use different neural circuits for song preception and production (Paton, Manogue and Nottebohm (1981); Kelley and Nottebohm 1979; see Konishi 1994 for review).

In oscine songbirds, auditory information reaches the song system via field L, which projects to regions contiguous with the nuclei of the song system such as HVC (higher vocal center) and RA (the robust nucleus of the archistriatum; Kelley and Nottebohm 1979; Fortune and Margoliash 1992; Vates et al. 1996). The oscine song system is composed of an anterior and a posterior pathway. The posterior motor pathway is required throughout life for song production and is composed of the HVC to RA projection. RA projects to the motor neurons of the hypoglossal nucleus that control the muscles of the syrinx (Bottjer and Arnold 1982; Vicario and Nottebohm 1987), and to motor neurons that control respiration. The anterior forebrain pathway is needed during song learning and is made up of HVC–nucleus X–DLM–LMAN–RA. The posterior pathway is the presumed site where the motor program underlying the bird's song is stored, while the anterior pathway contains neurons that respond to song stimuli (Bottjer et al. 1984; Margoliash 1983; Vicario 1994; Volman 1996; Doupe 1997).

Comparison of oscine and parrot song systems show that although field L provides auditory inputs to the oscine song system, it is not the main source of auditory information in the budgerigars, *Melopsittacus undulatus*. Instead, auditory inputs to the vocal control system appear to derive from nucleus basalis and the frontal neostriatum (Striedter 1994). These nuclei receive auditory input via ascending projections from the lemniscal nuclei (Arends and Ziegler 1986). Auditory nucleus basalis projects upon the overlying neostriatum, which in turn projects upon more caudal and lateral pallial regions more directly involved in motor control, as well as to field L (Striedter 1994; Farabaugh and Wild 1997). Despite independent evolution, the song control circuits of oscines and psittacines share many similar design features (Durand et al. 1997).

The descending projection from RA to the intercollicular area conforms to the pattern of projections from the archistriatum in all birds (Fig. 5.9; Bonke et al. 1979; Gurney 1981; Karten and Shimuzu 1989; Wild et al. 1993). Stimulation of the archistriatum and intercollicular nucleus results in calls in many vertebrates (Potash 1970; Brown 1971; Konishi 1973). Vocal control in oscine songbirds and parrots differs from other sauropsids, however, because the preexisting projection from the archistriatum to the intercollicular region appears to have been modified. In these birds RA projects directly to the motor nucleus of the hypoglossal nerve and respiratory control nuclei in the medulla, in addition to a region of the nucleus intercollicularis.

9. Summary

The auditory central nervous system is organized in a pattern common to sauropsids. The similarities between these groups are presumably due to the conserved nature of the auditory sense and their close phylogenetic relationship. The organization of the ascending auditory system shares a common plan (a morphotype). In the morphotype, the auditory nerve projects to two cochlear nuclei, the nucleus magnocellularis and the nucleus angularis. Nucleus magnocellularis projects to the second order nucleus laminaris that in turn projects to the superior olive, to the lemniscal nuclei, and to the central nucleus of the auditory midbrain (torus semicircularis, nucleus mesencephalicus lateralis dorsalis, inferior colliculus). The nucleus angularis projects to the superior olive, to the lemniscal nuclei, and to the central nucleus of the auditory midbrain. The parallel ascending projections of angularis and laminaris may or may not overlap with one another, and probably do overlap in the primitive condition. Hindbrain auditory connections are generally bilateral, although contralateral projections predominate. The lemniscal nuclei project to midbrain, thalamic, and forebrain targets. The central nucleus of the auditory midbrain projects bilaterally to its thalamic target (nucleus medialis or reuniens in reptiles, nucleus ovoidalis in birds). The auditory thalamus projects to the auditory region of the forebrain (medial dorsal ventricular ridge in reptiles, field L in birds). Field L projects to other forebrain nuclei that may be involved in the control of song and other vocalizations. Descending projections from the archistriatum to the intercollicular area (and directly to the hypoglossal nucleus in some) appear to mediate vocalization.

One central outstanding issue in auditory system evolution stems from the changes associated with the change of the stimulus from water to air. McCormick (1998) has shown that the ancestral pattern in amphibians and bony fish is characterized by first-order acoustic nuclei in a ventral octaval column. Primitive fish and non-anurans generally lack the dorsal octaval column that characterizes anurans and amniotes. There are many unanswered questions about how the dorsal octaval column originated, how many times dorsal octaval structures have evolved, and whether they were present in the amniote common ancestors (see McCormick 1998). Nevertheless, many modern fish and non-anurans possess dorsomedial first-order auditory neurons with lemniscal projections to the midbrain. These first-order neurons may or may not be homologous to the dorsal octaval nuclei with lemniscal projections in anurans and amniotes. Furthermore, the anuran primary auditory nuclei may or may not be homologous to the cochlear nuclei of sauropsids and mammals; understanding of these issues depends upon a clearer understanding of amniote evolution.

A second outstanding issue in auditory system evolution derives from the appearance of the many mammalian specializations for hearing. The well-known evolution of movable outer ears and the evolution of high-frequency hearing are both developments in the periphery that may have led to changes in the central nervous system (see Wilczynski 1984).

Specializations within the morphotype are easier to understand. Although there have been many independent developments within the conserved pattern, major differences among central auditory structures appear seldom. Those differences that have appeared predominate in the hindbrain or in the auditory forebrain. The selective forces driving these changes have been ascribed to the development of new end organs in the auditory periphery or the selective pressures resulting from the increased use of sound for communication (Wilczynski 1984). Many specializations may be understood within the framework of ideas presented on brain evolution (see, e.g., Ebbesson 1980; Aboitiz 1996). Careful analysis of the development of each cell type will be required to determine whether they might be homologous, or whether they have evolved in parallel.

Acknowledgments. The authors are indebted to the following for their generous gifts of time, helpful comments, and insightful discussions during the preparation of this manuscript: S. Brauth, S. Durand, U. Hausler, H. Karten, M. Kubke, L. Proctor, and M. Wild. Supported by NIH DC00436 to CEC, and by NIH DC02633 and Texas Woman's University's Research Enhancement Program to RAC.

References

Aboitiz F (1996) Does bigger mean better? Evolutionary determinants of brain size and structure. Brain Behav Evol 47:225–245.

Adams JC, Mroz EA, Sewell WF (1987) A possible neurotransmitter role for CGRP in a hair-cell sensory organ. Brain Res 419:347–351.

Adolphs R (1993) Acetylcholinesterase staining differentiates functionally distinct auditory pathways in the barn owl. J Comp Neurol 329:365–377.

Altschuler RA, Fex J (1986) Efferent neurotransmitters. In: Altschuler RA, Hoffman DW, Bobbin RP (eds) Neurobiology of Hearing: The Cochlea. New York: Raven Press, pp. 383–396.

Arends JJA (1981) Sensory and motor aspects of the trigeminal system in the mallard (*Anas platyrhonchos* L.) State Univ Leiden Netherlands.

Arends JJA, Zeigler HP (1986) Anatomical identification of an auditory pathway from a nucleus of the lateral lemniscal system to the frontal telencephalon (nucleus basalis) of the pigeon. Brain Res 398:375–381.

Balaban CD, Ulinski PS (1981) Organization of thalamic afferents to anterior dorsal ventricular ridge in turtles. I. Projections of thalamic nuclei. J Comp Neurol 200:95–129.

Barbas-Henry HA, Lohman AHM (1988) Primary projections and efferent cells of the VIIIth cranial nerve in the monitor lizard, *Varanus exanthematicus*. J Comp Neurol 277:234–249.

Belekhova MG, Zharskaja VD, Khachunys AS, Gaidaenko GV, Tumanova NL (1985) Connections of the mesencephalic, thalamic and telencephalic auditory centers in turtles. Some structural bases for audiosomatic interrelations. J Hirnforsch 26:127–152.

Biederman-Thorson M (1970) Auditory responses of units in the ovoid nucleus and cerebrum (field L) of the ring dove. Brain Res 24:247–256.

Bigalke-Kunz B, Rubsamen R, Dorrscheidt GJ (1987) Tonotopic organization and functional characterization of the auditory thalamus in a songbird, the European starling. J Comp Physiol 161:255–265.

Bonke BA, Bonke D, Scheich H (1979) Connectivity of the auditory forebrain nuclei in the guinea fowl (*Numida meleagris*). Cell Tissue Res 200:101–121.

Boord RL (1961) The efferent cochlear bundle in the caiman and pigeon. Exp Neurol 3:225–239.

Boord RL (1968) Ascending projections of the primary cochlear nuclei and nucleus laminaris in the pigeon. J Comp Neurol 133:523–542.

Boord RL, Rasmussen GL (1963) Projection of the cochlear and lagenar nerves on the cochlear nuclei of the pigeon. J Comp Neurol 120:462–475.

Bottjer SW, Arnold AP (1982) Afferent neurons in the hypoglossal nerve of the zebra finch (*Poephila guttata*): localization with horseradish peroxidase. J Comp Neurol 210:190–197.

Bottjer SW, Miesner EA, Arnold AP (1984) Forebrain lesions disrupt development but not maintainance of song in passerine birds. Science 224:901–903.

Brauth SE (1990) Investigation of central auditory nuclei in the budgerigar with cytochrome oxidase histochemistry. Brain Res 508:142–146.

Brauth SE, McHale CM, Brasher CA, Dooling RJ (1987) Auditory pathways in the budgerigar. Brain Behav Evol 30:174–199.

Brown JL (1971) An exploratory study of vocalization areas in the brain of the red-winged blackbird (*Agelaius phoeniceus*). Behavior 24:91–127.

Browner RH, Kennedy MC, Facelle T (1981) The cytoarchitecture of the torus semi-circularis in the red-eared turtle. J Morphol 169:207–223.

Browner RH, Marbey D (1988) Nucleus magnocellularis in the red-eared turtle, *Chrysemys scripta elegans*: eighth nerve endings and neuronal types. Hearing Res 33:257–272.

Bruce LL, Butler AB (1984) Telencephalic connections in lizards. II. Projections to anterior dorsal ventricular ridge. J Comp Neurol 229:602–615.

Carr CE (1992) Evolution of the central auditory system in reptiles and birds. In: Webster DB, Fay RR, Popper AN (eds) The Evolutionary Biology of Hearing New York: Springer-Verlag, pp. 511–543.

Carr CE, Boudreau RE (1991) The central projections of auditory nerve fibers in the barn owl. J Comp Neurol 314:306–318.

Carr CE, Boudreau RE (1993) Organization of nucleus magnocellularis and nucleus laminaris in the barn owl: encoding and measuring interaural time differences. J Comp Neurol 334:337–355.

Carr CE, Konishi M (1990) A circuit for detection of interaural time differences in the brainstem of the barn owl. J Neurosci 10:3227–3246.

Carr CE, Fujita I, Konishi M (1989) Distribution of GABAergic neurons and terminals in the auditory system of the barn owl. J Comp Neurol 286:190–207.

Carr CE, Amagai S, Kubke MF, Massoglia DP (1996) Evolution of time coding systems. In: Elsner N, Schnitzler U (eds) The Proceeding of the Goettingen Neurobiology Conference. Thieme, Stuttgart, Germany.

Chandler J (1984) Light and electron microscopic studies of the basilar papilla in the duck, *Anas platyrhyncos*. I. The hatchling. J Comp Neurol 222:506–522.

Code RA (1995) Efferent neurons to the macula lagena in the embryonic chick. Hear Res 82:26–30.

Code RA (1996) Chick auditory terminals contain dynorphin-like immunoreactivity. Neuroreport 7:2917–2920.

Code RA (1997) The avian cochlear efferent system. Poultry Avian Biol Rev 8:1–8.

Code RA, Carr CE (1994) Choline acetyltransferase-immunoreactive cochlear efferent neurons in the chick auditory brainstem. J Comp Neurol 340:161–173.

Code RA, Carr CE (1995) Enkephalin-like immunoreactivity in the chick brainstem: possible relation to the cochlear efferent system. Hear Res 87:69–83.

Code RA, Burd GD, Rubel EW (1989) Development of GABA immunoreactivity in brainstem auditory nuclei of the chick: ontogeny of gradients in terminal staining. J Comp Neurol 284:504–518.

Code RA, Darr MS, Carr CE (1996) Chick cochlear efferent neurons are not immunoreactive for calcitonin gene-related peptide. Hear Res 97:127–135.

Cohen YE, Knudsen EI (1994) Binaural tuning of auditory units in the forebrain archistriatal gaze fields of the barn owl basal ganglia. J Neurophysiol 72:285–298.

Cohen YE, Knudsen EI (1995) Auditory tuning for spatial cues in the barn owl: local organization but no space map. J Neurosci 15:5152–5168.

Cole KS, Gummer AW (1990) A double-label study of efferent projections to the cochlea of the chicken, *Gallus domesticus*. Exp Brain Res 82:585–588.

Coles RB, Aitkin LM (1979) The response properties of auditory neurones in the midbrain of the domestic fowl (*Gallus gallus*) to monaural and binaural stimuli. J Comp Physiol 134:241–251.

Coles RB, Guppy A (1988) Directional hearing in the barn owl (*Tyto alba*). J Comp Physiol 163:117–133.

Conlee JW, Parks TN (1986) Origin of ascending auditory projections to the nucleus mesencephalicus lateralis pars dorsalis in the chicken. Brain Res 367:96–113.

Cotanche DA, Lee KH, Stone JS, Picard DA (1994) Hair cell regeneration in the bird cochlea following noise damage or ototoxic drug damage. Anat Embryol (Berl) 189:1–18.

Cotanche DA, Hennig AK, Riedl AE, Messana EP (1997) Hair cell regeneration in the chick cochlea—where we stand after 10 years of work. In: Palmer AR, Rees

A, Summerfield AQ, Meddis A (eds) Psychophysical and Physiological Advances in Hearing: Proceedings of the 11th International Symposium on Hearing. London: Whurr, pp. 109–115.

Crawford AC, Fettiplace R (1980) The frequency selectivity of auditory nerve fibers and hair cells in the cochlea of the turtle. J Physiol 306:79–125.

Desmedt JE, Delwaide PJ (1963) Neural inhibition in a bird: effect of strychnine and picrotoxin. Nature 200:583–585.

Dezso A, Schwarz DWF, Schwarz IE (1993) A survey of the auditory midbrain, thalamus and forebrain in the chicken (*Gallus domesticus*) with cytochrome oxidase histochemistry. J Otolaryngol 22:391–396.

Diekamp B, Margoliash D (1991) Auditory responses in the nucleus ovoidalis are not so simple. Soc Neurosci Abstr 17:446.

Dooling RJ, Brown SD, Park TJ, Okanoya K (1990) Natural perceptual categories for vocal signals in budgerigars (*Melopsittacus undulatus*). In: Stebbins WC, Berkley MA (eds) Comparative Perception. Vol. II: Complex Signals. New York: John Wiley & Sons, pp. 345–374.

Doupe AJ (1997) Song- and order-selective neurons in the songbird anterior forebrain and their emergence during vocal development. J Neurosci 17:1147–1167.

Duckert LG, Rubel EW (1990) Ultrastructural observations on regenerating hair cells in the chick basilar papilla. Hear Res 48:161–182.

Duckert LG, Rubel EW (1993) Morphological correlates of functional recovery in the chicken inner ear after gentamycin treatment. J Comp Neurol 331:75–96.

Du Lac S, Knudsen EI (1990) Neural maps of head movement vector and speed in the optic tectum of the barn owl. J Neurophysiol 63:131–146.

Durand S, Tepper J, Cheng MF (1992) The shell region of the nucleus ovoidalis: a subdivision of the avian auditory thalamus. J Comp Neurol 323:495–518.

Durand SE, Zuo MX, Zhou SL, Cheng MF (1993) Avian auditory pathways show metenkephalin-like immunoreactivity Neuroreport 4:727–730.

Durand SE, Heaton JT, Amateau SK, Brauth SE (1997) Vocal control pathways through the anterior forebrain of a parrot (*Melopsittacus undulatus*). J Comp Neurol 877:179–206.

Ebbesson SO (1980) The parcellation theory and its relation to interspecific variability in brain organization, evolutionary and ontogenetic development, and neuronal plasticity. Cell Tissue Res 213:179–212.

Eybalin M (1993) Neurotransmitters and neuromodulators of the mammalian cochlea. Physiol Rev 73:309–373.

Farabaugh SM, Wild JM (1997) Reciprocal connections between primary and secondary auditory pathways in the telencephalon of the budgerigar (*Melopsittacus undulatus*). Brain Res 747:18–25.

Fay RR (1988) Hearing in Vertebrates: A Psychophysics Databook. Winnetka, IL: Hill-Fay Associates.

Firbas W, Muller G (1983) The efferent innervation of the avian cochlea. Hear Res 10:109–116.

Fischer FP (1992) Quantitative analysis of the innervation of the chicken basilar papilla. Hear Res 61:167–178.

Fischer FP (1994) Quantitative TEM analysis of the barn owl basilar papilla. Hear Res 73:1–15.

Fischer FP, Köppl C, Manley GA (1988) The basilar papilla of the barn owl *Tyto alba*: a quantitative morphological SEM analysis. Hear Res 34:87–101.

Fischer FP, Eisensamer B, Manley GA (1994) Cochlear and lagenar ganglia of the chicken. J Morphol 220:71–83.

Fortune ES, Margoliash D (1992) Cytoarchitectonic organization and morphology of cells of the field L complex in male zebra finches (*Taenopygia guttata*). J Comp Neurol 325:388–404.

Foster RJ, Hall WJ (1978) The organization of central auditory pathways in a reptile, *Iguana iguana*. J Comp Neurol 178:783–832.

Fritzsch B (1981) Efferent neurons to the labyrinth of *Salamandra salamandra* as revealed by retrograde transport of horseradish peroxidase. Neurosci Lett 26: 191–196.

Gleich O, Narins PM (1988) The phase response of primary auditory afferents in a songbird (*Sturnus vulgaris* L.). Hear Res 32:81–92.

Gurney M (1981) Hormonal control of cell form and number in the zebra finch song system. J Neurosci 1:658–673.

Hall WC, Ebner FF (1970) Thalamotelencephalic projections in the turtle (*Pseudemys scripta*). J Comp Neurol 140:101–127.

Hall WS, Cohen PL, Brauth SE (1993) Auditory projections to the anterior telencephalon in the budgerigar (*Melopsittacus undulatus*). Brain Behav Evol 41: 97–116.

Hartline PH, Campbell HW (1969) Auditory and vibratory responses in the midbrains of snakes. Science 163:1221–1223.

Hashimoto S, Kimura RS, Takasaka T (1990) Computer-aided three-dimensional reconstruction of the inner hair cells and their nerve endings in the guinea pig cochlea. Acta Otolaryngol (Stockh) 109:228–234.

Hausler U (1983) Histologische und elektrophysiologische Untersuchungen an einzelnen Neuronen des Nucleus ovoidalis im Zwischenhirn des Staren (*Sturnus vulgaris* L.) Diplomthesis, Faculty of Biology, Ruhr Universitaet Bochum, Germany.

Hausler U (1984) Neurophysiological and anatomical studies of nucleus ovoidalis neurons in the starling, *Sturnus vulgaris*. Verh Dtsch Zool Ges 77:291.

Hausler U (1988) Topography of the thalamotelencephalic projections in the auditory system of a songbird. In: Syka J, Masterton RB (eds) Auditory Pathway: Structure and Function. New York: Plenum Press, pp. 197–202.

Hausler U (1989) Die strukturelle und funktionelle Organisation der Hoerbahn im caudalen Vorderhirn des Staren (*Sturnus vulgaris*, L.) Ph.D. Thesis, Faculty of Biology, Technische Universitaet Muenchen, Germany.

Hausler UHL, Sullivan WE, Soares D, Carr CE (1999) A morphological study of the cochlear nuclei of the pigeon (*Columba livia*). Brain Behav Evol 54:290–302.

Heil P, Scheich H (1986) Effects of unilateral and bilateral cochlear removal on 2-deoxyglucose patterns in the chick auditory system. J Comp Neurol 252:279–301.

Hennig AK, Cotanche DA (1998) Regeneration of cochlear efferent nerve terminals after gentamycin damage. J Neurosci 18:3282–3296.

Hill KG, Stange G, Mo J (1989) Temporal synchronization in the primary auditory response in the pigeon. Hear Res 39:63–74.

Hotta T (1971) Unit responses from the nucleus angularis in the pigeon's medulla. Comp Biochem Physiol 40A:415–424.

Jeffress LA (1948) A place theory of sound localization. J Comp Physiol Psychol 41:35–39.

Jhaveri S, Morest DK (1982) Neuronal architecture in nucleus magnocellularis of the chicken auditory system with observations on nucleus laminaris: a light and electron microscope study. Neuroscience 7:809–836.

Johnson DH (1980) The relationship between spike rate and synchrony in responses of auditory nerve fibers to single tones. J Acoust Soc Am 68:1115–1122.

Joseph AW, Hyson RL (1993) Coincidence detection by binaural neurons in the chick brain stem. J Neurophysiol 69:1197–1211.

Kaiser A, Manley GA (1994) Physiology of single putative cochlear efferents in the chicken. J Neurophysiol 72:2966–2979.

Karten HJ (1967) The organization of the ascending auditory pathway in the pigeon (*Columba livia*) I. Diencephalic projections of the inferior colliculus (nucleus mesencephali lateralis, pars dorsalis). Brain Res 6:409–427.

Karten HJ (1968) The ascending auditory pathway in the pigeon (*Columba livia*) II. Telencephalic projections of the nucleus ovoidalis thalami. Brain Res 11: 134–153.

Karten HJ (1991) Homology and evolutionary origins of the 'neocortex.' Brain Behav Evol 38:264–272.

Karten HJ, Hodos W (1967) A stereotaxic atlas of the brain of the pigeon. Baltimore: Johns Hopkins University Press.

Karten HJ, Shimizu T (1989) The origins of neocortex: connections and lamination as distinct events in evolution. J Cogn Neurosci 1:291–301.

Katz LC, Gurney ME (1981) Auditory responses in the zebra finch's motor system for song. Brain Res 211:192–197.

Kelley DB, Nottebohm F (1979) Projections of a telencephalic auditory nucleus— field L—in the canary. J Comp Neurol 183:455–470.

Kennedy MC (1974) Auditory multiple-unit activity in the midbrain of the Tokay gecko (*Gekko gekko*, L.). Brain Behav Evol 10:257–264.

Kennedy MC (1975) Vocalization elicited in a lizard by electrical stimulation of the midbrain. Brain Res 91:321–325.

Kennedy MC, Browner RH (1981) The torus semicircularis in a gekkonid lizard. J Morphol 169:259–274.

Keppler C, Schermuly L, Klinke R (1994) The course and morphology of efferent nerve fibers in the papilla basilaris of the pigeon (*Columba livia*). Hear Res 74:259–264.

Kirsch M, Coles RB, Leppelsack H-J (1980) Unit recordings from a new auditory area in the frontal neostriatum of the awake starling (*Sturnus vulgaris*). Exp Brain Res 38:375–380.

Knudsen EI, Konishi M (1978) A Neural map of auditory space in the owl. Science 200:795–797.

Knudsen EI (1980) Sound localization in birds. In: Popper AN, Fay RR (eds) Comparative Studies of Hearing in Vertebrates. Berlin: Springer Verlag, pp. 287–322.

Knudsen EI (1983) Subdivisions of the inferior colliculus in the barn owl (*Tyto alba*). J Comp Neurol 218:174–186.

Knudsen EI, Knudsen PF (1983) Space-mapped auditory projections from the inferior colliculus to the optic tectum in the barn owl (*Tyto alba*). J Comp Neurol 218:187–196.

Knudsen EI, Konishi M, Pettigrew JD (1977) Receptive fields of auditory neurons in the owl. Science 198:1278–1280.

Knudsen EI, Knudsen PF, Masino T (1993) Parallel pathways mediating both sound localization and gaze control in the forebrain and midbrain of the barn owl. J Neurosci 13:2837–2852.

Konishi M (1970) Comparative neurophysiological studies of hearing and vocalization in songbirds. J Comp Physiol 66:257–272.

Konishi M (1973) How the owl tracks its prey. Am Sci 61:414–424.

Konishi M (1985) Birdsong: from behavior to neuron. Ann Rev Neurosci 8:125–170.

Konishi M (1986) Centrally synthesized maps of sensory space. Trends Neurosci 9:163–168.

Konishi M (1994) Pattern generation in birdsong. Current Opinion Neurobiol 4:827–831.

Konishi M, Sullivan WE, Takahashi T (1985) The owl's cochlear nuclei process different sound localization cues. J Acoust Soc Am 78:360–364.

Köppl C (1994) Auditory nerve terminals in the cochlear nucleus magnocellularis: differences between low and high frequencies. J Comp Neurol 339:438–446.

Köppl C (1997) Frequency tuning and spontaneous activity in the auditory nerve and cochlear nucleus magnocellularis of the barn owl *Tyto alba*. J Neurophysiol 77:364–377.

Köppl C, Manley GA (1990a) Peripheral auditory processing in the bobtail lizard *Tiliqua rugosa* II. Tonotopic organization and innervation pattern of the basilar papilla. J Comp Physiol A 167:101–112.

Köppl C, Manley GA (1990b) Peripheral auditory processing in the bobtail lizard *Tiliqua rugosa* III. Patterns of spontaneous and tone-evoked nerve-fibre activity. J Comp Physiol A 167:113–127.

Köppl C, Manley GA (1992) Functional consequences of morphological trends in the evolution of lizard hearing organs. In: Webster DB, Fay RR, Popper AN (eds) The Evolutionary Biology of Hearing. New York: Springer Verlag, pp. 489–510.

Köppl C, Carr CE (1997) A low-frequency pathway in the barn owl's auditory brainstem. J Comp Neurol 378:265–282.

Köppl C, Gleich O, Manley GA (1993) An auditory fovea in the barn owl cochlea. J Comp Physiol A 171:695–704.

Kreithen ML, Quine DB (1979) Infrasound detection by the homing pigeon: a behavioral audiogram. J Comp Physiol 129:1–4.

Kubke MF, Wild M, Carr CE (1998) Nucleus basalis of the barn owl contains both tonotopic and somatotopic maps. Int Soc Neuroethol Abstr.

Kunzle H (1986) Projections from the cochlear nuclear complex to rhombencephalic auditory centers and torus semicircularis in the turtle. Brain Res 379:307–319.

Lachica EA, Rubsamen R, Rubel EW (1994) GABAergic terminals in nucleus magnocellularis and laminaris originate from the superior olivary nucleus. J Comp Neurol 348:403–418.

Leake PA (1974) Central projections of the statoacoustic nerve in *Caiman crocodilus*. Brain Behav Evol 10:170–196.

Leibler LM (1975) Monaural and binaural pathways in the ascending auditory system of the pigeon. Ph.D. Thesis, Massachusetts Institute of Technology, Cambridge, MA.

Leppelsack HJ (1979) The increase of response selectivity within the avian auditory pathway. Exp Brain Res (suppl II):116–121.

Leppelsack H-J, Schwartzkopff J (1972) Eigenschaften von Aukutishen Neuronen im Kaudalen Neostriatum von Vogeln. J Comp Physiol 80:137–140.

Levin MD, Schneider M, Kubke M, Wenthold R, Carr CE (1997) Localization of glutamate receptors in the auditory brainstem of the barn owl. J Comp Neurol 378:239–253.

McCormick CA (1999) Anatomy of the central auditory pathways of fish and amphibians. In: Comparative Hearing: Fish and amphibians Fay RR, Popper AN (eds). New York: Springer-Verlag.

Maekawa M (1987) Auditory responses in the nucleus basalis of the pigeon. Hear Res 27:231–237.

Manley JA (1970a) Frequency sensitivity of auditory neurons in the Caiman cochlear nucleus. Z Vgl Physiol 66:251–256.

Manley GA (1970b) Comparative studies of auditory physiology in reptiles. Z Vgl Physiol 67:363–381.

Manley JA (1971) Single unit studies in the midbrain auditory area of Caiman. Z Vgl Physiol 71:255–261.

Manley GA (1974) Activity patterns of neurons in the peripheral auditory system of some reptiles. Brain Behav Evol 10:244–256.

Manley GA (1977) Response patterns and peripheral origin of auditory nerve fibres in the monitor lizard, *Varanus bengalensis*. J Comp Physiol 118:249–260.

Manley GA (1981) A review of the auditory physiology of reptiles. In: Autrum HE, Perl E, Schmidt RF (eds) Progress in Sensory Physiology. Berlin: Springer Verlag, pp. 49–134.

Manley GA, Köppl C, Konishi M (1988) A neural map of interaural intensity difference in the brainstem of the barn owl. J Neurosci 8:2665–2677.

Manley GA, Gleich O, Kaiser A, Brix J (1989) Functional differentiation of sensory cells in the avian auditory periphery. J Comp Physiol A 164:289–296.

Manley GA, Yates GK, Köppl C, Johnstone BM (1990) Peripheral auditory processing in the bobtail lizard *Tiliqua rugosa* IV. Phase locking of auditory-nerve fibres. J Comp Physiol A 167:129–138.

Marcellini DL (1978) The acoustic behavior of lizards. In: Greenberg N, MacLean PD (eds) Behavior and Neurology of Lizards. Rockville, MD: US Dept of Health, Education and Welfare, pp. 287–300.

Margoliash D (1983) Acoustic parameters underlying the responses of song specific in the white-crowned sparrow. J Neurosci 3:1039–1057.

Marin F, Puelles L (1995) Morphological fate of rhombomeres in quail/chick chimeras: a segmental analysis of hindbrain nuclei. Eur J Neurosci 7:1714–1738.

Masino T, Knudsen EI (1990) Horizontal and vertical components of head movement are controlled by distinct neural circuits in the barn owl. Nature 345: 434–437.

Mazer JA (1995) Integration of Parallel Processing Streams in the Inferior Colliculus of the Barn Owl. Ph.D. Thesis, California Institute of Technology, Pasadena, CA.

Mazer JA (1998) How the owl resolves auditory coding ambiguity. Proc Natl Acad Sci U S A 95:10932–10937.

Miller MR (1975) The cochlear nuclei of lizards. J Comp Neurol 159:375–406.

Miller MR (1980) The reptilian cochlear duct. In: Popper AN, Fay RR (eds) Comparative Studies of Hearing in Vertebrates. Berlin: Springer Verlag, pp. 169–204.

Miller MR, Beck J (1988) Auditory hair cell innervational patterns in lizards. J Comp Neurol 271:604–628.

Miller MR, Kasahara M (1979) The cochlear nuclei of some turtles. J Comp Neurol 185:221–236.

Mogdans J, Knudsen EI (1994a) Site of auditory plasticity in the brain stem (VLVp) of the owl revealed by early monaural occlusion. J Neurophysiol 72:2875–2891.

Mogdans J, Knudsen EI (1994b) Representation of interaural level difference in the VLVp, the first site of binaural comparison in the barn owl's auditory system. Hear Res 74:148–164.

Moiseff A (1989a) Bicoordinate sound localization by the barn owl. J Comp Physiol 164:637–644.

Moiseff A (1989b) Binaural disparity cues available to the barn owl for sound localization. J Comp Physiol 164:629–636.

Moiseff A, Konishi M (1983) Binaural characteristics of units in the owl's brainstem auditory pathway: precursors of restricted spatial receptive fields. J Neurosci 2:2553–2562.

Muller CM (1987) Gamma-aminobutyric acid immunoreactivity in brainstem auditory nuclei of the chicken. Neurosci Lett 77:272–276.

Nadol JB (1990) Synaptic morphology of inner and outer hair cells of the human organ of Corti. J Electron Microsc Tech 15:187–196.

Norberg RA (1978) Skull asymmetry, ear structure and function, and auditory localization in Tengmalm's owl, *Aegolius funerus* (Linne). Philos Trans R Soc Lond B Biol Sci 282:325–410.

Northcutt RG (1995) The forebrain of gnathostomes: in search of a morphotype. Brain Behav Evol 46:275–318.

Nottebohm F (1980) Brain pathways for vocal learning in birds: a review of the first 10 years. Prog Pyschobiol Physiol Psychol 9:85–124.

Ofsie MS, Hennig AK, Messana EP, Cotanche DA (1997) Sound damage and gentamicin treatment produce different patterns of damage to the efferent innervation of the chick cochlea. Hear Res 113:207–223.

Oertel D (1997) Encoding of timing in the brain stem auditory nuclei of vertebrates. Neuron 19:959–962.

Olsen JF, Knudsen EI, Esterly SD (1989) Neural maps of interaural time and intensity differences in the optic tectum of the barn owl. J Neurosci 9:2591–2605.

Overholt EM, Rubel EW, Hyson RL (1992) A circuit for coding interaural time differences in the chick brainstem. J Neurosci 12:1696–1706.

Parent A (1976) Striatal afferent connections in the turtle (*Chrysemys picta*) as revealed by retrograde axonal transport of horseradish peroxidase. Brain Res 108:25–36.

Parks TN, Rubel EW (1975) Organization and development of brain stem auditory nucleus of the chicken: organization of projections from N. magnocellularis to N. laminaris. J Comp Neurol 164:435–448.

Parks TN, Code RA, Taylor DA, Solum DA, Strauss KI, Jacobowitz DM, Winsky L (1997) Calretinin expression in the chick brainstem auditory nuclei develops and is maintained independently of cochlear nerve input. J Comp Neurol 383:112–121.

Peña JL, Viete S, Albeck Y, Konishi M (1996) Tolerance to sound intensity of binaural coincidence detection in the nucleus laminaris of the owl. J Neurosci 16:7046–7054.

Paton JA, Manogue KR, Nottebohm F (1981) Bilateral organization of the vocal control pathway in the budgerigar, *M. undulatus*. J Neurosci 1:1279–1288.

Potash LM (1970) Neuroanatomical regions relevant to production and analysis of vocalization within the avian torus semicircularis. Experientia 26:257–264.

Pritz MB (1974a) Ascending connections of a thalamic auditory area in a crocodile, *Caiman crocodilus*. J Comp Neurol 153:199–214.

Pritz MB (1974b) Ascending connections of a midbrain auditory area in a crocodile, *Caiman crocodilus*. J Comp Neurol 153:179–198.

Pritz MB, Stritzel ME (1992) A second auditory area in the non-cortical telencephalon of a reptile. Brain Res 569:146–151.

Proctor L (1997) The auditory thalamus of the barn owl: anatomy and physiological responses to sound localization cues. Ph.D. Thesis, California Institute of Technology, Pasadena, CA.

Proctor L, Konishi M (1997) Representation of sound localization cues in the auditory thalamus of the barn owl. Proc Natl Acad Sci U S A 94:10421–10425.

Puelles L, Robles C, Martinez-de-la-Torre M, Martinez S (1994) New subdivision schema for the avian torus semicircularis: neurochemical maps in the chick. J Comp Neurol 340:98–125.

Raman I, Trussell LO (1992) The kinetics of the responses to glutamate and kainate in neurons of the avian cochlear nucleus. Neuron 9:173–186.

Raman I, Zhang S, Trussell LO (1994) Pathway-specific variants of AMPA receptors and their contribution to neuronal signaling. J Neurosci 14:4998–5010.

Ramon y Cajal S (1908) Les ganglions terminaux du nerf acoustique des oiseaux. Trab Inst Cajal Invest Biol 6:195–225.

Reyes AD, Rubel EW, Spain WJ (1994) Membrane properties underlying the firing of neurons in the avian cochlear nucleus. J Neurosci 14:5352–5364.

Rieppel O, deBraga M (1996) Turtles as diapsid reptiles. Nature 384:453–455.

Roberts BL, Meredith GE (1992) The efferent innervation of the ear: variations on an enigma. In: Webster DB, Fay RR, Popper AN (eds) The Evolutionary Biology of Hearing. New York: Springer-Verlag, pp. 185–210.

Roberts BL, Maslam S, Los I, Van Der Jagt B (1994) Coexistence of calcitonin gene-related peptide and choline acetyltransferase in eel efferent neurons. Hear Res 74:231–237.

Rogers J (1989) Two calcium binding proteins mark many chick sensory neurons. Neuroscience 31:697–709.

Rubel EW, Parks TN (1975) Organization and development of brainstem auditory nuclei of the chicken: Tonotopic organization of N. Magnocellularis and N. Laminaris. J Comp Neurol 164:411–434.

Rubel EW, Parks TN (1988) Organization and development of the avian brainstem auditory system. In: Edelman GM, Gall WE, Cowan WM (eds) Auditory Function: Neurobiological Bases of Hearing. New York: John Wiley & Sons, pp. 3–92.

Rubsamen R, Dorrscheidt GJ (1986) Tonotopic organization of the auditory forebrain in a songbird, the European starling. J Comp Physiol 158:639–646.

Sachs MB, Sinnott JM (1978) Responses to tones of single cells in nucleus magnocellularis and nucleus angularis of the redwing blackbird (*Agelaius phoeniceus*). J Comp Physiol 126:347–361.

Sahley TL, Nodar RH, Musiek FE (1997) Efferent Auditory System Structure and Function. San Diego, CA: Singular Publishing Group, Inc.

Sams-Dodd F, Capranica RR (1994) Representation of acoustic signals in the eighth nerve of the Tokay gecko: I. Pure tones. Hear Res 76:16–30.

Sams-Dodd F, Capranica RR (1996) Representation of acoustic signals in the eighth nerve of the Tokay gecko: II. Masking of pure tones with noise. Hear Res 100:131–134.

Saunders JC, Adler HJ, Pugliano FA (1992) The structural and functional aspects of hair cell regeneration in the chick as a result of exposure to intense sound. Exp Neurol 115:13–17.

Scheich H, Langer G, Bonke D (1979) Responsiveness of units in the auditory neostriatum of the guinea fowl (*Numida meleagris*) to species specific calls and synthetic stimuli: II. Discrimination of iambus-like calls. J Comp Physiol 32:257–276.

Schermuly L, Klinke R (1990) Infrasound sensitive neurons in the pigeon cochlear ganglion. J Comp Physiol 166:355–363.

Schwarz DWF, Schwarz IE, Tomlinson RD (1978) Avian efferent vestibular neurons identified by axonal transport of [3H]adenosine and horseradish peroxidase. Brain Res 155:103–107.

Schwarz DWF, Schwarz IE, Frederickson JM, Landolt JP (1981) Efferent vestibular neurons: a study employing retrograde tracer methods in the pigeon (*Columba livia*). J Comp Neurol 196:1–12.

Schwarz DWF, Schwarz IE, Dezso A (1992) Cochlear efferent neurons projecting to both ears in the chicken, *Gallus domesticus*. Hear Res 60:110–114.

Smeets WZ, Gonzalez A (1994) Sensorimotor integration in the brain of reptiles. Eur J Morphol 32:307–310.

Smith CA (1985) Inner ear. In: King AS, McLelland J (eds) Form and Function in Birds, Vol 3. London: Academic Press, pp. 273–310.

Smith ZDJ, Rubel EW (1979) Organization and development of brain stem auditory nuclei of the chicken: dendritic gradients in nucleus laminaris. J Comp Neurol 186:213–239.

Smolders JWT, Klinke R (1986) Synchronized responses of primary auditory fibre-populations in *Caiman crocodilus* (L) to single tones and clicks. Hear Res 24:89–103.

Soares D, Simon J, Carr C (1999) The cochlear nuclei of the caiman. Soc Neurosci Abstr (in press).

Striedter GF (1994) The vocal control pathways in budgerigars differ from those in songbirds. J Comp Neurol 343:35–56.

Striedter GF (1997) The telencephalon of tetrapods in evolution. Brain Behav Evol 49:179–213.

Striedter GF, Marchant TA, Beydler S (1998) The "neostriatum" develops as part of the lateral pallium in birds. J Neurosci 18:5839–5849.

Strutz J (1981) The origin of centrifugal fibers to the inner ear in *Caiman crocodilus*: a horseradish peroxidase study. Neurosci Lett 27:95–100.

Strutz J (1982) The origin of efferent fibers to the inner ear in a turtle (*Terrapene ornata*). A horseradish peroxidase study. Brain Res 244:165–168.

Strutz J, Schmidt C (1982) Acoustic and vestibular efferent neurons in the chicken (*Gallus domesticus*). Acta Otolaryngol 94:45–51.

Strutz J, Schmidt CL, Sturmer C (1980) Origin of efferent fibers of the vestibular apparatus in goldfish. A horseradish peroxidase study. Neurosci Lett 18:5–9.

Sullivan WE (1985) Classification of response patterns in cochlear nucleus of barn owl: correlation with functional response properties. J Neurophysiol 53:201–216.

Sullivan WE, Konishi M (1984) Segregation of stimulus phase and intensity coding in the cochlear nucleus of the barn owl. J Neurosci 4:1787–1799.

Sullivan WE, Konishi M (1986) Neural map of interaural phase difference in the owl's brainstem. Proc Natl Acad Sci U S A 83:8400–8404.

Szpir MR, Sento S, Ryugo DK (1990) The central projections of the cochlear nerve fibers in the alligator lizard. J Comp Neurol 295:530–547.

Szpir MR, Wright DD, Ryugo DK (1995) Neuronal organization of the cochlear nuclei in alligator lizards: a light and electron microscopic investigation. J Comp Neurol 357:217–241.

Takahashi T, Konishi M (1986) Selectivity for interaural time difference in the owl's midbrain. J Neurosci 6:3413–3422.

Takahashi T, Konishi M (1988a) The projections of the cochlear nuclei and nucleus laminaris to the inferior colliculus of the barn owl. J Comp Neurol 274:190–211.

Takahashi T, Konishi M (1988b) Projections of nucleus angularis and nucleus laminaris to the lateral lemniscal nuclear complex of the barn owl. J Comp Neurol 274:212–238.

Takahashi T, Moiseff A, Konishi M (1984) Time and intensity cues are processed independently in the auditory system of the owl. J Neurosci 4:1781–1786.

Takahashi T, Carr CE, Brecha N, Konishi M (1987) Calcium binding protein-like immunoreactivity labels the terminal field of nucleus laminaris of the barn owl. J Neurosci 7:1843–1856.

Takahashi T, Wagner H, Konishi M (1988) The role of commissural projections in the representation of bilateral space in the barn owl's inferior colliculus. J Comp Neurol 281:545–554.

Takasaka T, Smith CA (1971) The structure and innervation of the pigeon's basilar papilla. J Ultrastruct Res 35:20–65.

Tanaka K, Smith CA (1978) Structure of the chicken's inner ear. SEM and TEM study. Am J Anat 153:251–272.

ten Donkelaar HJ, Bangma GC, Barbas-Henry HA, de Boer-van Huizen R, Wolters JG (1987) The brain stem in a lizard, *Varanus exanthematicus*. Adv Anat Embryol Cell Biol 103:56–60.

Theurich M, Langer G, Scheich H (1984) Infrasound responses in the midbrain of the guinea fowl. Neurosci Lett 49:81–86.

Trussell LO (1997) Cellular mechanisms for preservation of timing in central auditory pathways. Curr Opin Neurobiol 7:487–492.

Tucci DL, Rubel EW (1990) Physiological status of regenerated hair cells in the avian inner ear following aminoglycoside ototoxicity. Otolaryngol Head Neck Surg 103:443–450.

Vates GE, Broome BM, Mello CV, Nottebohm F (1996) Auditory pathways of caudal telencephalon and their relation to the song system of adult male zebra finches. J Comp Neurol 366:613–642.

Vicario DS (1994) Motor mechanisms relevant to auditory-vocal interactions in songbirds. Brain Behav Evol 44:265–278.

Vicario DS, Nottebohm F (1987) Organization of the zebra finch song control system: I. Representation of syringeal muscles in the hypoglossal nucleus. J Comp Neurol 271:346–354.

Volman SF (1996) Quantitative assessment of song-selectivity in the zebra finch higher vocal center. J Comp Physiol 178:849–862.

Volman SF, Konishi M (1990) Comparative physiology of sound localization in four species of owls. Brain Behav Evol 36:196–215.

von Bartheld CS, Code RA, Rubel EW (1989) GABAergic neurons in brainstem auditory nuclei of the chick: distribution, morphology, and connectivity. J Comp Neurol 287:470–483.

Wagner H (1993) Sound-localization deficits induced by lesions in the barn owl's auditory space map. J Neurosci 13:371–386.

Wagner H, Takahashi T, Konishi M (1987) Representation of interaural time difference in the central nucleus of the barn owl's inferior colliculus. J Neurosci 7:3105–3116.

Wang Y, Raphael Y (1996) Re-innervation patterns of chick auditory sensory epithelium after acoustic overstimulation. Hear Res 97:11–18.

Warchol ME, Dallos P (1989) Neural response to very low-frequency sound in the avian cochlear nucleus. J Comp Physiol 166:83–95.

Warchol ME, Dallos P (1990) Neural coding in the chick cochlear nucleus. J Comp Physiol 166:721–734.

Warr WB (1992) Organization of olivocochlear efferent systems in mammals. In: Webster DB, Popper AN, Fay RR (eds) The Mammalian Auditory Pathway: Neuroanatomy. New York: Springer-Verlag, pp. 410–448.

Weiss TF, Mulroy MJ, Turner RG, Pike CL (1976) Tuning of single fibers in the cochlear nerve of the alligator lizard: relation to receptor organ morphology. Brain Res 115:71–90.

Westerberg BD, Schwarz DWF (1995) Connections of the superior olive in the chicken. J Otolaryngol 24:20–30.

Wever EG (1978) The Reptile Ear. Princeton, NJ: Princeton University Press.

Whitehead MC, Morest DK (1978) Morphogenesis of synaptic endings of cochlear fibers in the chick basilar papilla. Soc Neurosci Abstr 4:397.

Whitehead MC, Morest DK (1981) Dual populations of efferent and afferent cochlear axons in the chicken. Neuroscience 6:2351–2365.

Wilczynski W (1984) Central neural systems subserving a homoplasous periphery. Am Zool 24:755–763.

Wild JM (1987) Nuclei of the lateral lemniscus project directly to the thalamic auditory nuclei in the pigeon. Brain Res 408:303–307.

Wild JM, Karten HJ, Frost BJ (1993) Connections of the auditory forebrain in the pigeon (Columba livia). J Comp Neurol 337:32–62.

Winter P, Schwartzkopf J (1961) Form und zellzahl der akustischen nervenzentren in der medulla oblongata von eulen (Striges). Experientia 17:515–516.

Woolf NK, Sachs MB (1977) Phase-locking to tones in avian auditory nerve fibers. J Acoust Soc Am 62:46.

Young SR, Rubel EW (1983) Frequency-specific projections of individual neurons in chick brainstem auditory nuclei. J Neurosci 7:1373–1378.

Young SR, Rubel EW (1986) Embryogenesis of arborization pattern and topography of individual axons in n. laminaris of the chicken brain stem. J Comp Neurol 254:425–459.

Zardoya R, Meyer A (1998) Complete mitochondrial genome suggests diapsid affinities of turtles. Proc Natl Acad Sci U S A 95:14226–14231.

Zaretsky MD, Konishi M (1976) Tonotopic organization in the avian telencephalon. Brain Res 111:167–171.

Zhang S, Trussell LO (1994) A characterization of excitatory postsynaptic potentials in the avian nucleus magnocellularis. J Neurophysiol 72:705–718.

Zidanic M, Fuchs PA (1995) Efferent innervation of the chick cochlea revealed by antibodies to choline acetyltransferase (ChAT) and synapsin. Assn Res Otolaryngol Abstr 18:193.

Zidanic M, Fuchs PA (1996) Synapsin-like immunoreactivity in the chick cochlea: specific labeling of efferent nerve terminals. Aud Neurosci 2:347–362.

6
Sound Localization in Birds

Georg M. Klump

1. Introduction

Without the ability to localize a sound, a bird's auditory world would be a cacophony of environmental sounds and vocalizations. When we can identify the location of a sound source, we can form auditory objects that help us to discern separate items in our environment (Bregman 1990). Given the parallels in the processing of auditory information in birds and mammals (e.g., humans), it can be assumed that this benefit of auditory localization will also accrue to birds. When a bird listens to the contact calls of members of its flock at a distance, for example, its directional hearing will help it to keep in touch with them. Through the mechanisms of sound localization it may be able to form auditory objects and thus could more easily separate its flock mates' calls from the acoustic background produced by other sources. A simple variation in the auditory sensitivity with direction may help the bird to increase the signal-to-noise ratio through a mechanism called "spatial release from masking" (Dent et al. 1997). When orienting its head in a way that it is more sensitive to the contact calls coming from one direction and less sensitive to the background noise arriving from other directions, a bird can considerably improve its signal detection.

In a territorial interaction between birds, a territory owner has a great advantage when it can accurately determine the location of a competitor that is singing. Its perception of the azimuth and the elevation of the angle of sound incidence helps to find out whether the singing neighbor that was identified by the song is still within the limits of its territory and thus no special threat, or whether it has moved beyond the limits of the territory (e.g., may be intruding into the recipient's territory). In territorial interactions, the ability to determine the azimuth and elevation of the sound source alone is not sufficient. A third measure, the distance of the sound source, is needed to determine precisely the signaling rival's location. Birds can determine all three parameters with some accuracy (e.g., see Nelson and Stoddard 1998).

Another behavioral context in which the ability to localize sounds plays a large role is predator-prey interaction. An avian prey producing sounds that are hard to localize by an avian predator will be at an advantage, since it can signal without the risk of becoming the focus of the predator's attention (Marler 1955). Alternatively, if a prey wants to signal a predator that it has been detected and the attack is unlikely to be successful (i.e., it would pay the predator to give up the chase), then the prey should use signals that are easily localizable (Klump and Shalter 1984). From the predator's viewpoint, good sound-localization ability is a prerequisite for success. This may explain why the highest accuracy in localizing sounds is found in a predatory bird, the barn owl (*Tyto alba*). This species has evolved multiple adaptations to achieve accurate localization performance (for reviews see Knudsen 1980; Konishi 1983, 1993a, 1993b; Konishi et al. 1988; Volman 1994).

Birds are not limited to the use of passive localization of sounds emitted from a distant source. Some species have evolved an active location system: they emit echolocation signals and analyze the reflections from surfaces around them. Such echolocation behavior is found, for example, in cave swiftlets (*Collocalia brevirostris unicolor*; Novick 1959) that emit clicks with a dominant frequency of 4–5 kHz at a rate of 10/s and in the oilbird (*Steatornis caripensis*; Griffin 1953; Konishi and Knudsen 1979) that emits click signals with a dominant frequency of 7 kHz at a rate of 300/s. Thus, the limited frequency response of the birds' middle ear has led them to evolve signals that are quite different from the ultrasonic echolocation sounds of bats. Compared to passive sound localization, however, little is known about echolocation in birds.

The examples given above emphasize the adaptive value of a well-developed ability to localize sound in a number of different behavioral contexts. Furthermore, they describe the basic problems a bird faces when locating a source. Birds solve the task of sound localization using various mechanisms for comparison of the information from their two ears. The most simple mechanism involves the lateralization of the sound source, that is, the animal only needs to be able to discern whether the sound originated from a source located left or right of the medial plane that is symmetrical to the position of the two ears. Lateralization does not enable the bird to pinpoint the direction from which the sound arrived using a single measurement of the incident sound, it only enables the bird to indicate whether the sound originated from the left or the right side. Information from sequential measurements taken by the animal while turning the head needs to be combined in direction finding. The direction of sound incidence might be estimated using a lateralization mechanism as the bearing halfway between the directions in which sound is perceived as originating from the left and from the right. A number of studies that are summarized below indicate, however, that birds use mechanisms of sound localization providing them with immediate directional information, that is, that involve direction sensitivity. By comparing the information obtained from the two ears, many species of birds can estimate the direction of sound incidence based on a single measurement. The mechanisms allowing the nervous system to compute the directional information from the interaural differences in the incident sound will be discussed below in detail. Finally, Quine and Kreithen (1981) have suggested the possibility of quite a different mechanism of sound localization for flying birds. They argued that the frequency shift resulting from the Doppler effect experienced by a flying pigeon (*Columba livia*) when it changes the direction of its flight path with respect to the angle of sound incidence could be used for localization. For a pigeon at a flight speed of 20 m/s, the maximum frequency shift could be 12%, which is much larger than the frequency-difference limen in pigeons and other birds (Quine and Kreithen 1981; see also Fay 1988 and Dooling et al., this volume). It is not known, however, whether pigeons or other birds apply this ability in actually locating sound sources.

Sound localization has been investigated in a large variety of bird species. The smallest species studied, zebra finch (*Poephila guttata*), canary (*Serinus canarius*), and great tit (*Parus major*), are members of songbird families. In total, data on the accuracy of sound localization in five different songbird species are known. The largest birds in which sound localization has been studied all belong to the order of the owls. Data on localization accuracy are available in the barn owl (*Tyto alba*), the great horned owl (*Bubo virginianus*), the short-eared owl (*Asio flammeus*), and the saw-whet owl (*Aegolius acadicus*). Besides studies in owls, data on the localization ability in three species of raptors, in the bobwhite quail (*Colinus virginianus*), in the pigeon (*Columba livia*), and the budgerigar (*Melopsittacus undulatus*) are available. Various physiological and physical aspects of binaural sound perception have been studied in some of these bird species and a number of additional species. This broad range of studies in birds provides the material for this review. It will be obvious that birds have developed some new solutions to solve similar problems faced by other terrestrial vertebrates in sound localization (see reviews by Wightman and Kistler 1993; Brown 1994).

Because of the lack of data on the localization ability in reptiles, this review will not discuss sound localization in this group. The morphology of the reptilian auditory system indicates that, for example, some lizards may be able to utilize similar sound localization mechanisms as some species of birds (i.e., a pressure-gradient system, see Lewis 1983).

2. Binaural Cues Available for Sound Localization

Most bird species have small heads and, therefore, a small interaural distance, which limits the size of binaural cues available in the sound field at the entrance of the left and right auditory meatus. In contrast to mammals with similar head size, birds use much lower frequencies in their communication signals. It will be shown below that this also limits the size of some of the interaural cues commonly used in terrestrial vertebrates for locating sound sources. Sounds that originate from a source that is not located in front of a bird's head but to the side will differ upon arrival at a bird's two ears in various measures. Each frequency component contained in the sound signal will be received with an amplitude that differs between the two ears, that is, with an interaural intensity difference (IID). The size of the IID depends on the frequency of the signal and on the angle of sound incidence. The IIDs of all frequency components in the signal can be combined into an interaural amplitude-difference spectrum that can be evaluated by the auditory system (see Knudsen 1980). Also, the signal phase of each frequency component will differ between the sounds arriving at the two ears providing the auditory system with an interaural phase-difference spectrum. Studies summarized below have shown that a measure derived from the interaural phase differences—the interaural time differences (ITD) reflected in the trains of action potentials encoding the phase—is used by the auditory system in the neural computation of the angle of sound incidence (e.g., Moiseff and Konishi 1981a; Konishi et al. 1985; Carr and Konishi 1988, 1990; Moiseff and Haresign 1992; Overholt et al. 1993). Besides ITDs related to the interaural phase difference of tonal signals, the time differences between amplitude peaks in the envelope of an ongoing sound signal (e.g., in a noise) arriving at the two ears also provide the auditory system with ITDs that are encoded in the auditory pathway (for examples in mammals, see Yin et al. 1984; Yin and Chan 1988; Batra et al. 1989). Finally, the time of arrival of the first wavefront of a signal, that is, its onset time, will differ between the two ears and may be used to compute the location of the sound source (for a review, see Blauert 1983). The availability of the various cues and their size in relation to the sound-signal parameters will be discussed in the following text in detail.

If IIDs are calculated based on the assumption that the bird's head acts like a sphere (e.g., see Rayleigh 1907; Stewart 1911; Schwarz 1943), it becomes obvious that the maximum IIDs between the entrance of the left and right auditory meatus of most bird species must be relatively small. As expected, measurements with probe microphones at the entrance of the meatus (Hill et al. 1980; Windt 1985; Lewald 1990a; Klump and Larsen 1992; Larsen and Dooling 1992, 1996, and unpublished data) in nonspecialized birds reveal small differences in interaural intensity cues for azimuth angles of sound incidence in the range from 60° left to 60° right of the bird's midline (see Fig. 6.1 for examples). Only if the sound source is presented in the quadrant in the back of the bird's head that is contralateral to the measured ear can a substantial drop in the amplitude be found for high-frequency signals (e.g., see the data for 8 kHz in Fig. 6.1). *T. alba*, with its specializations for auditory localization of prey (e.g., see Payne 1971; Konishi 1973a, 1973b), provides an exception from the pattern found in the nonspecialized birds with relatively small heads. As a consequence of its facial ruff that functions as an exponential horn with a gain of up to 20 dB (Konishi 1973b; Coles and Guppy 1988) and due to its asymmetries in the morphology of the external ears (for other owls with asymmetrical ears see reviews by Norberg 1977; Volman 1994), *T. alba*, compared to most other birds, experiences a high variation of sound intensity in relation to the angle of sound incidence. This variation is not prominent in the azimuth direction, however, but provides the owl with the main cue for determining the elevation of a sound source (e.g., Knudsen and Konishi 1979; Moiseff and Konishi 1983; Takahashi et al. 1984; Coles and Guppy 1988; Moiseff 1989a, 1989b; see Fig. 6.2). Birds without asymmetrical ears lack cues for elevation. This lack is reflected in the shape of the spatial receptive fields of neurons in the midbrain of these species that are limited in azimuth direction but show a large extension in elevation (e.g., Calford et al. 1985; Volman and Konishi 1989; for more details see below). So far, behavioral experiments that ask birds to determine the elevation of a sound source directly have failed in nonspecialized birds, whereas *T. alba* is clearly capable of doing so (Knudsen and Konishi 1979; Knudsen et al. 1979; Rice 1982).

According to model calculations, the interaural time differences at the meatus should also be relatively small. Woodworth (1962) estimated that the interaural time difference is proportional to the sine of the angle of sound incidence following the expression:

$$ITD = (2 \cdot a/c) \cdot \sin \theta \qquad (1)$$

where a = radius of the sphere, c = speed of sound, and θ = angle of sound incidence. Kuhn (1977) noted that for frequencies for which the product of the wave number (k) and the radius of the head (a) is smaller than a critical value in the range of between 1 and 2, it is more appropriate to use the expression

$$ITD = (3 \cdot a/c) \cdot \sin \theta \qquad (2)$$

for computing interaural time differences. This latter formula should be applied, for example, if the species under study has a head diameter of between 1 and 2 cm and if a stimulus frequency of 8,000 Hz or below is used (i.e., for 8,000-Hz signals $k \cdot a = 0.74$ and $k \cdot a = 1.48$ for 1 and 2 cm head diameter, respectively; k is directly proportional to frequency, i.e., k is halved if the frequency decreases by an octave). Applying Kuhn's calculations to a small songbird with a head diameter of 12 mm, a maximum interaural time difference at the meatus of 53 s would be predicted if the sound impinges on the head 90° from the side. For an angle of sound incidence of 30°, the interaural time difference would be reduced to only 27 µs. In birds with larger heads that localize mostly high-frequency sounds, such as *T. alba*, Woodworth's (1962) formula has been found to be more appropriate to describe the acoustic input at the meatus of the two ears (for 8,000-Hz signals $k \cdot a = 3.3$ in *T. alba*, see Moiseff and Konishi 1981a; Moiseff 1989a).

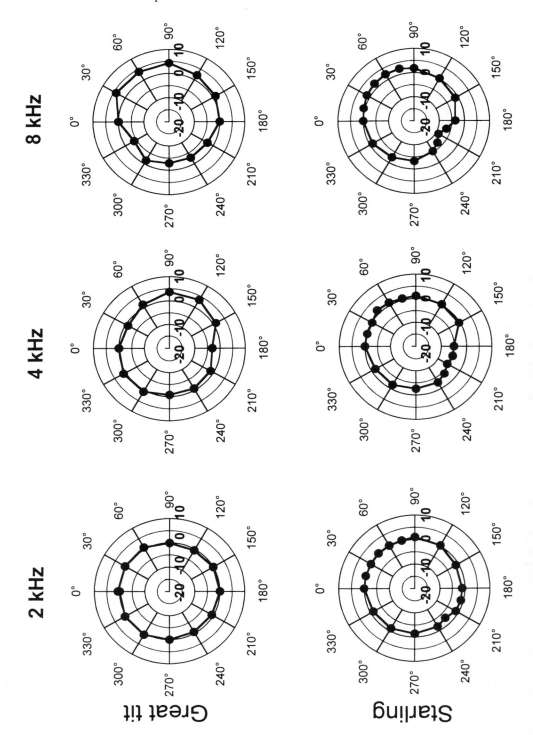

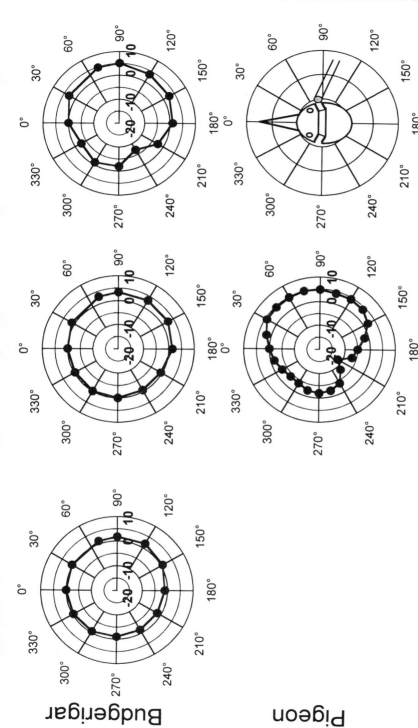

FIGURE 6.1. Sound pressure determined at the entrance of the right auditory meatus of four different bird species in relation to the azimuth angle of sound incidence (see sketch at bottom right). The sound pressure is expressed in decibels relative to the sound pressure measured with the sound source placed in front of the bird (i.e., at 0°). Since all species chosen for this figure have bilaterally symmetrical heads and ears, the interaural intensity difference that is available at the entrance of the ear canals can be calculated by subtracting the pattern shown from a pattern that is mirrored by the 0° to 180° axis. (*Sources:* great tit (*Parus major*): Windt 1985; starling (*Sturnus vulgaris*): Klump and Larsen 1992; budgerigar (*Melopsittacus undulatus*): Larsen and Dooling 1996; pigeon (*Columba livia*): Lewald 1990a.)

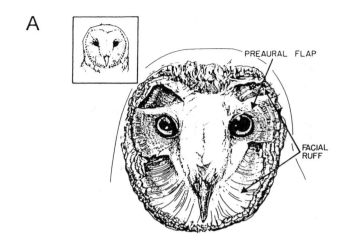

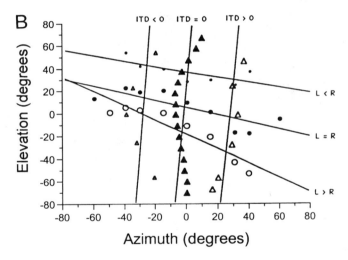

FIGURE 6.2. Generation of interaural time and intensity differences in the barn owl (*Tyto alba*): (**A**) The opening of the barn owl's left and right auditory meatus (from Knudsen 1980) lies behind the preaural flap at the bottom of an exponential horn formed by the facial ruff (see Coles and Guppy 1988). The left and right ear openings are shaped differently, which mainly leads to interaural intensity disparities that vary with the elevation of the sound source. (**B**) Interaural iso-disparity contours as determined for broadband noise pulses with small probe microphones inserted into the ear canals of an anesthetized owl (from Moiseff 1989a). The *small open, solid,* and *large open triangles* indicate speaker positions resulting in interaural time disparities of −100 μs, 0 μs, and +100 μs, respectively (negative values: left ear leads). The *small solid, large solid,* and *large open* circles indicate speaker positions resulting in interaural intensity disparities of −6 dB, 0 dB, and +6 dB, respectively. The *lines* show a linear fit to the data points.

The small ITD values at the entrance to the meatus have led some researchers to conclude that the head of most bird species is too small to provide reliable binaural cues for sound localization if the ears would sample the sound field independently as separate sound-pressure receivers (e.g., Coles et al. 1980; Hill et al. 1980; Lewis and Coles 1980; Lewis 1983; Calford 1988; Calford and Piddington 1988). As an alternative mechanism, these researchers suggested that the ears of birds are acoustically coupled through the interaural canal (see review in Kühne and Lewis 1985) and function as a pair of sound-pressure difference receivers (for the basic functional principles, see Olson 1946).

Sound-pressure difference receivers usually experience a larger variation of IID and ITD with a change in the angle of sound incidence than do sound-pressure receivers. This is illustrated in Figures 6.3 and 6.4. In ears that function as sound-pressure receivers, the vibration of the tympanum reflects the pressure changes in the auditory meatus. In ears that function as sound-pressure difference receivers, the tympanum is stimulated by the sum of two interfering waves: one driving the tympanum from the outside and one driving it from the inside after having traveled through the tympanum of the other ear and the interaural canal connecting the two middle ears (Fig. 6.3). The resultant of the two interfering waves, which is the adequate stimulus for a pressure-difference receiver, differs in amplitude and phase (i.e., in time) from the sound wave impinging on the outer side of the tympanum, which is the adequate single stimulus for a pressure receiver. A substantial variation in amplitude with changes in the angle of sound incidence can only be found in the pressure-difference receiver, if the sound-pressure level of the wave impinging on the tympanum from the inner side is not considerably smaller than the sound pressure of the wave impinging on the outer side. If, for example, the inner-side sound wave has a 10 dB lower amplitude than the outer-side sound wave, then the maximum change in amplitude that results from the interference of the two waves is 3.3 dB; for a sound wave 20 dB lower in amplitude, this change in amplitude would be reduced to a maximum of 1 dB. The maximum phase change that can be obtained with an inner-side sound wave having a 10 dB lower amplitude than the outer-side sound wave is 18.5°; and for an inner-side sound wave 20 dB lower in amplitude, this maximum phase change would be 5.7°. If the phase differences are transformed into time differences, then for a 1-kHz tone signal, the 18.5° and 5.7° phase differences would amount to time differences of 49 and 16 μs, respectively. At higher frequencies, as a result of the shorter period corresponding to one oscillatory cycle (i.e., a 360° phase difference), the time differences will be reduced in proportion to the increase in frequency. At a frequency of 4 kHz, for example, the 18.5° and 5.7° phase differences would amount to time differences of only 12.2 and 4.0 μs, respectively.

The combined effects of the pressure-difference mechanism at the left and right ear determine the IID and the ITD. In Figure 6.4 results from a

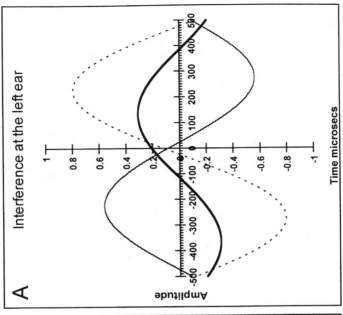

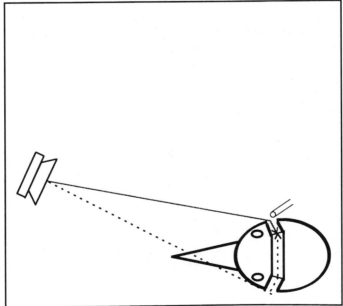

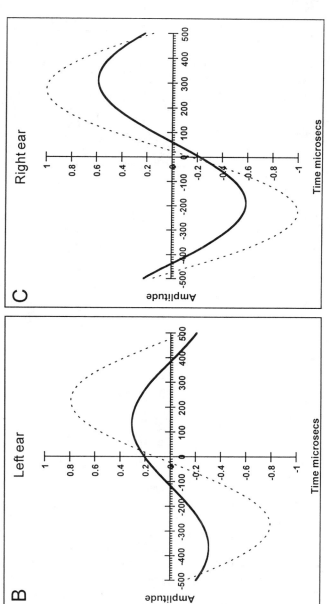

FIGURE 6.3. Example of a model calculation elucidating the role of the interaural canal in enhancing interaural time and intensity differences. A sound source is located to the right of the bird, and due to the difference in the path length the sound reaches the tympanum of the right ear earlier than that of the left ear. After passing the tympanum on one ear, the sound travels through the interaural canal to the inner surface of the tympanum of the other ear creating a pressure difference signal as is shown in (A). The example demonstrates the formation of the pressure-difference signal (*solid line*) by additive interference of the sound wave impinging on the outer surface of the tympanum (*dashed line*) and, after being conducted through the tympanum of the other ear and through the interaural canal, impinging on the inner surface of the tympanum (*solid gray line*). As can be seen by comparison of the patterns in (B) and (C) for the left and right ear, respectively, the interference enlarges the interaural time difference and intensity difference. The difference between the signals impinging on the outer surface of the tympanum of the left and right ear (*dashed lines*) is 45 μs and 1 dB whereas the difference between the resultant signals at both ears (*solid lines*) is 170 μs and 4dB.

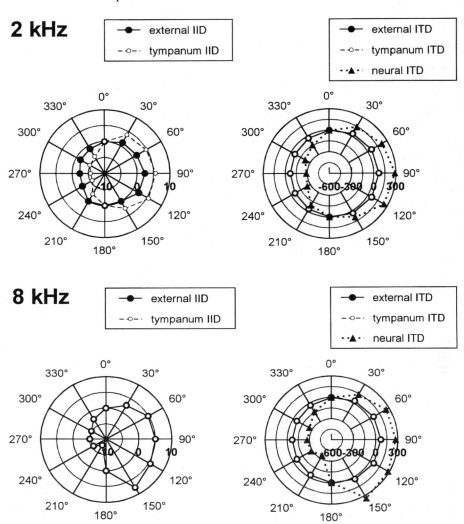

FIGURE 6.4. Interaural time differences (ITDs) and interaural intensity differences (IIDs) generated by the model calculations for a bird's head with the transmission characteristics of the European starling (*Sturnus vulgaris*). ITDs and IIDs at the outer surface of the tympana (*solid lines, filled circles*) of the left and right ears were computed on the basis of Kuhn's (1977) model or determined by measurement with a probe microphone (Klump and Larsen 1992), respectively. ITDs and IIDs were estimated for the tympanum vibration (*dashed lines, open circles*) of the left and right ear by computing the sound interference at the tympanum surface (see text). Potential neural ITDs (*dotted lines, filled triangles*) include the additional ITD that is predicted on the basis of the latency shift in the neural response of about 40 µs with a change in the sound pressure level at the ear of 1 dB (see Gleich and Narins 1988). In some conditions external and tympanum IID or ITD are so similar that the graphs lie on top of each other (for details see text).

model calculation of IID and ITD in relation to the angle of sound incidence in the European starling (*Sturnus vulgaris*) are shown for the frequencies of 2 and 8 kHz (based on data on head size, on the sound shadow
created by the head, and on the transmission characteristics of the interaural canal reported in Klump and Larsen 1992; for 8-kHz signals the transmission characteristics of the interaural canal was estimated based on
the data obtained for the frequency range from 1 to 4 kHz). At 1 kHz, the
interaural canal affects mainly the ITD. ITDs predicted by the pressure-
difference model are up to 51 μs larger than ITDs calculated for *S. vulgaris*
using Kuhn's (1977) formula (see above). At 1 kHz, the interaural canal has
little effect on the generation of IIDs in the starling. These findings are
similar to results obtained with the model presented by Calford and Piddington (1988) that did not incorporate the effect of the sound shadow
created by the head and any phase shifts in the transmission through the
interaural canal that occur on top of those dependent on the path length of
the sound traveling between the two tympana. At 2 kHz, the results of the
model presented here incorporating the actual values of the transfer functions that were experimentally determined for the interaural canal of *S. vulgaris* (for an overview see below) deviate from predictions obtained with
the model of Calford and Piddington (1988). At this frequency, the interaural transmission has little effect on the ITD found in the phase relationship of the tympanum vibration of the left and right ear. It affects, however,
the IID that is enhanced by up to 3.3 dB. Since sound intensity affects the
latency of the response of auditory-nerve fibers, this IID can help to considerably enhance the starling's neural ITD. Figure 6.4 shows the predicted
neural ITD that is calculated with a presumed latency shift of 40 μs/dB as
has been found in the cochlear ganglion cells of *S. vulgaris* (Gleich and
Narins 1988). The effect of the IID contributes more to the neural ITD than
the effect of the sound-wave interference on the phase relationship in the
tympanum vibration of the left and right ear. At 8 kHz, *S. vulgaris* ears
behave as a pair of pressure receivers. The attenuation of the interaural
canal presumably is so large that it does not affect the tympanum vibration
(a value of greater than 40 dB can be estimated by extrapolation of the
experimental data; see also below). Therefore, no additional ITD or IID is
produced by the pressure-difference mechanism. The sound shadow created
by the head of *S. vulgaris*, however, would be large enough to account for
an additional enhancement of neural ITDs of up to 380 μs. It is unlikely,
however, that *S. vulgaris* can utilize this large neural ITD in localizing pure
tones of 8 kHz, since it shows no significant phase locking at this frequency
(Gleich and Narins 1988). For noise-like high-frequency signals in which
the envelope varies over time, however, such large ITDs could be exploited
by analyzing the ITDs in the amplitude transients of the signal.

Despite an ongoing discussion over many years whether birds' ears function as pressure receivers or pressure-difference receivers, the debate is still
is not settled in some species. The variation between species and the dis-

crepancies between different measurements in the same species make it difficult to reach final conclusions as to the general importance of the two alternative mechanisms in the auditory periphery. The evidence for or against an involvement of the interaural canal in the generation of inter-aural time and intensity differences stems from different types of measurements. In a number of studies the ears were stimulated with a free-field sound source while the input to the inner ear was estimated either by means of laser vibrometry of the tympanum (Larsen et al. 1989; Klump and Larsen 1992; Larsen and Dooling 1992, 1996; Larsen et al. 1997) or by means of cochlear microphonic measurements (Schwartzkopff 1952, 1962a; Rosowski 1979; Coles et al. 1980; Moiseff and Konishi 1981a; Calford and Piddington 1988; Coles and Guppy 1988; Moiseff 1989a; Lewald 1990a; Hyson et al. 1994; Schmid et al. 1996) in anesthetized birds. Other studies used sound transducers that were directly coupled to the auditory meatus to determine the effects of the interaural canal using cochlear microphonics (anesthetized birds: Rosowski 1979; Rosowski and Saunders 1980; Counter and Borg 1982; chronic implants in awake birds: Klump and Larsen 1992) or neural thresholds at the brain stem level (anesthetized *T. alba*: Moiseff and Konishi 1981b). In some studies (e.g., Rosowski 1979; Rosowski and Saunders 1980; Moiseff and Konishi 1981b; Rangol and Plassmann 1993) the role of the interaural canal was evaluated by positioning probe microphones in the middle ear close to the inner surface of the tympanum to measure the pressure difference relative to that determined with a probe microphone positioned above the outer surface of the tympanum. These measurements, however, are very difficult to interpret since the microphone in the middle-ear cavity may modify the acoustic properties of the interaural connection, and large differences were observed between the transmission characteristics of the interaural canal determined by microphone and cochlear-microphonic measurements, respectively (e.g., see Rosowski 1979; Rosowski and Saunders 1980). Because of the problems inherent in the method, probe-microphone measurements of the transmission characteristics will not be discussed further.

Most neurophysiological measurements of transfer functions of the inter-aural canal with a closed sound system presenting the signals exclusively to one ear have revealed a lowpass characteristic (see Fig. 6.5). Using cochlear microphonics, Rosowski (1979) observed an attenuation of sounds transmitted through the interaural canal of the anaesthetized pigeon by 16 to 27 dB for frequencies in the range from 0.16 to 8 kHz. The phase difference between ipsilateral and contralateral stimulation was about 140° for frequencies below 0.5 kHz and increased from 180° at 0.5 to 580° at 8 kHz. Applying the same technique, Rosowski and Saunders (1980) measured phase and amplitude-transfer functions of the sound path including the tympanum on the side of acoustic stimulation and the interaural canal in anesthetized 3- to 10-day-old chickens (*Gallus gallus*). The attenuation of sounds transmitted through the interaural canal to the contralateral

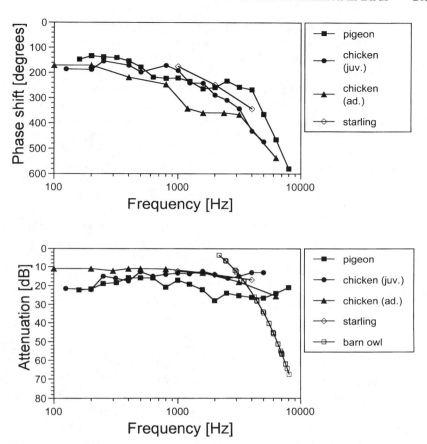

FIGURE 6.5. Transmission characteristics of the interaural sound path formed by one tympanum and the interaural canal in four bird species (pigeon (*Columba livia*): Rosowski 1979; juvenile chick (*Gallus gallus*): Rosowski and Saunders 1980; adult chick (*G. gallus*): Counter and Borg 1982; barn owl (*Tyto alba*): Moiseff and Konishi 1981b; starling (*Sturnus vulgaris*): Klump and Larsen 1992). The top and bottom graphs show the effect of the interaural canal on phase and amplitude, respectively.

tympanum was about 14 dB for frequencies between 0.25 and 5 kHz (this was much less than the acoustic transmission loss of about 30 dB that the authors determined with probe microphones). Sounds below 1 kHz traveling through the head of *G. gallus* to the inner side of the tympanum of the contralateral ear were received with a phase shift of 180° compared to sounds impinging on the tympanum from the outside, that is, the additional phase shift caused by the interaural transmission path was negligible. Above 1 kHz, the transmission through the tympanum and the interaural canal introduced an additional phase shift of about 110° to 150° per octave.

Measurements of transfer functions of the interaural canal by Counter and Borg (1982) using cochlear microphonics in anesthetized adult *G. gallus* revealed a similar pattern of the phase and amplitude-transfer functions (shown in Figure 6.5 for comparison). The attenuation for low-frequency sounds (<1 kHz) reaching the inner side of the tympanum through the interaural canal was about 11 dB. Above 1 kHz, the attenuation increased by between 4 and 7 dB per octave. As had been described by Rosowski and Saunders (1980) for the young *G. gallus* the sound reaching the inner side of the tympanum through the interaural canal was received with a phase shift of 180° at low frequencies (below 0.4 kHz), and the phase shift was increased on average by approximately 90° per octave at frequencies above 0.8 kHz. The data obtained in *S. vulgaris* by cochlear microphonic measurements in awake birds with chronic inner-ear electrode implants (Klump and Larsen 1992) were very similar to the data found in the anesthetized chicken. The attenuation of the sounds transmitted through the interaural canal to the inner side of the tympanum increased from 12 dB at 1 kHz to 17 dB at 4 kHz. At 1 kHz, the acoustic signal reaching the inside of the tympanum was 175° out of phase compared to the signal on the outside, and this phase difference increased with increasing frequency by about 70° to 90° per octave. Larsen et al. (1997) determined the interaural transmission gain in the anesthetized *M. undulatus* using laser vibrometry. At 2 kHz, the attenuation of sounds was similar to that determined in awake *S. vulgaris* but it was reduced to only 2 dB at a frequency of 0.5 kHz. At frequencies above 2 kHz, the attenuation was increased and as a whole the interaural transfer function showed a lowpass characteristic. Moiseff and Konishi (1981b) determined the attenuation of sound amplitude in the interaural canal of *T. alba* by analyzing the threshold differences of individual monaural neurons in the owl's cochlear nuclei. Neuronal thresholds at frequencies of 4 kHz and above differed considerably when sound stimulated the ipsi- and contralateral side indicating a lowpass characteristic of the interaural sound path in which transmission falls off with increasing frequency by about 35 dB per octave. Thus, in a frequency range in which *T. alba* shows the most accurate localization, the two ears are acoustically decoupled. At frequencies below 3 kHz, the thresholds of most neurons differed by about 6 dB or less suggesting only little attenuation at low frequencies on the internal sound path. Also cochlear-microphonic data from *B. virginianus* suggest that the two ears are acoustically decoupled at a frequency of 3 kHz (attenuation of 30 dB, see Volman and Konishi 1989).

In agreement with the lowpass characteristics of the interaural canal, studies of interaural directional cues using cochlear microphonic measurements combined with free-field stimulation have found an enhancement of interaural time differences at low frequencies compared to high frequencies (Fig. 6.6). The observed interaural time differences in most species were much larger than those predicted using the formulas suggested by Woodworth (1962) or Kuhn (1977). This has been interpreted as evidence

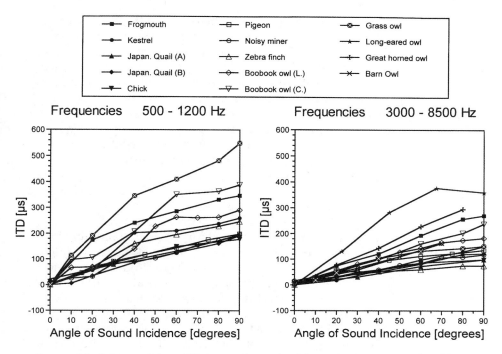

FIGURE 6.6. Interaural time differences determined from measurements of cochlear microphonics (data from Calford and Piddington 1988; Lewald 1990a; Hyson 1994; Larsen et al. 1989, and unpublished) or compound action potentials of the auditory nerve (long-eared owl: Schwartzkopff 1962). When data from two different frequency ranges are compared, a larger change of the interaural time difference with the angle of sound incidence can be found at low frequencies than at high frequencies.

for the enhancement of interaural time differences by the interaural canal, and Calford and Piddington (1988) calculated the amount of interaural sound transmission indirectly from the interaural time difference by modeling the tympanum as a pressure-difference receiver. Although the interpretation of the data on the interaural time differences seems to be straightforward, there are a number of inconsistencies in the literature that suggest there is no simple experimental approach to measuring the effect of the interaural canal. Measurements from different laboratories in the same species have yielded divergent results. Using the model proposed by Calford and Piddington (1988), Hyson and colleagues (1994) could fit theoretical functions to the chick's cochlear-microphonic data in the frequency range of 0.8–2 kHz assuming that the attenuation of the interaural canal is about 6 dB. The measurements of Rosowski and Saunders (1980), however, indicate an interaural attenuation of the connection of between 14 and

22 dB. Similar discrepancies exist between the results of the model calculations of the sound transmission through the interaural canal in *P. guttata* by Calford and Piddington (1988), which suggest a lowpass characteristic of the interaural canal and model calculations and cochlear microphonic measurements by Rangol and Plassmann (1993) and Schmid and co-workers (1996). Furthermore, Larsen and colleagues (1997) have shown that anesthesia may affect the intracranial air pressure in birds, which will change the transmission characteristics of the interaural canal. In the budgerigar, for example, especially at frequencies below 2 kHz, the attenuation of sounds traveling through the interaural canal may be increased by 15 to 20 dB if the anesthetized bird does not vent the intracranial air space by frequently opening the eustachian tube. Thus, artifacts resulting from the anesthesia may lead to erroneous conclusions.

It has been discussed above that, in general, the effects of the ITD and IID cannot be viewed separately. In the auditory periphery of birds there is a tradeoff between the neurons' response latency and the sound intensity of the stimulus. In *M. undulatus*, Schwartzkopff (1962b) reports a reduction in the response latency of auditory-nerve fibers of about 30 μs/dB for sound stimuli ranging in level from close to threshold to 40 dB above threshold. In *S. vulgaris*, Gleich and Narins (1988) concluded from a study of phase-coding by cochlear ganglion units that the response latency is shortened by 40 μs/dB increase of the signal level. Thus, in these two bird species with a relatively small interaural distance and small differences between the travel times of the signals to the two ears, even IIDs of only a few decibels may contribute profoundly to the generation of ITDs (see also Fig. 6.4). In owls a similar tradeoff between response latency and level was found that could enlarge ITDs that are available for analysis by the auditory system. In the long-eared owl, *Asio otus*, Schwartzkopff (1962a) reported a latency shift of the compound action potential of auditory nerve fibers of about −25 μs/dB. In the barn owl, Köppl (1997) estimated a shortening in the response delay of between 9 and 25 μs per dB increase in signal level when analyzing the phase-locked responses of auditory nerve fibers and of neurons in the cochlear nucleus magnocellularis. However, in more central neurons of the owls' auditory pathway (a description of the sound localization pathway in owls is provided below) this dependence of ITD on IID has been abolished. Volman and Konishi (1989, 1990) reported only a small dependence of best ITDs on the IID in neurons of the owls' external nucleus of the inferior colliculus—a nucleus that contains a map of auditory space (e.g., Knudsen and Konishi 1978a). In *B. virginianus* and in the burrowing owl, *Athene cunicularis*, the best ITDs varied by 1.5 μs/dB IID and 1.35 μs/dB IID, respectively. The shifts were in a direction opposite to that found in the auditory periphery of birds. Similar results have also been reported from auditory units in *T. alba* optic tectum (Olsen et al. 1989). In birds with asymmetrical outer ears, such as are found in *T. alba*, in which IIDs mainly encode the elevation and ITDs the azimuth of a sound source

(Moiseff 1989a, 1989b; see Fig. 6.2), the independence of best IIDs and ITDs in space-specific neurons may aid the processing of spatial information. It remains unclear why owls with symmetrical outer ears (e.g., *B. virginianus* or the *A. cuniculare*) have evolved this separation, although in these species the IIDs potentially would enhance the interaural temporal cues and could not provide information on elevation. Studies on the independence of best ITDs and IIDs of songbirds are still lacking.

Although birds have no movable pinnae that they could utilize to dynamically enhance the interaural differences of the sound signal, they may have means to spatially tune the physical response properties of their ears. Schwartzkopff (1962a) reported that *A. otus* can modify the shape and size of the opening of the auditory meatus by contraction of muscles, and he reported a large variation in the IIDs of *A. otus* that show movement of the external ears while coming out of anesthesia. Counter and Borg (1982) suggested that *G. gallus* may use contractions of the stapedius muscle to modify the binaural interaction through the interaural canal, since contraction of this muscle changed the phase and amplitude characteristics of the interaural transfer function and thus might effect the pressure-gradient system involved in the generation of spatial cues. Finally, Larsen and colleagues (1997) have demonstrated that *M. undulatus* potentially could affect the transmission characteristics of the interaural canal and consequently the ears' spatial sensitivity by controlling the air pressure in the middle-ear air space. Although Schwartzkopff (1962a) proposed to study the dynamics of spatial tuning of *A. otus* ears, however, neither in this nor in any other bird species has this suggestion led to experimental evidence for active spatial tuning of the auditory periphery in birds.

3. Behavioral Studies of Sound Localization

Behavioral observations in the field suggest that birds have a good ability to locate sounds. Oeming (cf. Payne 1971) watched great gray owls (*Strix nebulosa*) that were able to catch small rodents in deep snow, which they presumably could not see but had to locate by the sounds they made (for further examples of hunting without visual cues in owls, see Volman 1994). Rice (1982) played back sounds mimicking mouse squeaks from miniature speakers that he had hidden in a meadow beneath the grass and observed the responses of marsh hawks (*Circus cyaneus*) passing by. *C. cyaneus* were able to accurately strike the small speakers, even if no sound was played when the birds were closer than 3 to 4 m, indicating that they were able to locate the source of the squeak with an accuracy of between 1° and 2°. Observations of accurate sound localization in the field are not limited to watching predator behavior. Territorial songbirds indicate by their approach to a playback speaker presenting a rival song or alarm calls that they are able to locate the source. Windt (1985) played back a sample song or a mobbing call to *P. major* in their territory. When the targeted bird had approached the broadcasting speaker, he switched the playback to a second speaker located 25 m from the first speaker and estimated the initial angle of flight when the *P. major* took off from the perch now aiming to approach the new sound source (which, of course, was switched off instantly when the bird took off). The median deviation of this flight angle from the line connecting the starting point of the bird and the sound source was in the range of between 5° and 25°. Nelson and Stoddard (1998) studied sound localization in eastern towhees (*Pipilo erythrophthalmus*) in the field using the bird's approach to a speaker that had broadcast conspecific contact calls to measure the perceptual accuracy. The mean azimuth resolution in this species in the field was as low as 5°. Nelson and Stoddard also estimated the birds' auditory distance resolution. Their data suggest that eastern towhees may be able to resolve distance with an error of only 7% of the total distance.

Many of these examples of field studies on sound localization in birds have applied a mode of stimulus presentation that is typical for a closed-loop experiment. In this experiment, the sound is broadcast during the total orienting response and the subject can correct its response while orienting using the feedback from its sense organs (i.e., utilize a feedback loop in the orienting process). Closed-loop experiments will provide information on the best performance an animal is capable of (i.e., specify the minimum separabile for the angle of sound incidence), but only if the design of the experiments does not simply allow the animal to home in on the sound source. Closed-loop experiments do not reveal much information on the mechanism of sound localization. An animal that is only capable of determining whether the sound comes from the left or right instead of being able to perceive the true angle of sound incidence can do perfectly well in a closed-loop experiment. It only needs to orient itself until the feedback loop indicates no difference between the sensory input coming from the left and from the right ear. Open-loop experiments, which apply the second mode of stimulus presentation, provide much more information to the experimenter. In an open-loop experiment the signal is terminated before the subject starts its orienting response. The response of the animal is made without any additional sensory feedback. Therefore, it will reveal the angle of sound incidence that was perceived and allows a reliable estimate of the accuracy of perception. Thus, open-loop experiments provide more information regarding the possible mechanism of sound localization.

The bird can be presented with two types of localization tasks: absolute and relative (e.g., Lewald 1987a; Park and Dooling 1991). In an absolute localization task, the sound is presented from a single direction and originates from one source. The subject then has to decide from which direction the sound was presented, or from which of a number of possible sources located in different directions the sound originated. The experiment by Stevens and Newman (1936) is of this type, in which they required blindfolded humans to point to a sound source presented at some distance. The experiment conducted by Klump et al. (1986) with great tits tests the absolute localization ability. The birds had to indicate in a two-alternative forced choice procedure which of two speakers located in different directions had presented a single sound; the minimum angle between the two speakers as seen from the position of the subjects that allowed them to distinguish between the speakers was termed the minimum resolvable angle (MRA). In a relative localization task, the subject has to report the change in the direction from which a sound is received. First, one source broadcasts the signals, and then, during the ongoing experiment, the first source is switched off and an alternate source from a different direction is switched on that continues broadcasting the signal. The experiment by Mills (1958) is of this type, in which he required humans to indicate when the location of a sound source broadcasting a signal changed. Mills named the minimum separable angle for this task the minimum audible angle (MAA).

3.1. Studying Sound Localization by Using Natural Orientation Responses

Upon hearing sounds from potential prey, as a first response many avian predators turn their head so that the perceived location of the sound source casts an image onto the area of best visual acuity on the retina. This natural orienting response, which typically can be found in owls and hawks, has been exploited to measure the accuracy of sound localization (e.g., Knudsen et al. 1979; Knudsen and Konishi 1979). While the response occurs spontaneously, it helps to reinforce the head-turning with food to keep the subject's response propensity high rather than to habituate the response. In some of the early studies of sound localization using the head-turning response, researchers only determined whether *T. alba* (Shalter and Schleidt 1977) or goshawks (*Accipiter gentilis*) and pygmy owls (*Glaucidium passerinum*; Shalter 1978) oriented to the correct side if sounds were played back from one of two speakers placed at a large angle left or right of the direction to which the body axis of a bird sitting on a narrow perch was oriented (angles were 30° and 60°). These experiments were conducted under the premise that at least some of the signals presented to the subjects (among them a special type of aerial-predator call, the *seeet* call, see Marler 1955) presumably were difficult to localize, and thus no attempt was made to exactly measure the head-orientation angle to estimate the accuracy of localization.

Knudsen and colleagues (1979), in an elegant experiment aimed at determining accuracy of sound localization of *T. alba* in both open-loop and closed-loop conditions, adapted the search-coil technique that was first used in the study of eye movements in humans (Robinson 1963) for measuring head movements and orientation in behaving owls. The search coil is a small

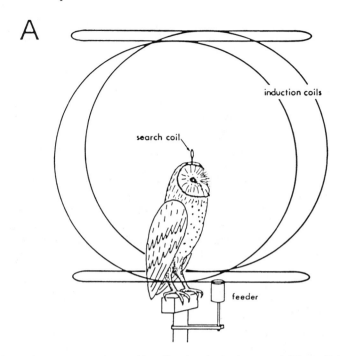

FIGURE 6.7. The accuracy of sound localization in the barn owl (*Tyto alba*) has been measured by means of the search coil technique during a behavioral head-turning response (from Knudsen and Konishi 1979). (**A**) The owl is positioned in the center of two orthogonal pairs of Helmholtz coils, each pair driven by an alternating current of a different frequency, which induce a voltage in the search coil depending on the orientation of the search coil in relation to the magnetic field created by the induction coils. (**B**) In this setup the head orientation in the horizontal plane (i.e., the azimuth orientation) and in the vertical plane (i.e., the elevation orientation) can be determined independently. (**C**) Examples of head-turning trajectories in two modes of stimulus presentation: in the open-loop sound presentation the signal ends before the owl starts its head movement; in the closed-loop sound presentation the signal continues throughout the head movement allowing the owl to correct its initial trajectory to realize a more accurate orientation on the target 70° right of the sound source used to which the owl oriented initially.

induction coil attached to the object whose position needs to be determined (e.g., to the bird's head, see Fig. 6.7A). It is placed in a spatially homogeneous magnetic field that is generated by a pair of Helmholtz coils driven by alternating current of a specific frequency. The amplitude of the voltage induced in the search coil depends on its orientation with respect to the orientation of the magnetic field lines. In a setup with at least two pairs of Helmholtz coils that are mounted perpendicular to each other, where each pair is driven by current of a different frequency, the orientation of the

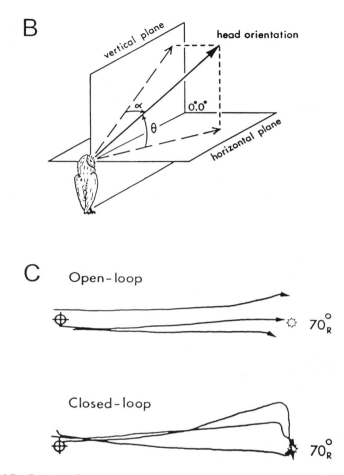

FIGURE 6.7. *Continued*

search coil (and thus of the object to which it is fixed) can be determined in three dimensions (Fig. 6.7B). The alternating voltage of the different frequencies that is induced in the search coil can be separated by frequency filtering to determine the search coil orientation with respect to each plane defined by a pair of Helmholtz coils. With two pairs of Helmholtz coils, Knudsen and Konishi (1979) measured the accuracy of the orienting response of *T. alba* both in azimuth and in elevation. By presenting sounds of either 75 ms or 1,000 ms duration, they could allow the owls to use only open-loop or both open- and closed-loop orientation mechanisms, respectively (the 75-ms sound was shorter than the owl's median latency to respond with a head turn). Although adjustment of the orientation trajectory could be observed in *T. alba* when sounds were presented using angles of sound incidence of 70° left or right (Fig. 6.7C), overall a remarkably small difference was observed between the localization accuracy determined using closed-loop and open-loop sound presentation (see also below for data obtained from a small songbird). There was no difference in the accuracy of localization in open-loop and closed-loop experiments up to azimuth angles of sound incidence of 30° from left or right. Only at larger azimuth angles of sound incidence, at which the error in the open-loop presentation was considerably increased, did the owls correct their head orientation and they performed better in the closed-loop experiments (Knudsen et al. 1979). These corrections were only made at the end of the owl's head saccade, and mostly concerned the elevational (i.e., vertical) component of the orienting response. The observations led Knudsen and co-workers (1979) to conclude that *T. alba* habitually uses the "open-loop strategy" in localizing sounds.

The total localization error for noise targets presented from a range of sound-incidence angles varying from −70° to +70° both in elevation and in azimuth that Knudsen and co-workers (1979) observed in the owl was about 3° to 5°. This is similar to the accuracy observed *T. alba* striking at speakers playing back rustling sounds to them in a quiet flight cage (Payne 1971; Konishi 1973a). Knudsen and colleagues (1979) found that for *T. alba* localization was most accurate for angles of azimuth and elevation of the sound source between −30° and +30°. This spatial range, in which the owl localizes best, correlates with the range of angles of sound incidence that are well represented in the midbrain (see below). If only the component of the error in azimuth or elevation was determined, the owl's accuracy of localization was even better (e.g., 2° in azimuth for noise bursts presented with small angles of sound incidence, for an overview see Table 6.1). For pure tones, *T. alba* orientation errors observed by Knudsen and Konishi (1979) were considerably larger. The smallest total orientation error of 12° was found for 6-kHz tones (again, no substantial difference in the owls' performance in open-loop and closed-loop experiments was observed), and in the frequency range from 4 to 8 kHz the owls exhibited a similarly accurate localization performance. Above and below this frequency range, the total orientation error rapidly increased. At low frequencies, this increase was mainly due to an enlargement of the elevational error component, whereas above 8 kHz both the azimuthal and the elevational error components grew considerably with increasing frequency.

T. alba computes the azimuth angle of sound incidence mainly using interaural time differences, whereas they compute elevation using interaural intensity differences in the spectral components of the sound (e.g., Knudsen and Konishi 1979; Moiseff and Konishi 1983). This separation of binaural ITD and IID cues may represent a special adaptation found in *T. alba* and some other owls with asymmetrical ears for creating different binaural cues for the two dimensions, azimuth and elevation (see review by Volman 1994, and above). In birds without morphological differences

TABLE 6.1. Minimum resolvable angles (MRA) in azimuth determined in absolute localization tasks in an echo-reduced laboratory environment[a]

Species	N	MRA	Task	Signal	Reference
Red-tailed hawk, *Buteo jamaicensis*	3	8–10	2AFC, multiple stimuli	noise 0.1–16 kHz	Rice 1982
American kestrel, *Falco sparverius*	3	10–12	2AFC, multiple stimuli	noise 0.1–16 kHz	Rice 1982
Short-eared owl, *Asio flammeus*	3	1–2	2AFC, multiple stimuli	noise 0.1–16 kHz	Rice 1982
Barn owl, *Tyto alba*	3	1–2	2AFC, multiple stimuli	noise 0.1–16 kHz	Rice 1982
	2	2–5	search coil closed loop	broadband noise	Knudsen, Blasdel, and Konishi 1979
	2	2–6	search coil open loop	broadband noise	Knudsen, Blasdel, and Konishi 1979
Marsh hawk, *Circus cyaneus*	3	2	2AFC, multiple stimuli	noise 0.1–16 kHz	Rice 1982
Great tit, *Parus major*	2	23	2AFC, single stimulus	broadband noise	Klump, Windt, and Curio 1986
Budgerigar, *Melopsittacus undulatus*	3	27	2AFC, single stimulus	broadband noise	Park and Dooling 1991
Zebra finch, *Poephila guttata*	3	101	2AFC, single stimulus	broadband noise	Park and Dooling 1991
Canary, *Serinus canarius*	3	29	2AFC, single stimulus	broadband noise	Park and Dooling 1991
Saw-whet owl, *Aegolius acadicus*	4	2	search coil closed loop	noise 2–7.5 kHz	Frost, Baldwin, and Csizy 1989
Sparrow hawk, *Accipiter nisus*	1	9–14	2AFC, multiple stimuli	bird calls 2–6 kHz	Klump and Kretzschmar, unpubl.
Great horned owl, *Bubo virginianus*	3	7	search coil closed loop	noise 0.2–7 kHz	Beitel 1991
Bobwhite quail, *Colinus virginianus*	4	~100	2AFC, multiple stimuli	broadband noise	Gatehouse and Shelton 1978

[a] In all experiments the broadband signals were played to the subject via a speaker (free-field). When not stated otherwise, in the two-alternative forced-choice (2AFC) experiments the sound was switched off when the bird approached the sound source so that the subject could not simply home in on the source.

between the left and right external ears (some owls may use movable skin folds that enable them to modify the shape of the entrance to the auditory meatus, e.g., see Schwartzkopff 1962a; Volman 1994), no substantial binaural cues are generated that could indicate the elevation of a sound source. The inability to train some birds of prey to localize sounds in elevation (e.g., red-tailed hawk, *Butes jamaicensis*, and American kestrel, *Falcs spaverius*) while other bird species tested in the same setup and with similar procedures (*T. alba*, *A. flammeus*, and *C. cyaneus*) were able to localize a broadband noise presented at 40° elevation suggests that there are species differences in the ability to assess the elevation of a sound source (Rice 1982). Payne (1971) reached a similar conclusion when failing to train screech owls (*O. asio*) and *B. virginianus* to capture prey in total darkness, which requires localization both in azimuth and elevation, while in his setup *T. alba* and *A. acadica* could easily learn the task.

Using the search-coil technique, Frost et al. (1989) measured the accuracy of *A. acadica* in localizing broadband mouse squeaks (500 ms duration, most energy in the range of 2–7 kHz) and cricket chirps (250 ms total duration, most energy in the range of 4–5.3 kHz) in the horizontal plane that were presented at azimuth angles in the range from 30° left to 30° of the owl. On average, the orientation error was 1.75°, which indicates an accuracy similar to that found *T. alba* (Knudsen et al. 1979). Matching the results in *T. alba*, the accuracy was higher when the targets were more frontal and was lower when the targets were more lateral. Furthermore, the elevational error was larger than the azimuthal error (all sounds were, however, presented from the same elevation). In *B. virginianus*, Beitel (1991) assessed the accuracy of the spontaneous head-orienting response to a broadband noise (200 Hz–7 kHz) by recording the behavior with a 16 mm film camera. In the anterior sound hemifield (i.e., in front of the owl from 90° to the left to 90° to the right), the azimuth-localization error of the great-horned owl was 7°. In the posterior sound hemifield, however, it was much larger (41°) when the owls made no subsequent corrective movements. Following corrective responses, which occurred mainly when long (1.6 s) stimuli were presented in the posterior sound hemifield, the owls were found to orient nearly as accurately as in the anterior sound hemifield. Beitel's (1991) finding that the accuracy of the orienting response of *B. virginianus* in the anterior sound hemifield was independent of stimulus duration suggests that, similar to *T. alba*, this species also habitually uses an open-loop strategy in acoustical orientation.

Studying the accuracy of sound localization outside in a deciduous forest rather than in a sound-proofed room, Brown (1982) reported data from videotape recordings of the head turning in three *B. virginianus* and two *B. jamaicensis* while intermittently presenting one of two different types of alarm calls that songbirds utter when mobbing a perching bird of prey or detecting an aerial predator on the wing (i.e., the mobbing call of a red-winged blackbird, *Agelaius phoeniceus*, and the *seeet* call—an 8-kHz tonal signal—of the American robin, *Turdus migratorius*), respectively. Compared to the data from laboratory studies (see Rice 1982; Beitel 1991), the localization errors reported by Brown were rather large (mean errors of orientation were 124.5° and 51.5° after presenting *seeet* and mobbing calls, respectively). While it may be more difficult for a bird in the natural environment to accurately locate sound signals (see also data from another bird of prey below), there are indications that the poor localization performance in *B. virginianus* for one of the signals, the *seeet* call, was probably due to its low level (65 dBA). This level of presentation probably was inaudible to *B. virginianus* (i.e., it was 20 dB below the behavioral threshold reported for 8 kHz tones by Trainer, 1946) and, therefore, not locatable. Brown (1982) also observed that his subjects turned their head on average by 45.5° per 3.33 s observation interval indicating that they scanned their surroundings now and then. Such orienting movements triggered by other stimuli may have increased the estimated error angles. Responding to a playback of calls and song elements, however, songbirds that have a much smaller head size than the owls or hawks apparently locate conspecific signals quite accurately (see above). More field experiments are needed to evaluate the accuracy of sound localization in the field.

3.2. Studying Sound Localization by Operant and Classical Conditioning

The sound localization of bird species that do not exhibit clear orienting responses in laboratory setups has been studied using operant and classical-conditioning paradigms. Mostly, the birds' absolute localization accuracy for azimuth angles was determined. In two studies, however, the ability to detect relative changes in the location of the sound source was also evaluated (Lewald 1987a; Park 1989)

The first songbirds in which the absolute localization accuracy was studied with operant techniques were the pine grosbeak (*Pinicola enucleator*) and the bullfinch (*Pyrrhula pyrrhula*). In *P. enucleator* the minimum resolvable angle between two locations at which Edelmann whistles were blown (only from one location in each trial) ranged from 20° to 23° (Granit 1941; no information regarding the frequencies is given in the paper that was posthumously published). Schwartzkopff (1950) used a two-alternative forced-choice procedure to determine the minimum resolvable angle in *P. pyrrhula*. He presented the birds with tones that were broadcast through one of two speakers, and the birds were rewarded with food when they flew from a starting position to a perch on the left or right side of the cage when the corresponding left or right speaker sounded. Since Schwartzkopff had no anechoic room, he placed the setup outdoors on the patio of the zoological institute. The smallest minimum resolvable angle between the speakers, indicating the most accurate localization, was 24° for tones of a frequency of 1500 Hz. Schwartzkopff (1950) observed a decrease of the localization accuracy at a lower frequency (850 Hz) and an only slightly larger minimum resolvable angle of 26° for 3-kHz tones (see Fig. 6.8B for data from other song birds). He also tried to answer the question whether *P. pyrrhula* utilizes time differences between transients in the sound reach-

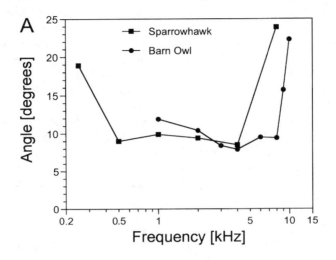

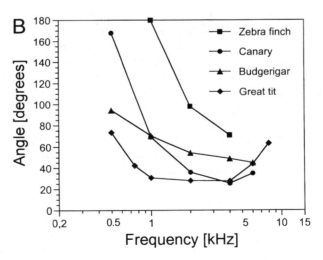

FIGURE 6.8. The accuracy of azimuth sound localization for pure-tone stimuli determined in an absolute localization task using behavioral response paradigms. (**A**) Data from two avian predators (Knudsen and Konishi 1979; Kretzschmar 1982). (**B**) Data from bird species with a small interaural distance (12–17 mm; Klump et al. 1986; Park and Dooling 1991).

ing the two ears by either presenting the birds with sounds which had a slow rise/fall (~300 ms) or with sounds that were switched on instantaneously without being ramped. The accuracy of localization of *P. pyrrhula* was not very different in the two stimulus conditions leading Schwartzkopff to conclude that interaural time differences probably were less important.

Studies of absolute azimuth sound localization for a wide range of pure-tone signals, broadband noise and various species-specific calls were conducted in *P. major*, *S. canarius*, and *P. guttata*, all small songbirds with an interaural distance of about 12 to 13 mm (Klump et al. 1986; Park and Dooling 1991). In addition, Park and Dooling (1991) tested *M. undulatus* (interaural distance 17 mm) in the same setup. Both studies used a two-alternative forced-choice procedure to determine the birds' ability to indicate which of two speakers had broadcast a signal while they were sitting on a waiting post. The distance between the speakers, and thus the angle between them as seen from the position of the subject on the waiting perch, was varied between sessions. In this way the minimum resolvable angle that still allowed the bird to identify the correct speaker could be determined. In both studies precautions were taken to prevent the birds from simply homing in on the sound source when taking off from the waiting perch. The whole setup was placed in a quiet and echo-reduced environment. With the exception of *P. guttata*, the angle between the two speakers that the subjects could resolve in this absolute task when a single pulse of broadband noise was broadcast was below 30° (for an overview, see Table 6.1). *P. guttata* could only discriminate between speakers if the angle between them on average was 101°. Also for tones and various bird calls, *P. guttata* showed much lower discrimination ability than the other species tested (see Fig. 6.8B). For pure tones, all four species exhibited a decrease in the angle between the speakers that could be resolved with the frequency increasing from 0.5 to 4 kHz. In *P. major*, the angle that could be discriminated clearly increased above 4 kHz with increasing frequency. The canary showed a similar trend. In all four species, the azimuthal localization accuracy shown for conspecific signals and other species' calls was as would be expected on the basis of their spectral content.

The predatory birds, in which azimuthal sound localization has been tested in a two-alternative, forced-choice procedure in a quiet room with a sound absorbing surface to reduce echoes, are the European sparrow hawk (*Accipiter nisus*, Kretzschmar 1982), *B. jamaicensis*, *F. spaverius*, *C. cyaneus*, *A. flammeus*, and *T. alba* (Rice 1982). Rice played broadband signals mimicking vole squeaks to the species of his study, and he found that the owls and *C. cyaneus* could discriminate between two speakers separated by 1° to 2° (at least 75% correct responses, sounds were presented repeatedly until the bird took off from a waiting perch). The accuracy of *T. alba* determined by Knudsen and Konishi (1979) measuring the head-orienting responses to broadband signals was similarly high. *B. jamaicensis* and *F. spaverius* were less accurate, needing an average angular separation between speakers of 9° and 11°, respectively. This is similar to the results obtained in *A. nisus*, which needed angles of between 9° and 14° to discriminate between two speakers playing back different broadband (2–6kHz) alarm calls from a tape until the hawk made a decision. Kretzschmar (1982) also tested the angular discrimination ability of *A. nisus* in azimuth for 300-ms pure tones in an anechoic room. Discrimination thresholds calculated from Kretzschmar's data for 75% correct responses were relatively constant at about 9° in the frequency range of 0.5–4kHz, and they were 19° and 24° at 0.25 and 8kHz, respectively (see Fig. 6.8A). The discrimination thresholds of *A. nisus*, when presented with multiple signals before making a decision, were similar to the azimuthal head-orienting error *T. alba* exhibited when being presented with a single signal (Knudsen and Konishi 1979). Kretzschmar (1982) also moved his testing equipment into the outdoor habitat of *A. nisus*, a deciduous temperate forest of mature trees devoid of leaves. In the natural environment, the performance of *A. nisus* significantly deteriorated. The minimum resolvable angles of *A. nisus* in the natural habitat were about two to three times larger than in the anechoic laboratory environment.

Studies on absolute sound localization in *C. livia* and *C. virginianus* have indicated a relatively inaccurate perception of the direction of the sound source in these species. Applying a threshold criterion of 75% correct responses to the data obtained in a two-alternative, forced-choice procedure, *C. virginianus* discriminated between sound sources broadcasting 1- or 2-kHz tones differing in the angle of sound incidence by 180° and 110°, respectively (Gatehouse and Shelton 1978). If a less rigorous response criterium would have been used (e.g., 65% correct responses), the minimum resolvable angle for 2-kHz tones would have improved considerably (to less than 45°). Sound sources playing broadband noise were only discriminated by *C. virginianus* if they were separated by an angle of about 100° as seen from the position of the bird (75% correct responses). In *C. livia*, the minimum angle that the birds could resolve in an absolute localization experiment was above 120° for tones with carrier frequencies between 0.5 and 2kHz (Jenkins and Masterton 1979). At low and high frequencies the performance of *C. livia* in the two-alternative, forced-choice experiments was much better, but no threshold values are available since the authors presented the birds with only one angular separation of speakers (i.e., 120°). It remains unclear whether the poor performance reflects an inaccurate perception of absolute sound direction, or whether they result from a tendency of some bird species to chose the two alternatives randomly in two-alternative, forced-choice procedures, which provide the bird with a reward with a probability of 50% (see Klump et al. 1986).

Relative sound localization in birds seems to be more accurate than absolute localization when determined in operant or classical-conditioning experiments. In a relative localization task, the bird does not need to determine the direction of sound incidence, it only needs to perceive a change in the sound with direction. If the change is due to cues typically used for sound localization, such as IID and ITD, the relative localization task gives an indication of directional resolution in the auditory system (Lewald 1987b). In *C. livia*, Lewald (1987a) measured MAAs for pure tones ranging in frequency from 0.25 to 6 kHz and for white noise using a classical conditioning paradigm. The smallest MAAs found in *C. livia* for frequencies up to 4 kHz varied between 4° and 6°. For 6 kHz tones the MAA was considerably higher. The median data of the five birds, however, showed a different pattern. For frequencies in the middle of the hearing range of *C. livia*, the median MAA was considerably larger than the smallest MAA (it was 11° and 25° for tones of 1 and 2 kHz, respectively). The pigeon's MAA for white noise was 4°, and there was little variation between the subjects. The small MAAs contrast with the large absolute localization errors observed by Jenkins and Masterton (1979) in *C. livia*, which was greater than 120° for 1 and 2 kHz tones (see above). Park (1989) determined the performance of *S. canarius*, *M. undulatus*, and *P. guttata* in a relative sound localization task using a go/nogo operant paradigm. He presented various pulsed-noise stimuli and calls from speakers that differed in position from a reference speaker in azimuth, elevation, or both (then the speaker array pointed 45° upwards). The smallest angle of change that was presented was 3°. The performance of the birds in relative sound localization appeared to be extremely accurate. *P. guttata* could detect a change of the broadcast location in azimuth of 2° (compared to 101° in an absolute task), *M. undulatus* seemed to be as accurate (3° in this task versus 27° in an absolute task), and in *S. canarius* the apparent localization accuracy improved from 29° in the absolute task to an MAA of 15° in the relative task. It remains unclear what cues the birds used to differentiate between speaker locations, but the small heads of these species with the short interaural distance make it unlikely that location cues such as IIDs and ITDs could be used to achieve this high accuracy. Unless extreme precautions are taken to assure that no alternative cues other than IIDs and ITDs can be exploited by the birds (e.g., a change in the spectrum of the sound arriving at the subject that is related to the change of speaker location), data obtained in a relative localization task may not provide reliable information on directional resolution.

4. Neural Mechanisms of Sound Localization

Although the phylogenetic variety of bird species in which the neural substrate for sound localization has been studied is now nearly as broad as the species diversity in behavioral studies, most of our knowledge stems from one very specialized bird species, *T. alba*. In this species, which has developed into a model organism for the study of sound localization mechanisms, more than 20 years of research provide a detailed picture of the neuronal processing involved in the evaluation of localization cues. Various aspects of the *T. alba* studies have been reviewed in depth (e.g., Knudsen 1980, 1984b; Konishi et al. 1988; Takahashi 1989; Konishi 1993a, 1993b) so that this chapter can provide a rather brief account of the neural substrate of *T. alba* for sound localization. It will highlight, however, some new studies on how *T. alba* solves more complex problems in sound localization such as localizing sounds in an environment with clutter echoes and processing the localization cues of nonstationary sound sources. For studies on the development of the parts of the neural system involved in sound localization, the reader is referred to the reviews by Knudsen and coworkers (Knudsen 1988, 1994; Knudsen and Brainard 1995).

4.1. The Pathways Involved in Sound Localization

The behavioral studies and the studies on the acoustic input to the ears indicated that two acoustical cues are available to the birds: IIDs (i.e., spectral differences) and ITDs. These spectral and temporal cues are represented separately in two parallel afferent pathways of the bird auditory system that are started by branches of the auditory nerve projecting to the nucleus angularis and the nucleus magnocellularis in the brain stem, respectively (e.g., Takahashi et al. 1984). Figure 6.9 shows the sound localization pathways of *T. alba* as an example, for details on the morphology and physiology of the auditory pathway in different birds see the chapters by Köppl and Carr and the review by Carr 1992). Both pathways converge again in the external part of the inferior colliculus (IC) in the midbrain. The IC of birds, which was previously named MLd, is homologous to the mammalian colliculus inferior (see the discussion in Carr 1992).

FIGURE 6.9. Known connections in the neural sound localization pathways of the barn owl (*Tyto alba*). (After data summarized by Carr 1992 and by Cohen et al. 1998). For sake of clarity, the left and right set of nuclei are shown, but the connections are drawn only on one side (the full set of connections can be obtained by adding the mirror image of the connections shown). Nuclei marked with a *gray background* form part of the brain stem to midbrain time-coding pathway that provides information on interaural time differences (ITDs) used in the neural computation of space-specific receptive fields in the external nucleus of the inferior colliculus (ICx). Nuclei in the pathway to the inferior colliculus (IC) that have a *white background* form part of the brain stem to midbrain spectral-coding pathway that provides information on interaural intensity differences (IIDs) used in the neural computation in ICx. Besides in the ICx, space-specific neurons are also found in the optic tectum (as a result of a projection from ICx) and in the auditory archistriatum (which are not a consequence of a projection from ICx). There are additional efferent connections from the auditory archistriatum to the midbrain tegmental nuclei not shown in this scheme (e.g., to the nucleus Darkshewitsch, the interstitial nucleus of Cajal, the nucleus ruber, and the mesencephalic reticular formation; see Cohen et al. 1998).

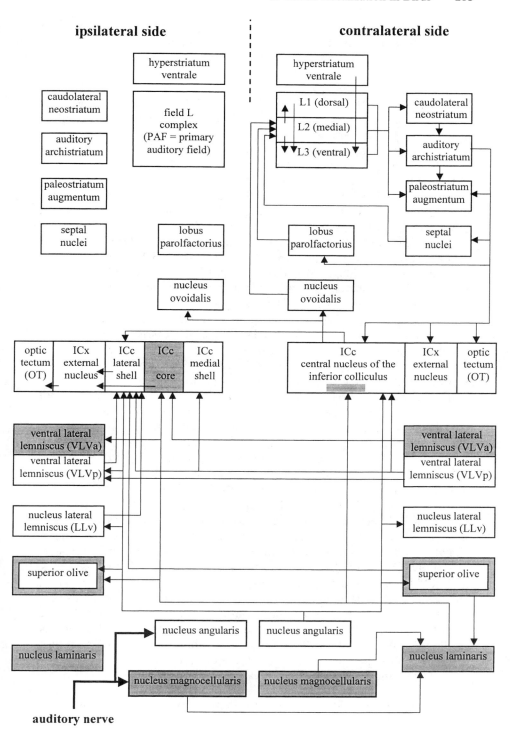

Typical for neurons in the temporal-feature pathway encoding the ITDs are specializations for accurate processing of temporal information in the spike train. The synapses of auditory nerve fibers on the neurons of the nucleus magnocellularis have a cup-like form and cover large parts of the soma of the postsynaptic cell that only have short dendrites (Carr and Boudreau 1991, 1993; Carr 1993). Studies in *G. gallus* have shown that nucleus magnocellularis neurons express a set of ion channels that result in an accurate transmission of the first spike in a series of action potentials suppressing subsequent responses (Koyano et al. 1996). The nucleus magnocellularis neurons project bilaterally to the nucleus laminaris that provides for the first stage in the time-difference pathway, at which binaural comparisons are made. It is most developed in *T. alba*, with a thick layer of cells, and in less specialized bird species it is composed of a single layer of bipolar neurons (e.g., see Rubel and Parks 1975; Young and Rubel 1983; Carr and Konishi 1990). Neurophysiological studies in *T. alba*, *G. gallus*, and other bird species (e.g., Sullivan and Konishi 1984, 1986; Konishi et al. 1985; Carr and Konishi 1988, 1990; Overholt et al. 1993; Amagai et al. 1995; Viete et al. 1997) suggest that the nucleus laminaris functions as a neural cross-correlator as was proposed by Jeffress (1948) in a neural circuit model. Axons projecting into the nucleus laminaris from the ipsi- and contralateral nucleus magnocellularis form delay lines that are organized into a map representing a range of interaural time differences. The nucleus laminaris neurons themselves operate as coincidence detectors that fire when inputs from the ipsi- and contralateral sides arrive simultaneously (e.g., see Carr and Konishi 1988, 1990; Reyes et al. 1996). The axons of the laminaris neurons project contralaterally to the superior olive (which may regulate the activity of neurons in the nucleus laminaris via an efferent projection, e.g. see Brückner and Hyson 1998) and to the anterior part of the nucleus of the ventral lateral lemniscus, and bilaterally to the core of the IC (ICc). ICc receives an additional indirect input from the nucleus laminaris via the anterior part of the nucleus of the ventral lateral lemniscus. Since spikes that are phase-locked to the signal frequency provide the input to the coincidence-detection circuit, the responses of the laminaris neurons to interaural time differences exhibit a periodicity that corresponds to the signal period. In behavioral experiments using narrow-band stimuli, Saberi, et al. (1998) reported that *T. alba* exhibits a cyclic orientation response just as would be predicted from the neuronal response to narrow-band sounds. In the external nucleus of the IC (ICx), the ambiguities inherent in this periodicity are resolved by convergence from ICc neurons that are tuned to different frequencies but to the same interaural time difference onto neurons in the ICx (Wagner et al. 1987). GABAergic inhibition is involved in creating a response pattern in which an ICx neuron fires most intensely to one ITD (Fujita and Konishi 1991; Albeck 1997; Mori 1997). As a result of the nonlinear interactions between converging neurons, in ICx a map of ITDs is generated that encodes broadband stimuli unambiguously (Wagner et al.

1987; Takahashi 1989; Brainard et al. 1992; Mazer 1998). The neuronal elements of the map in the ICx express a response pattern resembling various features of a cross-correlator (Albeck and Konishi 1995). In other bird species besides *T. alba*, broad tuning of ICx neurons in comparison to ICc neurons and their preferred response to broadband stimuli can be viewed as evidence for the convergence of the ICc neurons in the ICx (e.g., Calford et al. 1985; Pettigrew and Larsen 1990; Volman and Konishi 1990), although a study by Lewald (1990b) failed to demonstrate a functional subdivision of the IC typically found in owls (see review by Volman 1994).

The spectral intensity pathway encoding IIDs that originates in the cochlear nucleus angularis and converges in the IC with the time-difference pathway (see Fig. 6.9) has been studied in much less detail. The neurons of the nucleus angularis (NA) connect binaurally *en passant* to the superior olive, to the nucleus of the lateral lemniscus (LLv), and to the lateral shell of the ICc. NA also projects to the posterior nucleus of the ventral lateral lemniscus (VLVp) on the contralateral side. In *T. alba*, VLVp was shown to project to the lateral shell of the ICc on the ipsi- and contralateral sides and to the medial shell on the contralateral side (see Adolphs 1993; and review by Carr 1992). The VLVp on both sides are reciprocally connected (Takahashi and Keller 1992b; Adolphs 1993). Neural computations in this pathway result in a map of characteristic IIDs in the lateral shell of the ICc. With the exception of studies in *T. alba* (Moiseff and Konishi 1983; Manley et al. 1988; Takahashi and Keller 1992b; Adolphs 1993), no detailed investigations on the processing of binaural intensity cues in birds have been made below the level of the IC. Manley and colleagues (1988) have demonstrated that VLVp already contains a map of IIDs. Neurons in this nucleus respond to a specific IID in a step-like mode, that is, they show a sigmoidal transition between a low discharge rate below the threshold IID and a high discharge rate above it. Cells responding to an IID with a greater contralateral sound level were found more dorsal in the nucleus, whereas cells responding to an IID with a greater ipsilateral sound level were found more ventral in the nucleus (Manley et al. 1988). This distribution of responses is likely to result from a gradient of inhibitory connections between ipsi- and contralateral VLVp (Manley et al. 1988; Takahashi and Keller 1992b; Adolphs 1993). The output of the VLVp from both sides projects to the medial and lateral shell of the ICc and may route the information represented in the map of IIDs to the ICc shell. In the lateral shell of the ICc neurons possibly become tuned to specific IIDs (Adolphs 1993). These tuned IID responses and the tuned ITD responses in the ICc core lead to the spatially restricted receptive fields that are found in the ICx of *T. alba* Adolphs (1993) reported more afferent connections to the lateral shell of the ICc with so far unknown function (e.g., originating from the ipsilateral LLv and from the contralateral superior olive).

Most comparative neurophysiological data from different bird species on the coding of auditory spatial information are available from the IC. *T. alba* has developed spatial receptive fields in the ICx that are restricted both in azimuth and elevation (Knudsen and Konishi 1978a, 1978b, 1978c; Fig. 6.11A). These spatial receptive fields are organized topographically into a map representing the auditory space (Fig. 6.12). Each ICx neuron receives input from a number of neurons in the core of the IC that are all tuned to the same interaural time difference but to different frequencies (e.g., Knudsen 1983b, 1984b; Wagner et al. 1987; Fujita and Konishi 1991). In the core of the ICc, the neurons exhibit preferences for multiple interaural time differences that are separated by a time period corresponding to the reciprocal of the characteristic frequency of the neuron (examples are provided in Fig. 6.10). Core neurons tuned to a specific interaural time difference but

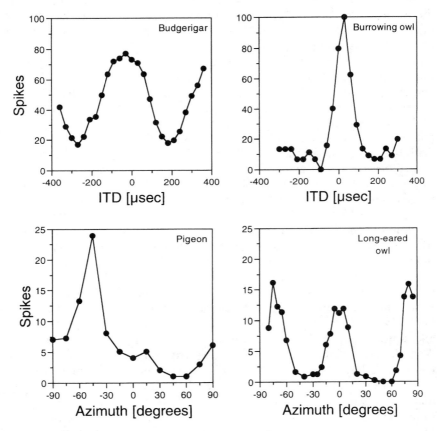

FIGURE 6.10. Typical interaural time difference (ITD)-response curves and free-field azimuth response curves recorded in the nucleus laminaris (NL) or in the inferior colliculus (IC) of various bird species. NL of budgerigar (*Melopsittacus undutatus*): Amagai et al., unpublished data; IC of burrowing owl (*Athene cunicularis*): Volman and Konishi 1990; IC of pigeon (*Columba livia*): Lewald 1990b; IC of long-eared owl (*Asio otus*): Volman and Konishi 1989.

to different frequencies thus have an excitatory peak in common encoding the interaural time difference, but they differ in their side peaks. Nonlinear interactions between these neurons projecting to the ICx that are mediated by inhibition result in a single excitatory response peak corresponding to the common interaural time difference that represents a specific azimuth angle of sound incidence (Wagner et al. 1987; Fujita and Konishi 1991). In the core of the left ICc of *T. alba*, the left spatial hemifield is mapped and vice versa in the right ICc (Takahashi et al. 1989). Commissural projections from the core of the left ICc to the lateral shell of the right ICc and vice versa enable the combination of the two ITD maps into an undivided representation of the frontal auditory space in the ICx. This frontal space, in which *T. alba* also shows its highest localization accuracy (Knudsen and Konishi 1979), is more extensively represented in the ICx than the lateral space (Knudsen and Konishi 1978a; Knudsen 1980). Maps of auditory space in the ICx of other owls also indicate a strong representation of the frontal and frontal-lateral directions (Volman and Konishi 1990). In *T. alba*, with its asymmetrical outer ears, elevation is encoded by ICx neurons responding to differences between the spectral profiles of sound intensity at the left and right ears. The IID tuning of these neurons is mediated by the responses of neurons in the VLVp and in the lateral shell of the ICc (see above). Their spatial receptive fields result from a non linear interaction between the IID and ITD tuning provided by the projections to the ICx (i.e., they respond only if stimulated with sound having both the appropriate IID and ITD; e.g., see Moiseff and Konishi 1981a; Takahashi et al. 1984).

A study in the ICx of *B. virginianus*, which does not possess asymmetrical outer ears like *T. alba*, revealed neurons with spatial receptive fields that were narrowly tuned in azimuth direction but broadly tuned in elevation (Volman and Konishi 1989; see Fig. 6.11B). Conforming to the lack of systematic changes of the IID in relation to the elevation of the sound source, in this owl IID tuning contributed little to the limitation of the spatial receptive fields in the vertical direction. The map of auditory space observed in the ICx of *B. virginianus* reflected mainly the ITD tuning, and the ITD tuning of ICx neurons was little affected by changes in the IID (Volman and Konishi 1989). Studies of *A. otus* and *A. cunicularis* (Volman and Konishi 1990; see also the review by Volman 1994), the former species having asymmetrical and the latter species having symmetrical outer ears, fit in this general picture. Whereas no elevation tuning was evident in *A. cunicularis*, in *A. otus* 54% of the units in ICx exhibited a response that was limited in elevation. The elevation tuning, however, was not a sharp as in *T. alba* (Volman and Konishi 1990). The receptive field widths of ICx units in azimuth in *B. virginianus* was as narrow as in *T. alba* (16° and 18°, respectively), and it was increased to 29° and 44° in *A. otus* and *A. cunicularis*, respectively (receptive field widths were measured at the 50% of maximum response threshold).

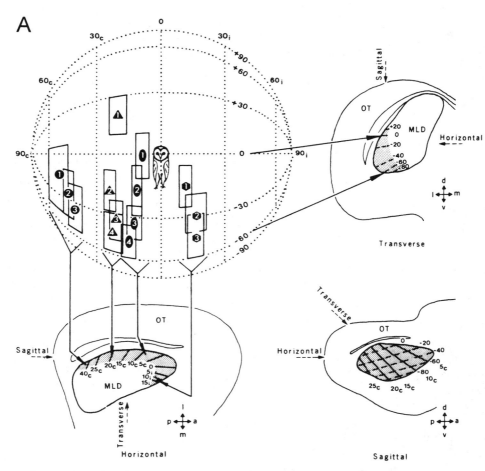

FIGURE 6.11. Spatial receptive fields of units recorded in the external nucleus of the inferior colliculus (ICx) of (**A**) the barn owl (*Tyto alba*; from Knudsen and Konishi 1978a) and (**B**) the great horned owl (*Bubo virginianus*; from Volman and Konishi 1989). In both owl species, the space-specific neurons of ICx are assembled into a map of auditory space. In *T. alba*, the receptive fields of space-specific neurons (boxes in the upper left graph; similar symbols numbered from 1 to 4 indicate units encountered in one electrode penetration) are limited both in azimuth and elevation, and both azimuth and elevation are represented in the map in ICx in different directions (see the two schematic drawings left and right of the globe on which the units' receptive fields are projected). In *B. virginianus*, that possesses no asymmetrical external ears as *Tyto alba* does, the receptive fields of space-specific neurons in the ICx are primarily limited in the azimuth direction (**B**).

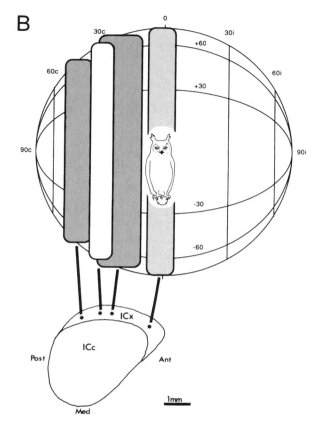

FIGURE 6.11. *Continued*

Although diurnal raptors of the genera *Accipiter* and *Circus* have shown to be capable of accurate sound localization in behavioral tests (Kretzschmar 1982; Rice 1982; see above), neurophysiological recordings from cells in the IC have not revealed a narrow spatial tuning of the responses in species of these genera (Calford et al. 1985). Similar to findings in *T. alba*, Calford and co-workers (1985) could define a central and a lateral subdivision in the IC of the diurnal raptors (they studied the brown falcon, *Falco berigora*, brown goshawk, *Accipiter fasciatus*, and the swamp harrier, *Circus aeruginosus*) that differed in the response characteristics of auditory neurons. The lateral region of the IC (probably corresponding to the owls' ICx) contained neurons that were sensitive to stimulus position, whereas there was no clearly defined spatial tuning in the central region of the IC (probably corresponding to the owls' ICc). Calford and colleagues (1985) did not describe the quasi periodic responses of neurons in relation to the azimuth angle of sound incidence that is typical for owls and other species

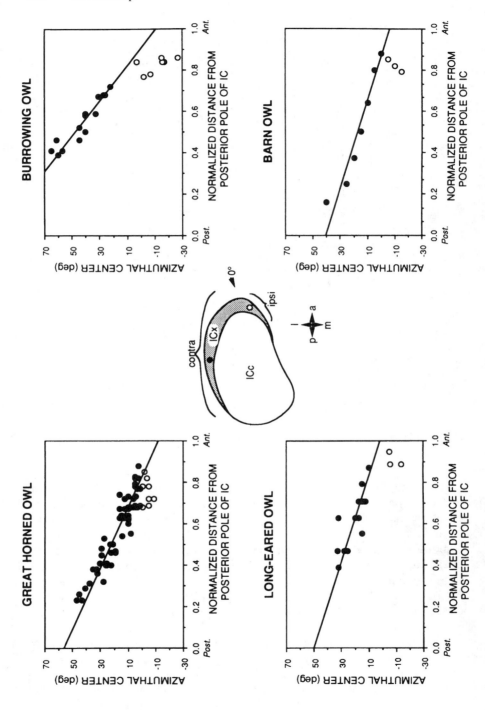

and results from phase ambiguity of the input to the ears (e.g., see Fig. 6.10). Few units in the raptors' lateral region of the IC showed a restriction of the response in elevation (one in *F. berigora* and one in *C. aeruginosus*). Most of the space-selective neurons showed a hemifield response characteristics that was only affected by variation in the azimuth angle of sound incidence. The cells fired when stimulated with a sound source in the free field that was presented from the contralateral side showing a clearly defined medial border in their space-specific response. Few of the hemifield units exhibited both a medial and a lateral border. The obvious discrepancy between the large spatial-receptive fields in diurnal raptors and the ability to accurately locate sound sources remains unexplained. It is not known whether broadband signals, which were not presented in the study by Calford et al. (1985), would have resulted in spatially more restricted receptive fields.

When studying the neural coding of the azimuth angle of sound incidence in the IC of *C. livia*, Lewald (1988, 1990b) also reported that most units responded best to contralateral stimulation (53%). These units, which were not tuned to specific directions but responded in a threshold-like fashion to the stimulus azimuth, probably are similar to the hemifield units described by Calford and colleagues (1985). In a small number of cells in the IC of *C. livia* (6%) the best response areas were on the ipsilateral side and had only a clear medial border (i.e., their response pattern was a mirror image of that in contralateral units). About a fifth of the neurons (18%) in *C. livia* IC exhibited a response that was space-tuned to frontal directions with clear borders on the medial and lateral sides. The width of the receptive fields in these neurons was on average 61° (measured at the 75% of maximum response threshold), which was much wider than that found in the owls (see above). In a few units in the IC of *C. livia* (3%), Lewald (1990b) reported response patterns with multiple excitatory areas that were separated by inhibitory areas. The frequency of the stimulating sounds had little effect on the width of the frontal response areas, which was contrary to the expectation based on the observation of seemingly poor directional sensitivity at frequencies of about 1–2 kHz in behavioral tests (Jenkins and Masterton 1979; Lewald 1987a). At least some *C. livia* studied in the behavioral tests by Lewald (1987a), however, did not show a reduced localization accuracy at 1 or 2 kHz, indicating that this may not be a general deficit based on poor processing mechanisms that can be found in all individuals. The

◀ ────────────────────────────────────

FIGURE 6.12. Spatial maps in the external nucleus of the inferior colliculus (ICx) of four species of owls (from Volman and Konishi 1990) suggest an orderly representation of interaural time differences. The location of the centers of the units' receptive fields is plotted in relation to the position of their recording site. *Filled* and *open circles* indicate data points obtained with stimulation of the ear contra or ipsilateral to the recording site, respectively. The regression lines are computed using data points from contralateral stimulation only.

only other bird species in which the directional response of IC neurons has been studied in the free field is the plains-wanderer (*Pedionomus torquatus*; Pettigrew and Larsen 1990). In the lateral part of the IC of this species, neurons with a broad frequency tuning and a space-specific response were found, whereas in the medial part of the IC the neurons were tuned to specific frequencies and did not show a clear directional response pattern.

The mechanisms that form the basis for creating space-specific receptive fields in the birds' IC have been studied using dichotic stimulation. In *T. alba*, it has been demonstrated by combining free-field and dichotic stimulation that the ITDs provide the cue for the azimuth angle of sound incidence and the IIDs provide the cue for the elevation angle of sound incidence (e.g., Moiseff and Konishi 1983). In other owls that do not show the asymmetry of the outer ears leading to the differentiation of the role of the two cues, ITDs and IIDs may both encode the stimulus azimuth. In these birds, the exchangeability of the two cues has yet to be demonstrated in behavioral tests. Interestingly, on the level of the owls' IC no substantial tradeoff between the IID of the stimulus and the ITD has been found, although this tradeoff would enhance the ITD in a free-field stimulation (the observed shift of $1.3–1.5\,\mu s/dB$ in IC neurons even was in the direction opposite to that deducted from the tradeoff between time and intensity; see Volman and Konishi 1989, 1990, and above). Best IID and ITD, however, were observed to be closely correlated in *A. cunicularis*, a species in which both cues encode the azimuth of the sound source (Volman and Konishi 1990). This correlation, which was found despite a lack of time-intensity trading, suggests that IID and ITD provide independent cues for the stimulus azimuth in nonspecialized birds that do not exploit ITDs and IIDs as separate cues for azimuth and elevation, respectively. The study by Lewald (1990b) in *C. livia* comparing responses of IC neurons obtained with free-field and dichotic stimulation similarly demonstrates that ITD or IID tuning, or both, can explain the neurons' firing rate in relation to the azimuth angle of sound incidence. The findings relate to the duplex theory of sound localization (Rayleigh 1907) that was proposed for perception of sound direction in humans. Rayleigh (1907) suggested that ITDs and IIDs will be used interchangeably in the perception of the azimuth angle of sound incidence, the former at low frequencies, at which the head provides no sound shadow and thus no substantial IIDs, and the latter at high frequencies. In birds that show a species-specific variation in the physical properties of the acoustic input to the ears (e.g., due to differences in head size, morphology of the external ears, and in the transfer function of the interaural canal, see above) and thus vary in the availability of IID and ITD cues, studies in different species may lead to quite different results. In some species, ITDs may be the predominant cue for azimuth localization of a sound source (e.g., as suggested by Klump et al. 1986 for *P. major*, a small song bird), whereas in other species both ITD and IID cues may contribute substantially to azimuth sound localization (either concurrently or exchangeably; e.g., Volman and Konishi 1989, 1990; Lewald 1990b), and it is even possible that in some species IIDs may provide the most important cue (e.g., as suggested for *G. gallus* Coles and Aitkin 1979).

Only in *T. alba* has neuronal encoding of sound localization been studied beyond the level of the IC in some detail. Space-specific neurons have been found in areas to which the IC either projects directly (e.g., in the optic tectum that receives a direct projection from the ICx, see Knudsen 1982, 1984a; Olsen et al. 1989) or indirectly via the thalamic nucleus ovoidalis (e.g., to forebrain areas, see Fig. 6.9). The auditory space map in the optic tectum that is relayed from the ICx interacts with the visual space map in the development of sound localization, matching the coordinates of the visual and auditory worlds (e.g., Knudsen 1983a; Knudsen et al. 1994; and see reviews by Knudsen 1988, 1994; and by Rucci et al. 1997). This alignment assures that multimodal cells are tuned to the same location in space when the owl turns the head to bring the sound source into the region of the visual fovea.

The observation by Knudsen et al. (1993) that lesions in the optic tectum alone were not sufficient to completely abolish the sound localization behavior of *T. alba*, and the fast recovery of the owls' localization ability that was observed by Wagner (1993) when lesioning the ICx both emphasize the significance of the forebrain pathway for sound localization. Only when Knudsen and colleagues (1993) deactivated the thalamic nucleus ovoidalis was the owls' acoustic orientation behavior no longer reliably directed toward the sound source. These findings and the evidence from the tracing study by Cohen et al. (1998) suggest that the space representation in the auditory forebrain is created independently from the map of auditory space found in the ICx. A majority of neurons in the thalamic nucleus ovoidalis are tuned both to IID and ITD (Proctor and Konishi 1997), but ITD tuning often is multipeaked, similar to the response pattern of cells found in the lateral shell of the ICc. Some neurons in the nucleus ovoidalis also show side-peak suppression in their ITD responses like cells in the ICx, but no systematic map of spatially tuned neurons is found at this level of the sound localization pathway. In the projection area of the nucleus ovoidalis, the field L complex (called primary auditory field by Cohen and Knudsen 1998), the tuning of the cells to binaural cues and the lack of a spatial map was similar to the situation found in the owls' auditory thalamus. In projection areas of the primary auditory field (i.e., in the auditory archistriatum and in the paleostriatum augmentum, see Fig. 6.9), however, the typical neurons' response showed an enhanced side-peak suppression of ITDs that together with the IID tuning and broad frequency tuning now resembled more the response pattern of space-specific cells in the ICx. Contrary to the organization in the ICx, however, no space map was formed in the forebrain areas, but units with similar space-specific responses were clustered together (Cohen and Knudsen 1994, 1995, 1998).

Cohen and co-workers (1998) speculate that the sound localization areas in the forebrain serve different functions. Behavioral experiments in *T. alba* indicated that the auditory archistriatum and adjacent areas are involved in directing the gaze of the owl to the location of the sound source (Knudsen et al. 1993; Knudsen and Knudsen 1996a). In contrast to the role of the ICx-optic tectum projection, that provides the basis for fast reflex-like head saccades aligning the visual and auditory maps for multimodal processing, the area of the auditory archistriatum and its surround may be more involved in voluntary acoustically guided orientation behavior. In a recent study, for example, Knudsen and Knudsen (1996b) demonstrated that this area plays a role in sound localization tasks that require the formation of an auditory spatial memory. Based on evidence from studies in songbird species, Cohen et al. (1998) propose that the additional indirect projection from the primary auditory field to the auditory archistriatum via the caudolateral neostriatum may serve mainly the recognition of auditory signals at specific locations. However, no experimental studies have been conducted that directly address this proposed function.

In comparison to the ascending projections, little is known about the role of descending projections in the sound localization pathway of birds. The superior olive was found to provide GABAergic inhibition to the nucleus laminaris, and it also projects to the nucleus magnocellularis and the nucleus angularis (e.g., Yang et al. 1999). Recent studies in *T. alba* (Pena et al. 1996) and *G. gallus* (Brückner and Hyson 1998; Yang et al. 1999) suggest, that the inhibitory projection may help to adjust the operating point of the auditory brain stem nuclei by preventing the responses from being driven into saturation at high sound-pressure levels.

4.2. Mechanisms for Auditory Detection of Motion

Evidence from behavioral observations in hunting owls suggest that they can predict the trajectory of a moving prey using acoustic cues. When striking the prey, the rectangle that the owls form with their talons is oriented in the direction of the trajectory defined by the prey movement (Payne 1971). In the owl's auditory system, a correlate of this ability to evaluate the prey movement can be found on the midbrain level. About 25% of the neurons in the IC and optic tectum of *T. alba* differ in their response when stimulated with sounds simulating different directions of motion in the azimuth plane (Wagner and Takahashi 1990, 1992; Wagner et al. 1994; Kautz and Wagner 1998; for a recent review see Wagner et al. 1997). Motion at various speeds (apparent velocities ranged from 125 to 1,200°/s) was simulated by presenting pulses of noise sequentially through seven speakers placed at 30° intervals from −90° to +90° azimuth in the horizontal plane 1 m in front of the owl (Fig. 6.13A). The sounds had a rise/fall time of 5 ms, and in some of the experiments they just overlapped (i.e., while one stimulus was switched off, the next stimulus was switched on), whereas in others

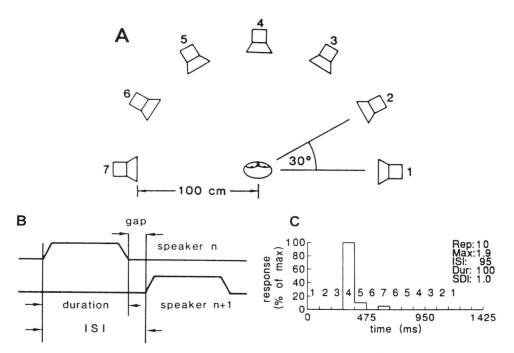

FIGURE 6.13. Acoustic motion-direction sensitivity of auditory neurons in the barn owl (*Tyto alba*; after Wagner and Takahashi 1992). (**A**) Motion was simulated by means of an array of seven speakers positioned in the horizontal plane. (**B**) Sounds of a certain duration were first broadcast from one speaker (*n*) and then from the adjacent speaker (*n* + 1) to simulate a sound source moving during one inter-stimulus interval (ISI; if the ISI was larger than the sound duration, an additional silent gap preceded the next sound) by 30° from the position of speaker *n* to speaker *n* + 1. (**C**) Averaged response (10 repetitions) of a motion-direction sensitive neuron in the optic tectum responding if sounds were presented from the speaker at position 4 and the simulated motion direction was counterclockwise, and not respond-ing during the clockwise motion of the source (Max: number of spikes per stimulus presentation at most effective speaker; SDI: signed directionality index—a value of −1 indicates an exclusive response for counterclockwise motion; Dur: duration). (**D**) Model of an acoustic-motion detector based on the inhibitory Barlow-Levick scheme (for details see Wagner and Takahashi 1992). Two space-specific auditory units (R1 and R2) pointing to locations separated by an angle of Δφ are stimulated sequentially by the moving sound source. The two units feed their response time-shifted by Δt into a nonlinear neural circuit (NL) thus producing a motion-direction dependent output by means of inhibition (direction-dependent inhibition, DDI, shown in **E**). An additional direction-independent inhibition (DII) affecting the response (Sum) may lead to a further reduction in response strength increasing the motion-direction sensitivity. (**E**) Response strength of a hypothetical neuron at the output of the nonlinear circuit in relation to the location of the speakers (numbers 1 to 7) presenting the sounds sequentially in time (DI: directionality index). (**F**) The response strength of the hypothetical neuron is decreased but the directionality of the response is increased if an additional direction-independent inhibition is added.

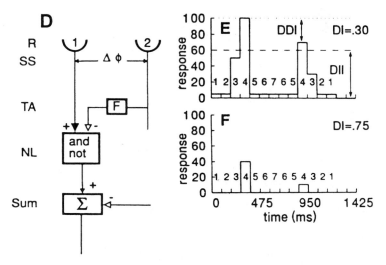

FIGURE 6.13. *Continued*

they were separated by a gap (Fig. 6.13B). Neurons that were directionally sensitive responded preferentially when sounds were presented sequentially from the speakers either in the clockwise or the counterclockwise direction (Fig. 6.13C). The strength of the preference was described by a directionality index computed as the ratio of the number of action potentials in percent that were observed while stimulating in opposite directions. The directionality index was relatively independent of the simulated speed of motion (on average the best response was found at a simulated speed of 310°/s. For sounds presented at 20 dB or more above the level of the units' response threshold, the average direction-sensitive response was constant and independent of the level of the sound (directionality index of about 80%). This level independence matches the level independence observed in the spatial receptive field of units in the owl's ICx and optic tectum (see above). At levels between 0 and 20 dB above the response units' threshold the directionality index increased linearly with the level in decibels (Wagner et al. 1994). When the sound level was expressed in the form of a signal-to-noise ratio (Wagner et al. 1994 added background noise in most of the experiments), a relative level independence was already observed above a signal-to-noise ratio of 5–10 dB. At equal levels of the signal and the background noise, the average directionality index was 42%.

Simulated motion has also been generated using dichotic stimulation with two continuous sine waves of slightly differing frequencies (e.g., Takahashi and Keller 1992a). This binaural stimulus shows a continuously varying interaural phase difference (Moiseff and Konishi 1981a; Moiseff and Haresign 1992). When using simulated angular speeds in azimuth of mostly between 52°/s and 258°/s (the maximum speed tested was 1,360°/s), which correspond to the speeds that could be encountered by owls in nature when flying, Takahashi and Keller (1992a) found a directional selectivity in about 10% of the cells in the ICc, which is somewhat less than in the other studies using free-field stimulation with noise bursts. Spatial selectivity, as derived by Takahashi and Keller (1992a) from the vector strength of the histograms, was enhanced when the tones were presented in moderate levels (rms signal-to-noise ratios of about -10 to $0\,dB$) of "indifferent" background noise (i.e., binaurally uncorrelated background noise added to the tone signals presented to the left and right ear). They interpreted the higher vector strength in the histograms obtained with stimuli simulating motion compared to simulated stationary stimuli as an indication of a more sharply tuned spatial selectivity in azimuth of the neurons in the case of motion stimuli (a simple amplitude modulation of the stimulus in the neuron's spatial receptive field did not enhance spatial tuning). Wagner and co-workers (1994) similarly present data that indicate a better detectability of a stimulus with simulated motion compared to a stationary stimulus presented in background noise ($p < 0.03$, McNemar test for the significance of change, data from figure 10 in Wagner et al., 1994). These results from the owl's auditory system resemble findings in the visual processing that motion enhances the detectability of a signal (see references in Takahashi and Keller 1992a).

Two alternative mechanisms have been suggested for auditory motion detection in humans: the snapshot hypothesis and a hypothesis invoking a specialized motion-sensitive system (for a review see Middlebrooks and Green 1991). In the snapshot hypothesis it is assumed that the auditory system compares the perceived spatial location of a sound source at two distinct times and computes the movement trajectory. In the motion-sensitive system hypothesis it is assumed that there are neurons that are sensitive to dynamic aspects of the localization cues. The computational mechanism proposed by Wagner and Takahashi (1992) for motion detection in the midbrain of *T. alba* belongs to the latter mechanisms, and it has been developed by the authors in parallel to a mechanism suggested by Barlow and Levick (1965) for visual motion processing (Fig. 6.13D). Space-specific neurons tuned to different locations (see above), which are stimulated sequentially by a sound passing through the different neurons' receptive fields, form the input to the motion detector. The time delay between the responses from different locations is passed through a temporal lowpass filter F (Wagner and Takahashi 1992; Wagner et al. 1997) producing an output in the form of a transient response. The nonlinear interaction NL in the next stage creates a direction-dependent inhibition (DDI; Fig. 6.13E) that suppresses the response in the "null direction." Experimental evidence suggests that this inhibition is due to GABAergic neurons in the IC (Kautz and Wagner 1998). Finally, a direction-independent inhibition (DII; Fig. 6.13F) lowers the overall response thereby increasing the relative difference. Wagner and Takahashi (1992) found, indeed, that size of the directionality index was inversely related to the number of discharges that they observed.

Psychoacoustic studies on the analysis of auditory motion in birds are lacking. In psychoacoustic studies in humans, the sensitivity in motion detection has been described in the form of the minimum audible movement angle (MAMA; see review by Middlebrooks and Green 1991; and by Wagner et al. 1997). In such studies, listeners are, for example, asked to discriminate between directions of motion or to discriminate between stationary and moving sounds. Thresholds are determined for the change in the angle of sound incidence that can be perceived (i.e., the MAA). Similar studies would be feasible in animals like *T. alba*, since simulated motion could be used to conduct the experiments.

4.3. Mechanisms for Reducing the Effect of Clutter Echoes: the Precedence Effect

In contrast to the echo-reducing environments in which studies on sound localization in physiological laboratories are conducted, the real world provides a much more complex acoustic environment in which sounds travel on multiple paths from the source to the receiver. First the sound traveling on the shortest direct path arrives, which is rapidly followed by numerous echoes originating from reflecting surfaces. Provided that the sound arriving on the direct path and the echoes are separated by at least 1 ms, humans locate only the first sound and not the trailing echoes. This effect has been termed the "law of the first wavefront" or the "precedence effect" (for reviews see Blauert 1983 and Zurek 1987). Recent behavioral and neurophysiological experiments by Keller and Takahashi (1996a) suggest that the precedence effect also operates in *T. alba*. When they presented naive *T. alba* in a behavioral experiment with two identical 3-ms bursts of noise from two speakers in front of the bird separated by an angle of 40° (as seen from the position of the owl), the owls consistently oriented toward the leading speaker, if the delay between the signals emitted from the two speakers ranged between 2 and 10 ms. At longer delays (e.g., 15–40 ms) the owls' head turns were less consistent, and at a number of trials the owls first oriented toward one speaker and then rapidly afterward to the other speaker. This may indicate that the owls perceived two separate sources in those conditions that they fixated sequentially. In ICx of *T. alba* containing a retina-like map of auditory space, Keller and Takahashi (1996a) presented space-specific neurons with the same stimuli as those used in the behavioral experiments. One speaker was placed in the neurons' space-specific receptive field while the other speaker was separated from the first one so that the angle of sound incidence from the two speakers differed, for example, by 40°. With such a free-field stimulation, the neurons' responses to trailing sounds presented in the excitatory center of the spatial-receptive field were substantially suppressed when the trailing sounds were presented with a delay of up to 5 ms (see the examples in Figs. 6.14A and 6.14C). When the sound was presented simultaneously by both speakers and the receptive field of the unit was located midway between them, the neurons responded as would be expected as a result of summing localization (Fig. 6.14B, see

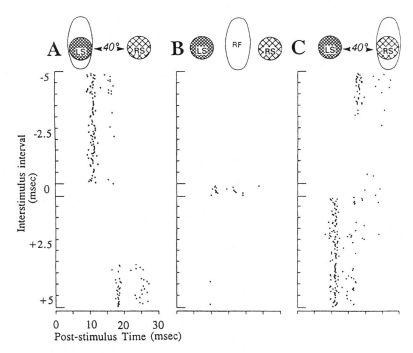

FIGURE 6.14. Neural response pattern of an external nucleus of inferior colliculus (ICx) unit in the barn owl (*Tyto alba*) showing suppression of simulated echoes as may be indicative of the precedence effect (from Keller and Takahashi 1996a). Identical 3-ms bursts of noise were presented from each of two speakers with an interstimulus interval (ISI) varying in steps of 0.5 ms (negative times indicate that the left speaker (LS) leads the right speaker (RS), positive times vice versa). The receptive-field (RF) center of a space-specific ICx neuron either was located at one of the speaker positions (**A, C**), or the speakers were positioned left and right of the receptive-field center (**B**). If the speaker in the center of the unit's RF was stimulated first, the neuron responded with action potentials (negative ISI in A, positive ISI in C); if it lagged the other speaker by a small time delay, its response was suppressed (positive ISI in A, negative ISI in C). With increasing time delay between the sounds broadcast from the two speakers the response recovered. In (**B**) it is shown that the neuron also responded to a phantom target located midway between the two speakers. In (**A**) and (**C**) the location of the phantom target produced by an ISI of 0 ms is located outside the neuron's spatial receptive field and thus no response occured.

also Takahashi and Keller 1994; Keller and Takahashi 1996b). When delays longer than 5 ms were used, the neurons responded to both the leading and the trailing sound. At larger angular speaker separations, the suppression of the response was smaller and neurons were capable of responding to both successive sounds with short delays. This indicates that the observed suppression was not due to a refractory period during which the ICx neurons could not respond. Keller and Takahashi (1996a) suggest that the pattern of neuronal responses is caused by inhibitory input to the ICx. When a site in the space map of the ICx is excited, it is surrounded by an annulus of inhibition that lasts for 5–10 ms and extends over a 20° to 60° radius from the center of excitation. The authors suggest that this pattern may be a manifestation of the center-surround organization shown in individual space-specific neurons (see Knudsen and Konishi 1978b) that leads to an enhanced temporal contrast in the space-map of the ICx.

5. Concluding Remarks

The comparison of the physical properties of the ear, the behavioral sound localization performance, and the physiological data from various bird species has made clear that quite large interspecific differences may exist in the mechanisms of sound localization. The studies in *T. alba* provide the foundation for our understanding of the physiological mechanisms underlying the sound localization ability. It appears to be the most parsimonious hypothesis that the computational principles and the structure of the auditory pathway, with its parallel processing of binaural differences of the temporal and spectral information, provide the basis for sound localization in the other bird species as well. *T. alba*, however, shows a number of specializations that have evolved to serve its nocturnal hunting by acoustic cues. Therefore, it would be desirable to study other less specialized birds from various taxonomic groups in more detail and see whether the same rules apply to the neural mechanisms in the different species. As has been demonstrated in this chapter, this seems to be the case in the inferior colliculus of other owls and in the brain stem nuclei of the sound localization pathway of various birds. Contrary to the computational mechanisms in *T. alba*, however, on the level of the inferior colliculus of some bird species, ITD and IID cues seem to jointly encode stimulus azimuth. Since this aspect is characteristic for sound localization in mammals, it would be very interesting to understand the integration of the two different cues into one coherent spatial auditory representation. Furthermore, a much better understanding of the generality of the computational principles in birds would be achieved if an attempt would be made in a less specialized bird species to describe the neural responses on all known levels of the sound localization pathway and relate these to the behavioral performance.

The behavioral studies of sound localization in birds have so far mainly addressed the question of the accuracy of sound localization in relation to stimulus characteristics, and even in this area we only have a basic knowledge of the sensory capabilities of birds. Nothing is known, for example, about the localization of moving sound sources. In addition, there are many more interesting aspects to sound localization that are crucial for understanding the birds' sensory analysis of their acoustic environment. The study of Nelson and Stoddard (1998), for example, demonstrated a remarkable capability of birds to judge not only the direction from which a sound originated but also the distance of the sound source. The ability of birds to segregate the different voices in the cacophony of a dawn chorus, that is, to form auditory objects with the help of location cues, is another area of research that has not been taken up. If we understand these examples of perception of the location of sounds by birds in the real world, we will gain insights that are of relevance for understanding the perception of the environmental cues by all vertebrates including us.

Acknowledgments. During the preparation of this chapter the author was funded by grants from the Deutsche Forschungsgemeinschaft. E. Futterer and K. Senft helped in the preparation of the figures. I thank S. Amagai, C.E. Carr, M. Dent, R.J. Dooling, and O.N. Larsen for providing unpublished data C.E. Carr, R.J. Dooling, and A.N. Popper critically read a previous version of the manuscript and provided helpful feedback.

References

Adolphs R (1993) Bilateral inhibition generates neuronal responses tuned to interaural level differences in the auditory brainstem of the barn owl. J Neurosci 13:3647–3668.

Albeck Y (1997) Inhibition sensitive to interaural time difference in the barn owl's inferior colliculus. Hear Res 109:102–108.

Albeck Y, Konishi M (1995) Responses of neurons in the auditory pathway of the barn owl to partially correlated binaural signals. J Neurophysiol 74:1689–1700.

Amagai S, Carr CE, Dooling RJ (1995) Physiology and anatomy of brainstem auditory nuclei of several small birds. In: Burrows M, Matheson T, Newland PL, Schuppe H (eds) Nervous Systems and Behaviour. Thieme: Stuttgart, p. 314.

Barlow HB, Levick WR (1965) The mechanisms of directionally selective units in rabbit's retina. J Physiol (Lond) 178:477–504.

Batra R, Kuwada S, Stanford TR (1989) Temporal coding of envelopes and their interaural delays in the inferior colliculus of the unanesthetized rabbit. J Neurophysiol 61:257–268.

Beitel RE (1991) Localization of azimuthal sound direction by the great horned owl. J Acoust Soc Am 90:2843–2846.

Blauert J (1983) Spatial Hearing. Cambridge MA: MIT Press, 427 pp.

Brainard MS, Knudsen EI, Esterly SD (1992) Neural derivation of sound localization: resolution of spatial ambiguities in binaural cues. J Acoust Soc Am 91:1015–1017.

Bregman AS (1990) Auditory Scene Analysis. Cambridge, MA: MIT Press.

Brown CH (1982) Ventroloquial and locatable vocalizations in birds. Z Tierpsychol 59:338–350.

Brown CH (1994) Sound localization. In: Fay RR, Popper AN (eds) Comparative Hearing: Mammals. New York: Springer, pp. 57–96.

Brückner S, Hyson RL (1998) Effect of GABA on the processing of interaural time differences in nucleus laminaris neurons in the chick. Eur J Neurosci 10: 3438–3450.

Calford MB (1988) Constraints an the coding of sound frequency imposed by the avian interaural canal. J Comp Physiol A 162:491–502.

Calford MB, Piddington RW (1988) Avian interaural delay enhances interaural delay. J Comp Physiol A 162:503–510.

Calford MB, Wise LZ, Pettigrew JD (1985) Coding of sound localition and frequency in the auditory midbrain of diurnal birds of prey, families Accipitridae and Falconidae. J Comp Physiol A 157:149–160.

Carr CE (1992) Evolution of the central auditory system in reptiles and birds. In: Webster DB, Fay RR, Popper AN (eds) The Evolutionary Biology of Hearing. New York: Springer, pp. 511–543.

Carr CE (1993) Processing of temporal information in the brain. Ann Rev Neurosci 16:223–243.

Carr CE, Boudreau RE (1991) Central projections of auditory nerve fibers in the barn owl. J Comp Neurol 314:306–318.

Carr CE, Boudreau RE (1993) Organization of the nucleus magnocellularis and the nucleus laminaris in the barn owl: encoding and measuring interaural time differences. J Comp Neurol 334:337–355.

Carr CE, Konishi M (1988) Axonal delay lines for time measurement in the owl's brainstem. Proc Natl Acad Sci U S A 85:8311–8315.

Carr CE, Konishi M (1990) A circuit for detection of interaural time differences in the brain stem of the barn owl. J Neurosci 10:3227–3246.

Carr CE, Fujita I, Konishi M (1989) Distribution of GABAergic neurons and terminals in the auditory system of the barn owl. J Comp Neurol 286:190–207.

Cohen YE, Knudsen EI (1994) Auditory tuning for spatial cues in the barn owl basal ganglia. J Neurophysiol 72:285–298.

Cohen YE, Knudsen EI (1995) Binaural tuning of auditory units in the forebrain archistratal gaze fields of the barn owl: local organization but no space map. J Neurosci 15:5152–5168.

Cohen YE, Knudsen EI (1998) The representation of binaural spatial cues in the primary auditory field of the barn owl forebrain. J Neurophysiol 79:879–890.

Cohen YE, Miller GL, Knudsen EI (1998) A forebrain pathway for auditory space processing in the barn owl. J Neurophysiol 79:891–902.

Coles RB, Aitkin LM (1979) The response properties in the midbrain of the domestic fowl (Gallus gallus) to monaural and binaural stimuli. J Comp Physiol 134:241–251.

Coles RB, Guppy (1988) Directional hearing in the barn owl (Tyto alba). J Comp Physiol A 163:117–133.

Coles RB, Lewis DB, Hill KG, Hutchings ME, Gower DM (1980) Directional hearing in the Japanese quail (*Coturnix coturnix japonica*). II. Cochlear physiology. J Exp Biol 86:153–170.

Counter SA, Borg E (1982) The avian stapedius muscle. Acta Otolaryngol 94:267–274.

Dent ML, Larsen ON, Dooling RJ (1997) Free-field binaural unmasking in budgerigars (*Melopsittacus undulatus*). Behav Neurosci 111:590–598.

Fay RR (1988) Hearing in Vertebrates. Winnetka, IL: Hill-Fay Associates.

Frost BJ, Baldwin PJ, Csizy ML (1989) Auditory localization in the northern saw-whet owl, *Aegolius acadicus*. Can J Zool 67:1955–1959.

Fujita I, Konishi M (1991) The role of GABAergic inhibition in processing of interaural time difference in the owl's auditory system. J Neurosci 11:722–739.

Gatehouse RW, Shelton BR (1978) Sound localization in the bobwhite quail (*Colinus virginianus*). Behav Biol 22:533–540.

Gleich O, Narins PM (1988) The phase response of primary auditory afferents in a songbird (*Sturnus vulgaris* L.). Hearing Res 32:81–92.

Granit O (1941) Beiträge zur Kenntnis des Gehörsinns der Vögel. Ornis fennica 18:49–71.

Griffin DR (1953) Acoustic orientation in the oil bird, *Steatornis*. Proc Natl Acad Sci 39:884–893.

Hill KG, Lewis DB, Hutchings ME, Coles RB (1980) Directional hearing in the Japanese quail (*Coturnix coturnix japonica*). I. Acoustic properties of the auditory system. J Exp Biol 86:135–151.

Hyson RL, Overholt EM, Lippe WR (1994) Cochlear microphonic measurements of interaural time differences in the chick. Hear Res 81:109–118.

Jeffress LA (1948) A place theory of sound localization. J Comp Physiol Psychol 41:35–39.

Jenkins WM, Masterton RB (1979) Sound localization in the pigeon (*Columba livia*). J Comp Physiol Psychol 93:403–413.

Kautz D, Wagner H (1998) GABAergic inhibition influences auditory motion-direction sensitivity in barn owls. J Neurophysiol 80:172–185.

Keller CH, Takahashi TT (1996a) Responses to simulated echoes by neurons in the barn owl's auditory space map. J Comp Physiol A 178:499–512.

Keller CH, Takahashi TT (1996b) Binaural cross-correlation predicts the responses of neurons in the owl'a auditory space map under conditions simulating summing localization. J Neurosci 16:4300–4309.

Klump GM, Larsen ON (1992) Azimuth sound localization in the European starling (*Sturnus vulgaris*): physical binaural cues. J Comp Physiol A 170:243–251.

Klump GM, Shalter MD (1984) Acoustic behaviour of birds and mammals in the predator context. Z Tierpsychol 66:189–226.

Klump GM, Windt W, Curio E (1986) The great tit's (*Parus major*) auditory resolution in azimuth. J Comp Physiol 158:383–390.

Knudsen EI (1980) Sound localization in birds. In: Popper AN, Fay RR (eds) Comparative Studies of Hearing in Vertebrates. New York: Spinger, pp. 289–322.

Knudsen EI (1982) Auditory and visual maps of space in the optic tectum of the barn owl. J Neurosci 2:1177–1194.

Knudsen EI (1983a) Early auditory experience aligns the auditory map of space in the optic tectum of the barn owl. Science 222:939–942.

Knudsen EI (1983b) Subdivisions of the inferior colliculus in the barn owl (*Tyto alba*). J Comp Neurol 218:174–186.

Knudsen EI (1984a) Auditory properties of space-tuned units in the owl's tectum. J Neurophysiol 53:709–723.

Knudsen EI (1984b) Synthesis of a neural map of auditory space in the owl. In: Edelman GM, Cowan WM, Gall WE (eds) Dynamic Aspects of Neocortical Functioning. New York: Wiley, pp. 375–396.

Knudsen EI (1988) Experience shapes sound localization and auditory unit properties during development in the barn owl. In: Edelman GM, Gall WE, Cowan WM (eds) Auditory Function: Neurobiological Bases of Hearing. New York: Wiley, pp. 137–149.

Knudsen EI (1994) Supervised learning in the brain. J Neurosci 14:3985–3997.

Knudsen EI, Brainard MS (1995) Creating a unified representation of visual and auditory space in the brain. Annu Rev Neurosci 18:19–43.

Knudsen EI, Knudsen PF (1996a) Contribution of the forebrain archistratal gaze fields to auditory orienting behavior in the barn owl. Exp Brain Res 108:23–32.

Knudsen EI, Knudsen PF (1996b) Disruption of auditory spatial working memory by inactivation of the forebrain archistriatum in barn owls. Nature 383:428–431.

Knudsen EI, Konishi M (1978a) A neural map of auditory space in the owl. Science 200:795–797.

Knudsen EI, Konishi M (1978b) Center-surround organization of auditory receptive fields in the owl. Science 202:778–780.

Knudsen EI, Konishi M (1978c) Space and frequency are represented separately in the auditory midbrain of the owl. J Neurophysiol 41:870–884.

Knudsen EI, Konishi M (1979) Mechanisms of sound localization in the barn owl (*Tyto alba*). J Comp Physiol 133:13–21.

Knudsen EI, Konishi M, Pettigrew JD (1977) Receptive fields of auditory neurons in the owl. Science 198:1278–1280.

Knudsen EI, Blasdel GG, Konishi M (1979) Sound localization by the barn owl (*Tyto alba*) measured with the search coil technique. J Comp Physiol 133:1–11.

Knudsen EI, Knudsen PF, Masino T (1993) Parallel pathways mediating both sound localization and gaze control in the forebrain and midbrain of the barn owl. J Neurosci 13:2837–2852.

Knudsen EI, Esterly SD, Olsen JF (1994) Adaptive plasticity of the auditory space map in the optic tectum of adult and baby barn owls in response to external ear modification. J Neurophysiol 71:79–94.

Köppl C (1997) Phase locking to high frequencies in the auditory nerve and cochlear nucleus magnocellularis of the barn owl, *Tyto alba*. J Neurosci 17:3312–3321.

Konishi M (1973a) Locatable and non-locatable acoustic signals for barn owls. Am Nat 107:775–785.

Konishi M (1973b) How the owl tracks its prey. Am Sci 61:414–424.

Konishi M (1983) Neuroethology of acoustic prey localization in the barn owl. In: Huber F, Markl H (eds) Neuroethology and Behavioral Physiology. Berlin: Springer, pp. 303–317.

Konishi M (1993a) Listening with two ears. Sci Am 286:66–73.

Konishi M (1993b) Neuroethology of sound localization in the owl. J Comp Physiol A 173:3–7.

Konishi M, Knudsen EI (1979) The oilbird: hearing and echolocation. Science 204:425–427.

Konishi M, Sullivan WE, Takahashi TT (1985) The owl's cochlear nuclei process different sound localization cues. J Acoust Soc Am 78:360–364.

Konishi M, Takahashi TT, Wagner H, Sullivan WE, Carr CE (1988) Neurophysiological and anatomical substrates of sound localization in the owl. In: Edelman GM, Gall WE, Cowan WM (eds) Auditory Function: Neurobiological Bases of Hearing. New York: Wiley, pp. 721–745.

Koyano K, Funabiki K, Ohmori H (1996) Voltage-gated currects and their roles in timing coding in auditory neurons of the nucleus magnocellularis of the chick. Neurosci Res 26:29–45.

Kretzschmar E (1982) Wie hört ein Sperber (*Accipiter nisus* L.) Alarmrufe seiner Beutevögel? Thesis, Ruhr-University Bochum.

Kuhn GF (1977) Model for interaural time differences in the azimuthal plane. J Acoust Soc Am 62:157–167.

Kühne R, Lewis B (1985) External and middle ears. In: King AS, McLelland J (eds) Form and Function in Birds, Vol. 3. London: Academic Press, pp. 227–271.

Larsen ON, Dooling RJ (1992) Binaural physical cues available for sound localization in the free sound field by small birds. Proceedings 3rd International Congress Neuroethology, Montreal 1992, p. 338.

Larsen ON, Dooling RJ (1996) Intracranial pressure modifies hearing in birds. Abstracts 19th Midwinter Meeting Association for Research in Otolanryngology, Feb. 4–8, 1996 St. Petersburg Beach.

Larsen ON, Tweedale R, Calford MB (1989) Binaural cues for directional hearing in acoustically non-specialized birds. Soc Neurosci Abs 15:114.

Larsen ON, Dooling RJ, Ryals BM (1997) Roles of intracranial air pressure in bird audition. In: Lewis ER, Long GR, Lyon RF, Narins PM, Steele CR (eds) Diversity in Auditory Mechanics. Singapore: World Scientific Publishing, pp. 253–259.

Lewald J (1987a) The acuity of sound localization in the pigeon (*Columba livia*). Naturwiss 74:296–297.

Lewald J (1987b) Interaural time and intensity difference thresholds of the pigeon (*Columba livia*). Naturwiss 74:449–451.

Lewald J (1988) Neuronal coding of azimuthal sound direction in the auditory midbrain of the pigeon. Naturwiss 75:470–472.

Lewald J (1990a) The directionality of the ear of the pigeon (*Columba livia*). J Comp Physiol A 167:533–543.

Lewald J (1990b) Neural mechanism of directional hearing in the pigeon. Exp Brain Res 82:423–436.

Lewis B (1983) Directional cues for auditory localization. In: Lewis B (ed) Bioacoustics: A Comparative Approach. London: Academic Press, pp. 233–257.

Lewis B, Coles R (1980) Sound localization in birds. TINS 3:102–105.

Manley GA, Köppl C, Konishi M (1988) A neural map of interaural intensity difference in the brainstem of the barn owl. J Neurosci 8:2665–2677.

Marler P (1955) Characteristics of some animal calls. Nature 176:6–8.

Mazer JA (1998) How the owl resolves auditory coding ambiguity. Proc Natl Acad Sci U S A 95:10932–10937.

Middlebrooks JC, Green DM (1991) Sound localization by human listeners. Annu Rev Psychol 42:135–159.

Mills AW (1958) On the minimum audible angle. J Acoust Soc Am 30:237–246.

Moiseff A (1989a) Binaural disparity cues available to the barn owl for sound localization. J Comp Physiol A 164:629–636.

Moiseff A (1989b) Bi-coordinate sound localization by the barn owl. J Comp Physiol A 164:637–644.

Moiseff A, Haresign T (1992) Responses of auditory units in the barn owl's inferior colliculus to continuously varying interaural phase differences. J Neurophysiol 67:1428–1436.

Moiseff A, Konishi M (1981a) Neuronal and behavioral sensitivity to binaural time differences in the owl. J Neurosci 1:40–48.

Moiseff A, Konishi M (1981b) The owl's interaural pathway is not involved in sound localization. J Comp Physiol A 144:299–304.

Moiseff A, Konishi M (1983) Binaural characteristics of units in the owl's brainstem auditory pathway: precursors of restricted spatial receptive fields. J Neurosci 3:2553–2562.

Mori K (1997) Across-frequency nonlinear inhibition by GABA in processing of interaural time difference. Hear Res 111:22–30.

Nelson BS, Stoddard PK (1998) Accuracy of auditory perception of distance and azimuth by a passerine bird. Anim Behav 56:467–477.

Norberg RÅ (1977) Occurrence and independent evolution of bilateral ear asymmetry in owls and implications on owl taxonomy. Phil Trans R Soc London B Biol Sci 282:375–408.

Novick A (1959) Acoustic orientation in the cave swiftlet. Biol Bull 117:497–503.

Olsen JF, Knudsen EI, Esterly SD (1989) Neural maps of interaural time and intensity differences in the optic tectum of the barn owl. J Neurosci 9:2591–2605.

Olson HF (1946) Gradient microphones. J Acoust Soc Am 17:192–198.

Overholt EM, Rubel EW, Hyson RL (1993) A circuit for coding interaural time differences in the chick brain stem. J Neurosci 12:1698–1708.

Park TJ (1989) Sound localization in small birds. Dissertation, University of Maryland, MD, College Park.

Park TJ, Dooling RJ (1991) Sound localization in small birds: absolute localization in azimuth. J Comp Psychol 105:125–133.

Payne RS (1971) Acoustic location of prey by barn owls (Tyto alba). J Exp Biol 54:535–573.

Pena JL, Viete S, Albeck Y, Konishi M (1996) Tolerance to sound intensity of binaural coincidence detection in the nucleus laminaris of the owl. J Neurosci 16:7046–7054.

Pettigrew JD, Larsen ON (1990) Directional hearing in the plains wanderer Pedionomus torquatus. In: Rowe M, Aitkin L (eds) Information Processing in Mammalian Auditory and Tactile Systems. New York: Alan R. Liss, pp. 179–190.

Proctor L, Konishi M (1997) Representation of sound localization cues in the auditory thalamus of the barn owl. Proc Natl Acad Sci U S A 94:10421–10425.

Quine DB, Kreithen ML (1981) Frequency shift discrimination: can homing pigeons locate infrasounds by doppler shifts? J Comp Physiol 141:153–155.

Rangol H-P, Plassmann W (1993) The middle-ear of the zebrafinch behaves like a pressure-gradient receiver. In: Elsner N, Heisenberg M (eds) Gene—Brain—Behaviour. Stuttgart: Thieme, p. 250.

Rayleigh Lord [Strutt JW] (1907) On our perception of sound direction. Philos Mag 13:214–232.

Reyes AD, Rubel EW, Spain WJ (1996) In vitro analysis of optimal stimuli for phase-locking and time-delayed modulation of firing in avian nucleus laminaris neurons. J Neurosci 16:993–1007.

Rice WR (1982) Acoustical location of prey by the marsh hawk: adaptation to concealed prey. Auk 99:403–413.

Robinson DA (1963) A method of measuring eye movement using a scleral search coil in a magnetic field. IEEE Trans Biomed Electron Eng 10:137–145.

Rosowski JJ (1979) The interaural pathway of the pigeon and sound localization: does the pigeon ear act as a differential pressure transducer? Dissertation, University of Pennsylvania, Philalelphia, PA.

Rosowski JJ, Saunders JC (1980) Sound transmission through the avian interaural pathways. J Comp Physiol A 136:183–190.

Rubel EW, Parks TN (1975) Organization of development of brain stem auditory nuclei of the chicken: tonotopic organization of n. magnocellularis and n. laminaris. J Comp Neurol 164:411–434.

Rucci M, Tononi G, Edelman GM (1997) Registration of neural maps through value-dependent learning: modeling the alignment of auditory and visual maps in the barn owl's optic tectum. J Neurosci 17:334–352.

Saberi K, Farahbod H, Konishi M (1998) How do owls localize interaurally phase-ambiguous signals? Proc Natl Acad Sci U S A 95:6465–6468.

Schmid O, Rangol H-P, Plassmann W (1996) Directional hearing in zebra finches—analysis on the level of CM-potential recordings. In: Elsner N, Schnitzler H-U (eds) Brain and Evolution. Stuttgart: Thieme, p. 188.

Schwartzkopff (1950) Beitrag zum Problem des Richtungshörens bei Vögeln. Z vergl Physiol 32:319–327.

Schwartzkopff J (1952) Untersuchungen über die Arbeitsweise des Mittelohres und das Richtungshören der Singvögel unter Verwendung von Cochlea-Potentialen. Z Vergl Physiol 34:46–68.

Schwartzkopff J (1962a) Zur Frage des Richtungshörens von Eulen (*Striges*). Z Vergl Physiol 45:570–580.

Schwartzkopff J (1962b) Die akustische Lokalisation bei Tieren. Ergebn Biol 24: 136–176.

Schwarz L (1943) Zur Theorie der Beugung einer ebenen Schallwelle an der Kugel. Akust Z 8:91–117.

Shalter MD (1978) Localization of passerine seeet and mobbing calls by goshawks and pygmy owls. Z Tierpsychol 46:260–267.

Shalter MD, Schleidt WM (1977) The ability of barn owls *Tyto alba* to discriminate and locaize avian alarm calls. Ibis 119:22–27.

Stevens SS, Newman EB (1936) The localization of actual sources of sound. Am J Psychol 48:297–306.

Stewart GW (1911) The acoustic shadow of a rigid sphere with certain applications in architectural acoustics and audition. Phys Rev 33:467–479.

Sullivan WE, Konishi M (1984) Segregation of stimulus phase and intensity coding in the cochlear nucleus of the barn owl. J Neurosci 4:1787–1799.

Sullivan WE, Konishi M (1986) Neural map of interaural phase difference in the owl's brainstem. Proc Natl Acad Sci U S A 83:8400–8404.

Takahashi TT (1989) The neural coding of auditory space. J Exp Biol 146:307–322.

Takahashi TT, Keller CH (1992a) Simulated motion enhances neuronal selectivity for a sound localization cue in background noise. J Neurosci 12:4381–4390.

Takahashi TT, Keller CH (1992b) Commissural connections mediate inhibition for the computation of interaural level difference in the barn owl. J Comp Physiol A 170:161–169.

Takahashi TT, Keller CH (1994) Representation of multiple sound sources in the owl's auditory space map. J Neurosci 14:4780–4793.

Takahashi TT, Konishi M (1986) Selectivity for interaural time difference in the owl's midbrain. J Neurosci 6:3413–3422.

Takahashi TT, Moiseff A, Konishi M (1984) Time and intensity cues are processed independently in the auditory system of the owl. J Neurosci 4:1781–1786.

Takahashi TT, Wagner H, Konishi M (1989) Role of commissural projections in the representation of bilateral auditory space in the barn owl's inferior colliculus. J Comp Neurol 281:545–554.

Trainer JE (1946) The auditory acuity of certain birds. Doctoral dissertation, Cornell University, Ithaca, NY.

Viete S, Pena JL, Konishi M (1997) Effects of interaural intensity difference on the processing of interaural time difference in the owl's nucleus laminaris. J Neurosci 17:1815–1824.

Volman SF (1994) Directional hearing in owls: neurobiology, behavior and evolution. In: Davies MNO, Green PR (eds) Perception and Motor Control in Birds. Berlin: Springer, pp. 292–314.

Volman SF, Konishi M (1989) Spatial selectivity and binaural responses in the inferior colliculus of the great horned owl. J Neurosci 9:3083–3096.

Volman SF, Konishi M (1990) Comparative physiology of sound localization in four species of owls. Brain Behav Evol 36:196–215.

Wagner H (1993) Sound-localization deficits induced by lesions in the barn owls's auditory space map. J Neurosci 13:371–386.

Wagner H, Takahashi TT (1990) Neurons in the midbrain of the barn owl are sensitive to the direction of apparent acoustic motion. Naturwiss 77:439–442.

Wagner H, Takahashi TT (1992) Influence of temporal cues on acoustic motion-direction sensitivity of auditory neurons in the owl. J Neurophysiol 68:2063–2076.

Wagner H, Takahashi TT, Konishi M (1987) Representation of interaural time difference in the central nucleus of the barn owl's inferior colliculus. J Neurosci 7:3105–3116.

Wagner H, Trinath T, Kautz D (1994) Influence of stimulus level on acoustic motion-direction sensitivity in barn owl midbrain neurons. J Neurophysiol 71:1907–1916.

Wagner H, Kautz D, Poganiatz I (1997) Principles of acoustic motion detection in animals and man. Trends Neurosci 20:583–588.

Wightman FL, Kistler DJ (1993) Sound localization. In: Yost WA, Popper AN, Fay RR (eds) Human Psychophysics. New York: Springer, pp. 155–192.

Windt W (1985) Lokalisation von Kunst- und Naturlauten durch Kohlmeisen (*Parus major*). Diplom Thesis, Ruhr-Universität Bochum.

Wise LZ, Frost BJ, Shaver SW (1988) The representation of sound frequency and space in the midbrain of the saw-whet owl. Soc Neurosci Abstr 14:1095.

Woodworth (1962) Experimental Psychology. New York: Holt, Rinehart & Winston.

Yang L, Monsivais P, Rubel EW (1999) The superior olivary nucleus and its influence on nucleus laminaris: a source of inhibitory feedback for coincidence detection in the avian auditory brainstem. J Neurosci 19:2313–2325.

Yin TCT, Chan JCK (1988) Neural mechanisms underlying interaural time sensitivity to tones and noise. In: Edelman GM, Gall WE, Cowan WM (eds) Auditory Function: Neurobiological Bases of Hearing. New York: Wiley, pp. 385–430.

Yin TCT, Kuwada S, Sujaku Y (1984) Interaural time sensitivity of high-frequency neurons in the inferior colliculus. J Acoust Soc Am 76:1401–1410.

Young SR, Rubel EW (1983) Frequency-specific projections of individual neurons in chick brainstem auditory nuclei. J Neurosci 3:1373–1378.

Zurek PM (1987) The precedence effect. In: Yost WA, Gourevitch G (eds) Directional Hearing. New York: Springer, pp. 85–105.

7
Hearing in Birds and Reptiles

Robert J. Dooling, Bernard Lohr, and Micheal L. Dent

1. Introduction

1.1. Birds and Reptiles

The comparative hearing of birds and reptiles should always be considered together. It is clear from the vertebrate fossil record that birds and reptiles split over 200 million years ago from the diapsid reptiles of the early Triassic period (Fedducia 1980; Carroll 1987). Because of this common ancestry, there is considerable similarity between the hearing organs of modern day birds and reptiles, especially the Crocodilia (Manley and Gleich 1991; Manley, Chapter 4). However, comparisons between reptiles and birds are difficult for a number of reasons. In reptiles, the auditory anatomy is extraordinarily diverse. While this presents investigators with excellent opportunities to understand the relation between form and function, direct data on the behavior of hearing in reptiles are almost nonexistent. This leaves our understanding of hearing in this group of vertebrates entirely based on indirect measures from anatomy and physiology. Thus, any comparison of hearing between reptiles and birds is somewhat unbalanced because it also involves a comparison across methodologies: hearing estimates from anatomical and physiological data in the case of reptiles along with behavioral estimates of hearing capabilities in birds.

In spite of the considerable behavioral data on hearing in birds, complementing a considerable array of anatomical and physiological data on the auditory system, there are still some problems. Because birds are not well represented in the fossil record, it has proven more difficult to understand the taxonomic relationship among different groups of birds. Modern treatments consider birds to be in four groups. One group, the ratites (large flightless birds) represented by the emus (*Dromaius novaehollandiae*), ostriches (*Struthio camelus*), and rheas (Rheidae) retain an ancient paleognathous palate (with an extensive vomer-pterygoid joint, a caudal process of the palatine, and a movable joint between the pterygoid and the base of the braincase).The other three groups share a neognathus palate (a modern condition consisting of reduced or absent vomers, the presence of a joint between pterygoid and palatine, and the absence of a joint between the pterygoid and braincase). These three groups are divided into a primitive and more advanced land bird assemblage and a water bird assemblage (Brodkorb 1971; Fedducia 1980; Carroll 1987; Manley and Gleich 1991). Birds such as pigeons (*Columba livia*), chickens (*Gallus gallus*), mallard ducks (*Anas platyrhynchos*), and budgerigars (*Melopsittacus undulatus*) belong to the primitive subdivision of the land bird assemblage while birds such as songbirds (Passeriformes) and owls (Strigiformes) are more recently evolved (Fedducia 1980; Carroll 1987).

Birds—and reptiles to a lesser degree—have undergone an extensive and complex adaptive radiation into a wide range of habitats (Brodkorb 1971). For both vertebrate groups, acoustic information transmitted through the habitat can play a crucial role in a variety of behaviors including individual and species recognition, mate selection, and territorial defense. For these reasons, the study of hearing capabilities in birds and reptiles provides a fertile window on the evolution and adaptation of auditory systems to different environments and the mechanisms involved in sound detection and discrimination.

1.2. The Study of Hearing

The sense of hearing in animals may be studied using anatomical, physiological, and behavioral approaches and each method has advantages and disadvantages. Hearing is traditionally and most generally defined as the behavioral response to sound involving the whole, awake organism. For this reason, the present review focuses on behavioral or psychoacoustic measures of hearing that are the most direct and usually the most appropriate as assessments of an animal's hearing capabilities. Because there are only two known behavioral studies of hearing in reptiles, this chapter considers data from two kinds of studies. First, behavioral studies are described that meet the criteria of rigorous animal psychophysics, including good regulation and definition of the acoustic stimulus, use of a reliable response measure, and incontrovertible evidence of stimulus control (Stebbins 1970; Klump et al. 1995). Second, data are considered from physiological approaches where there is strong evidence that the particular methods used typically correlate well with other (behavioral) estimates of absolute sensitivity and hearing range. Examples include hearing estimates obtained from the distribution of thresholds across frequency from single-unit recordings, evoked potentials from either peripheral or central nervous system structures, or cochlear microphonics. Incorporating these data in the present review strengthens comparisons among different species or groups of animals and between reptiles and birds.

2. Audibility Curves—Absolute Thresholds and Bandwidths

The minimum audible sound pressure at frequencies throughout an organism's range of hearing defines the audibility curve and provides the starting point for understanding hearing in any organism as well as for comparative statements of auditory capability. Over the past 25 years, a number of behavioral audibility curves have become available for birds. Behavioral audibility curves are now available for 38 species of birds, and this database can be extended by another 10 species of birds by including data from physiological recordings. Consideration of similar physiological data for the reptiles (Wever 1978) allows at least a crude estimate of the hearing of over 200 species of reptiles. In characterizing avian and reptilian audibility curves for each species, polynomial functions were fitted to published behavioral or physiological data so that threshold estimates at the same frequencies throughout each species' hearing range can be compared. In this way, average audibility curves can be generated and additional quantitative comparisons among phylogenetic groups are possible.

2.1. Birds

Short of comparing each individual species, there are useful comparisons of the hearing of different groups of birds based on phylogeny and lifestyle as has appeared in earlier reviews (Dooling 1980, 1992a). One such comparison involves the most recently evolved order of birds, the songbirds or Passeriformes (Corvidae, Estrildidae, Emberizidae, Fringillidae, Muscicapidae, Paridae, Passeridae, Ploceidae, Sturnidae, and Turdidae in this sample), the evolutionarily older orders of birds constituting many of the non-Passeriformes (Anseriformes, Caprimulgiformes, Casuariformes, Charadriiformes, Columbiformes, Falconiformes, Galliformes, and Psittaciformes in this sample), and a rather special group of birds that are nocturnal predators, the Strigiformes (Tytonidae and Strigidae in this sample). Roughly equal numbers of species have been tested behaviorally in each of these three broad groups of birds.

The median audibility curves for these three groups of birds are shown in Figure 7.1. The scientific name, common name, and the references appropriate for the audibility curves of each species are given in Table 7.1. These curves illustrate the general trends reported in earlier reviews (Dooling 1980, 1982, 1992a). Birds hear best at frequencies between about 1 and 5 kHz with absolute sensitivity often approaching 0–10 dB SPL at the most sensitive frequency which is usually in the region of 2–3 kHz (Dooling 1980, 1982, 1992a). Nocturnal predators in general hear better than either songbirds or non-songbirds over their entire range of hearing. Songbirds tend to hear better at high frequencies than non-songbirds and non-songbirds

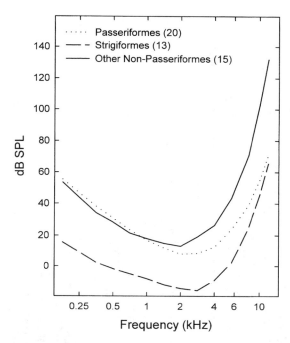

FIGURE 7.1. Median audibility curve based on all of the birds for which data are available (See Table 7.1 for references). Number of species is listed in parentheses.

tend to hear better at low frequencies than songbirds. On average, the limits of "auditory space" available to a bird for vocal communication extends from about 0.5 kHz to 6.0 kHz (the frequency range or bandwidth 30 dB above the most sensitive region of the typical audibility curve). The long-term average power spectrum of most bird vocalizations falls well within this frequency region and there tends to be a correlation between hearing sensitivity at high frequencies and the highest frequencies contained in the species' vocalizations (Dooling 1980, 1982).

This is not to say there are no exceptions to this homogeneous picture of avian hearing. Common pigeons (*C. livia*) may have an unusual auditory sensitivity to very low frequency sounds (Quine 1978; Yodlowski 1980). By some estimates, they may be almost 50 dB more sensitive than humans in the frequency region of 1–10 Hz (Kreithen and Quine 1979). The absolute auditory sensitivity of nocturnal predators, such as barn owls (*Tyto alba*) and great horned owls (*Bubo virginianus*), is probably driven more by their predatory lifestyle and less by correlation with the distribution of energy in their vocal repertoire (Konishi 1973a, 1973b; Van Dijk 1973; Dyson et al. 1998). On the other hand, echolocating birds such as the oilbird (*Steatornis caripensis*) appear to have an audibility curve that is similar to other birds

TABLE 7.1. References for individual audibility curves listed by order and including common name, genus, and species for each bird. Where several references are listed, average audibility curves were compiled

Order	Common Name	Genus and Species	References
Anseriformes	Mallard duck	*Anas platyrhynchos*	Trainer 1946
Apodiformes	Australian grey swiftlet	*Collocalia spodiopygia*	Coles, Konishi, and Pettigrew 1987
Caprimulgiformes	Oilbird	*Steatornis caripensis*	Konishi and Knudsen 1979
Casuariiformes	Emu	*Dromaius novaehollandiae*	Manley, Köppl, and Yates 1997
Charadriiformes	Plains wanderer	*Pedionomus torquatus*	Pettigrew and Larsen 1990
Columbiformes	Pigeon	*Columba livia*	Trainer 1946; Heise 1953; Stebbins 1970; Harrison and Furumoto 1971; Hienz, Sinnott, and Sachs 1977; Goerdel-Leich and Schwartzkopff 1984
Falconiformes	American kestrel	*Falco sparverius*	Trainer 1946
	European sparrowhawk	*Accipiter nisus*	Trainer 1946; Klump, Kretzschmar, and Curio 1986
Galliformes	Bobwhite quail	*Colinus virginianus*	Barton, Bailey, and Gatehouse 1984
	Chicken	*Gallus gallus*	Saunders and Salvi 1993; Gray and Rubel 1985
	Japanese quail	*Coturnix coturnix japonica*	Linzenbold, Dooling, and Ryals 1993; Niemiec, Raphael, and Moody 1994
	Turkey	*Meleagris gallopavo*	Maiorana and Schleidt 1972
Passeriformes	American robin	*Turdus migratorius*	Konishi 1970
	Blue jay	*Cyanocitta cristata*	Cohen, Stebbins, and Moody 1978
	Brown-headed cowbird	*Molothrus ater*	Hienz, Sinnott, and Sachs 1977
	Bullfinch	*Pyrrhula pyrrhula*	Schwartzkopff 1949
	Chipping sparrow	*Spizella passerina*	Konishi 1970
	Common canary	*Serinus canarius*	Okanoya and Dooling 1985, 1987
	Common crow	*Corvus brachyrhynchos*	Trainer 1946
	European starling	*Sturnus vulgaris*	Kuhn et al. 1982; Dooling et al. 1986; Trainer 1946; Konishi 1970
	Field sparrow	*Spizella pusilla*	Dooling, Peters, and Searcy 1979
	Red-billed firefinch	*Lagonosticta senegala*	Lohr and Dooling 1999
	Great tit	*Parus major*	Langemann, Gauger, and Klump 1998; Klump, Kretzschmar, and Curio 1986

Common name	Scientific name	Reference
House finch	*Carpodacus mexicanus*	Dooling, Zoloth, and Baylis 1978
House sparrow	*Passer domesticus*	Konishi 1970
Pied flycatcher	*Ficedula hypoleuca (sp.)*	Aleksandrov and Dmitrieva 1992
Red-winged blackbird	*Agelaius phoeniceus*	Hienz, Sinnot (sp.) and Sachs 1977
Slate-colored junco	*Junco hyemalis*	Konishi 1970
Song sparrow	*Melospiza melodia*	Okanoya and Dooling 1987, 1988
Swamp sparrow	*Melospiza georgiana*	Okanoya and Dooling 1987, 1988
Western meadowlark	*Sturnella neglecta*	Konishi 1970
Zebra finch	*Taeniopygia guttata*	Okanoya and Dooling 1987; Hashino and Okanoya 1989
Psittaciformes		
Bourke's parrot	*Neophema bourkii*	Dooling et al., unpublished data
Budgerigar	*Melopsittacus undulatus*	Saunders and Dooling 1974; Dooling and Saunders 1975a; Saunders, Rintelmann, and Bock 1979; Saunders and Pallone 1980; Okanoya and Dooling 1987; Hashino, Sokabe, and Miyamoto 1988; Hashino and Sokabe 1989
Strigiformes		
Cockatiel	*Nymphicus hollandicus*	Okanoya and Dooling 1987
African wood owl	*Strix woodfordii*	Nieboer and Van der Paardt 1977
Barn owl	*Tyto alba*	Konishi 1973; Dyson, Klump, and Ganger (sp.) 1998
Brown fish owl	*Ketupa zeylonensis*	Van Dijk 1973
Eagle owl	*Bubo bubo*	Van Dijk 1973
Forest eagle owl	*Bubo nipalensis*	Van Dijk 1973
Great horned owl	*Bubo virginianus*	Trainer 1946
Long-eared owl	*Asio otus*	Van Dijk 1973
Mottled owl	*Strix virgata*	Van Dijk 1973
Scops owl	*Otus scops*	Van Dijk 1973
Snowy owl	*Nyctea scandiaca*	Van Dijk 1973
Spotted wood owl	*Strix seloputo*	Van Dijk 1973
Tawny owl	*Strix aluco*	Van Dijk 1973
White-faced scops owl	*Otus leucotis*	Van Dijk 1973

(Griffin 1954; Konishi and Knudsen 1979). Perhaps this is because energy in the echolocation pulses of this species falls in the frequency range of 1.5–2.5 kHz, which is near the region of maximum auditory sensitivity of most birds.

Birds may be distinguished among vertebrates by the remarkable consistency in their auditory structures and in their basic hearing capabilities, such as absolute thresholds and range of hearing. It is intriguing to consider whether the characteristics of the audibility curves of different orders of birds are related to other biological parameters such as a bird's size. Even a cursory inspection of the data suggests that larger birds hear better at low frequencies and poorer at high frequencies than smaller birds. Exploring the relationship between biological parameters such as weight and overall length and various characteristics of audibility curves in this data set such as bandwidth, best frequency, best intensity, low-frequency cutoff, and high-frequency cutoff (Rosowski and Graybeal 1991) reveal that both center-frequency and high-frequency cutoff are significantly and inversely correlated with a bird's size and weight. In other words, small birds (such as many songbirds) hear better at high frequencies and worse at low frequencies than larger birds while larger birds, such as turkeys (*Meleagris gallopavo*) and chickens, hear better at low frequencies and worse at high frequencies than small birds. Perhaps the simplest explanation for these trends is that body size puts a constraint on the low-frequency sensitivity of small birds.

2.2. *Reptiles*

In contrast to the apparent consistency in gross auditory anatomy and behavioral sensitivity among birds, the diversity in reptiles provides a wealth of opportunity for exploring the relationship between auditory structures and hearing capabilities. This same diversity, however, presents quite a challenge in providing a meaningful summary of hearing in reptiles. This problem is further exaggerated when direct behavioral measures of hearing are available for only two species of reptiles (Patterson 1966; Manley, Chapter 4). Thus, what is known about hearing in reptiles comes almost exclusively from cochlear microphonic potential, auditory evoked potential, and auditory nerve single unit data (Wever 1978; Manley 1990). With these methods, particularly the cochlear microphonic methods, hearing estimates are available for a remarkable number of species in the reptile orders Testudines, Rhynchocephalia, Squamata (including the suborders Amphisbaenia, Serpentes, and Lacertilia), and Crocodilia (Wever 1978; Manley 1990; Köppl and Manley 1992).

The problems with estimating what an animal hears from physiological data are well known (see for example, Wever 1978; Manley 1990; Köppl and Manley 1992). Nevertheless, these data constitute the best understanding of hearing in reptiles available and the correlation between physiological and behavioral measures of hearing in reptiles, as in birds, is probably reasonable. As an example, a behavioral audibility curve obtained using a head-retraction response and negative reinforcement in *Trachemys scripta* (the red-eared turtle), shows best sensitivity between 200–600 Hz at a level of about 35–40 dB SPL (Patterson 1966). Cochlear microphonic data are in reasonable agreement with these findings, though absolute sensitivity is higher (best absolute sensitivity 50–60 dB SPL) and the range of hearing is broader and extends further into the high frequencies (Wever and Vernon 1956; Gulick and Zwick 1966; Wever 1978). On the other hand, auditory nerve data from *T. scripta* by Fettiplace and his colleagues (Crawford and Fettiplace 1980; Art and Fettiplace 1984) show an excellent agreement with the audibility curves obtained behaviorally.

Figure 7.2 shows the median audibility curves for the six major orders and suborders of reptiles summarized from Wever (1978) and references therein. These are divided into two groups—species without an external ear opening (Fig. 7.2a) and species with an external ear opening (Fig. 7.2b). As a rule, reptiles without external ears show absolute sensitivities roughly 20–30 dB higher than reptiles with external ears. It is also clear that lizards and crocodilians, two relatively derived groups, exhibit better hearing at high frequencies than other groups of reptiles. A restriction to low-frequency hearing in the more ancestral groups is generally associated with electrical tuning of the hair cells. The extension to higher frequencies in the derived condition may involve several different strategies of mechanical resonance systems (Manley, Chapter 4).

2.2.1. Reptile Orders and Hearing

An anatomical comparison suggests that the hearing organ in turtles and the tuatara (*Sphenodon punctatus*), the only living member of the order Rhynchocephalia, both represent a primitive stage of development, despite an early split in the radiation of reptiles (Starck 1978; Manley 1990). The few data on hearing in the tuatara show hearing sensitivity that is similar to that of turtles (Gans and Wever 1976). Indeed the physiological audibility curves for the tuatara and turtles are very similar except for a 30-dB shift likely due to the lack of an external ear in the tuatara (Fig. 7.2). The similarity in hearing sensitivity between the tuatara and turtles accords reasonably well with the traditional phylogeny of living reptiles (Romer 1966; Carroll 1987). This phylogeny is based primarily on morphological characters and suggests that the descendants of turtles branched first from the reptile lineage during the late Paleozoic or early Mesozoic, about 250 million years ago (Romer 1966). Other extant reptiles fall into two groups: the Archosauria, which includes crocodilians and birds, and the Lepidosauria, which includes the tuatara and squamates (lizards, snakes, and amphisbaenians). Both the turtles, as ancestral reptiles, and the tuataras, as ancestral lepidosaurs, may never have developed specialized hearing

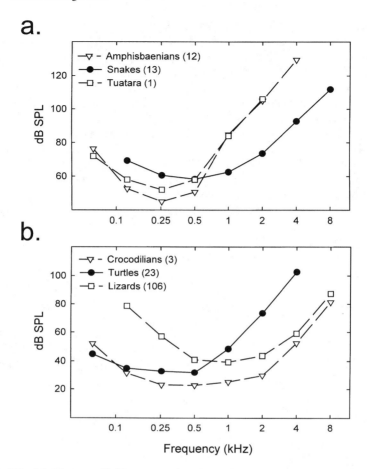

FIGURE 7.2. (a) Mean audibility curves for the major orders and suborders of reptiles without external ear openings. (b) Mean audibility curves for the major orders and suborders of reptiles having external ear openings. Number of species is listed in parentheses. Curves are averages based on the cochlear microphonic data from Wever 1978 and references therein.

abilities. Recent molecular data, however, have thrown these relationships into question (Hedges and Poling 1999), placing the turtles closer to the crocodilians—two groups with considerably different hearing abilities based on physiological data.

The order Squamata is divided into the suborders Amphisbaenia (an ancestral group of fossorial animals), Serpentes (the snakes), and Lacertilia (the lizards) (Starck 1978; Carroll 1987). What few data exist on hearing in amphisbaenians are limited to cochlear microphonic recordings from a dozen species (Wever 1978). Amphisbaenians have no external ear opening, and anatomically the inner ear exhibits extreme uniformity within the group (Gans and Wever 1972). The average cochlear microphonic audibility curves for Amphisbaenia (Fig. 7.2a) suggests relatively poor hearing with best absolute sensitivity at approximately 60 dB SPL and best frequencies between 300–700 Hz. Thus, amphisbaenian hearing also is limited to relatively low frequencies.

Snakes, like amphisbaenians and the tuatara, lack external ear openings, but they are not deaf as is popularly thought. There are cochlear potential data from 19 species of snakes from six families (Wever and Vernon 1960; Wever 1978). The average snake audibility curve based on cochlear microphonic (CM) recordings shows best frequencies in the range of 100–500 Hz with the absolute sensitivity of most species falling in the range of 25–55 dB SPL at best frequency. Hartline and colleagues (Hartline and Campbell 1969; Hartline 1971a, 1971b) recorded auditory evoked and single-unit potentials from the midbrain of several species of snakes in three taxonomic families and showed absolute sensitivities that generally agreed well with data from cochlear potentials for the same species. Estimates of the bandwidth of hearing from midbrain recordings, however, were substantially narrower than those obtained with CM recordings. Bandwidth estimates at 30 dB up from the point of maximum sensitivity ranged from 150 to 450 Hz for the midbrain single-unit data versus 60 to 600 Hz for the cochlear microphonic data in a species of the Viperidae, *Crotalus viridis*. A similar comparison of data sets in a colubrid, *Pituophis melanoleucus*, showed the same trend.

Figure 7.2b shows an average audibility curve for lizards based on the cochlear microphonic data from Wever (1978). Table 7.2 lists the families and number of species within each family that were used to construct this average audibility curve. Lizards tend to be more sensitive at higher frequencies than most other reptiles. In part, because lizards are by far the most diverse group of reptiles with respect to cochlear anatomy (Wever 1978; Manley 1990; Köppl and Manley 1992), care should be taken that

TABLE 7.2. Families of lizards used to produce average lizard curve in Figure 7.2b

Family	Number of Species
Iguanidae	38
Agamidae	7
Anguidae	5
Varanidae	3
Teiidae	7
Lacertidae	9
Gerrhosauridae	6
Scincidae	16
Cordylidae	10

these average data are not over-interpreted. This is particularly true of hearing thresholds at high frequencies. Hair cell regions of opposite polarity in the high-center frequency (CF) segment of the lizard basilar papilla could produce cancellation effects in cochlear microphonic data (Köppl and Manley 1992; Manley, Chapter 4). Manley and his colleagues (Manley 1990; Köppl and Manley 1992) describe several other specific criticisms of the cochlear microphonic data of Wever (1978) with regard to estimating hearing thresholds. Nevertheless, taken as a whole, newer data concentrating on hair cell or single nerve-fiber activity broadly conform to the earlier cochlear microphonic data presented in Wever (1978), particularly with respect to regions of best frequency.

Most recent physiological studies of reptile hearing have concentrated on a few species from different families. These include the granite spiny lizard (Iguanidae (*Sceloporus orcutti*)), alligator lizard (Anguidae (*Gerrhonotus multicarinatus*)), two *Podarcis* species (Lacertidae), the bengal monitor (Varanidae (*Varanus bengalensis*)), tokay gecko (Gekkonidae (*Gekko gecko*)), and bobtail skink (Scincidae (*Tiliqua rugosa*)) (Köppl and Manley 1992). Generally, absolute threshold estimates based on cochlear microphonic potentials are about 20 dB higher and the hearing range estimates are broader when compared with estimates based on other physiological measures.

Because the data for lizards are so diverse, several families were intentionally omitted from the average audibility curve summarized in Figure 7.2b. These highly variable families fell into two groups. Summary audibility curves for one group, the geckos, are shown in Figure 7.3a. Since the geckos represent the single most diverse family of reptiles from the perspective of cochlear morphology and physiology, the overall shape and sensitivity of audibility functions based on cochlear microphonic data should probably not be taken as representative of any particular species (Manley 1990). Audibility for a second group of families with clearly divergent middle and inner ear structure are shown in Figure 7.3b. The divergent features exhibited in these families range from the absence of tympanic membranes to more dramatic modifications such as the loss of the round window in the Chamaelionidae and Anniellidae (Wever 1978). Although hearing in the lizards is quite variable, some species probably do hear quite well both in terms of absolute sensitivity and in terms of range of hearing.

The data for the geckos *Gekko gecko* and *Coleonyx variegatus* stand as exceptions to the general similarity, at least in shape, of the different physiological data sets for lizards. Auditory threshold estimates for *G. gecko* are summarized in Manley (1972) and Eatock et al. (1981), and are quite variable depending on the recording technique used and the particular study. The data here suggest that cochlear microphonic thresholds overestimate thresholds based on earlier auditory nerve fiber and cochlear nucleus unit recordings at high frequencies (above about 800 Hz) by about 20–30 dB. Recent eighth nerve recordings, however, suggest a pattern more similar to that observed with cochlear microphonic audibility curves (Sams-Dodd and

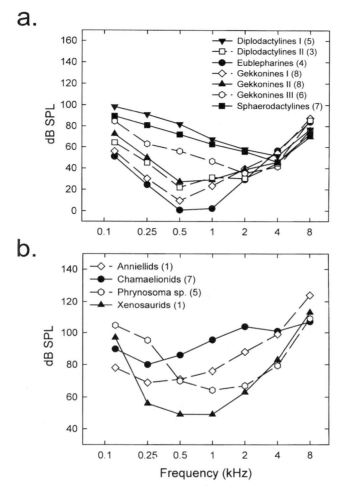

FIGURE 7.3. (**a**) Average audibility curves of Gekkonid subgroupings according to Wever 1978. (**b**) Average audibility curves for lizard families having divergent middle ear types and/or no external ear (Wever 1978). Number of species is listed in parentheses. Curves are averages based on cochlear microphonic data.

Capranica 1994). Still, sensitivities at some frequencies may have been affected by the methods used in this study (see Manley, Chapter 4). Similarly, single-unit thresholds from neurons in the brain stem of *C. variegatus* (Suga and Campbell 1967) showed less sensitivity at the best frequency (between 800–2000 Hz) but were considerably more sensitive at high frequencies (above 2 kHz) compared with cochlear microphonic thresholds (Wever 1978). This comparison reinforces the caveat that cochlear microphonic data may greatly overestimate thresholds especially at high frequencies for species in this group.

Finally, anatomical data from studies of the inner ear also support a close evolutionary relationship between crocodilians and birds (Manley 1990). The crocodilian-bird lineage is thought to have evolved independently from other groups of reptiles for over 200 million years. Relatively few studies exist on their hearing ability, however, and only three species have been examined in this regard. Cochlear microphonic data exist for *Caiman crocodilus*, *Alligator mississippiensis*, and *Crocodylus acutus* (Wever and Vernon 1957; Wever 1971) and audibility curves from these data are very similar for all three species, showing both good absolute sensitivity and a relatively broad range of hearing. Best frequencies are between 200–1000 Hz with a level of sensitivity approaching that observed in most birds (5–35 dB SPL). Primary auditory fibers of *C. crocodilus* were recorded by Manley (1970) and by Klinke and Pause (1980) and these results accord relatively well with those of Wever (1971), both in terms of absolute sensitivity and frequency range. Absolute sensitivity at best frequency differed by no more than 15 dB between these studies, while audibility curves based on eighth nerve recordings were slightly narrower than those based on the cochlear microphonic data. The caiman audibility curve becomes even more like that of a bird when one factors in the higher ambient temperatures characteristic of its natural habitat (see below).

2.2.2. Effects of Temperature on Hearing in Reptiles

Ambient temperature has a strong influence on the hearing physiology of reptiles (see also Manley, Chapter 4). Campbell (1969) measured auditory evoked potentials in several species of lizards and found an overall decrease in sensitivity as temperatures were decreased from those considered optimal for the animal. There were no consistent changes in the frequency of best sensitivity with a change in temperature, however. Wever (1978) presented several cases in which cochlear microphonics were recorded at different temperatures and obtained generally similar results, though changes in sensitivity were often less pronounced. Thresholds based on compound evoked potentials and single units in the snake midbrain also showed a decrease in sensitivity with suboptimal temperatures. In addition, the characteristic frequency of single units decreased in a linear fashion with a decrease in temperature. Similarly, auditory nerve afferents from *C. crocodilus* (Smolders and Klinke 1984) and *G. gecko* (Eatock and Manley 1981) showed a decrease in sensitivity and characteristic frequency as temperature was decreased from optimal. Interestingly, sharpness of tuning, as measured with Q_{10dB} was not dramatically affected. While the change in characteristic frequency in *Caiman* was linear, it was not in *Gekko*, with temperature changes nearer the species optimum having less of an influence on a single unit's characteristic frequency. In a related study, pigeon (*C. livia*) auditory neurons also showed a decrease in sensitivity and characteristic frequency with a temperature decrease (Schermuly and Klinke

1985), demonstrating that this effect is not limited to ectotherms. Generally, individual cells in the auditory system show a decrease in characteristic frequency and a decrease in sensitivity as temperature is decreased from the species optimum. The implication for such changes to an animal's hearing ability is considerable, given the changes in temperature and metabolic rate that many reptiles undergo on a daily basis.

3. Masking and Frequency Analysis in Birds

While the audibility curve defines the active acoustic space within which birds and reptiles must operate, hearing is obviously more complicated than simply detecting sound. In the case of birds, a large number of creative psychophysical tests over the years have explored how birds discriminate both simple and complex sounds. These tests are sometimes purposely designed to parallel the kinds of acoustic problems birds face in their natural habitat, including the need to locate conspecifics, to discriminate among potential mates, to defend their territories, to avoid predators, and to locate prey. These auditory discrimination tasks involve very complex sounds and, certainly in nature, these tasks are performed in a noisy environment. For this reason, the ability to detect whether or how a sound has changed against a background of noise is critically important. Results from masking studies, in particular, have direct relevance for investigations of vocal transmission distance in natural or noisy environments. Some of these masking studies provide insights into frequency analysis by the avian auditory system and are worth considering in relation to other measures of frequency analysis.

3.1. Critical Ratios

The simplest masking experiments to conduct in animals are those measuring the threshold for a pure tone in the presence of a known level of background noise. The ratio between the power in the pure tone at threshold and the power per Hertz (spectrum level) of the background noise is called the critical ratio. Critical ratios have obvious implications for hearing under natural conditions including the perception of vocal signals, the range over which vocal signals may be heard, and the evolution of detection mechanisms. In a more ecological context, critical ratios might aid in understanding the selective pressures driving territory sizes and the extent to which anthropogenic noises may interfere with the acoustic communication of birds in their natural habitats.

Critical ratio data have now been obtained for 13 species of birds including songbirds, non-songbirds, and even nocturnal predators. Typically, there is an orderly increase of about 3 dB in critical ratio with each doubling of frequency over a frequency range of two to three octaves. In mammals, this phenomenon is related to the mechanics of the peripheral auditory system and the logarithmic organization of traveling wave maximum displacement along the basilar membrane (Békésy 1960; Greenwood 1961a, 1961b; Buus et al. 1995). The same may be true of birds, although frequency maps from single-unit recordings suggest a more complicated story (Gleich and Manley, Chapter 3). Figure 7.4 shows the critical ratio functions for 13 species of birds. Ten of these species, shown as an average curve, follow quite closely the typical pattern of approximately a 2–3 dB/octave increase in signal-to-noise ratio. This is the pattern characteristic of critical ratio functions in mammals. Figure 7.4 also shows critical ratio functions shown separately for three avian species that depart from this general trend: the

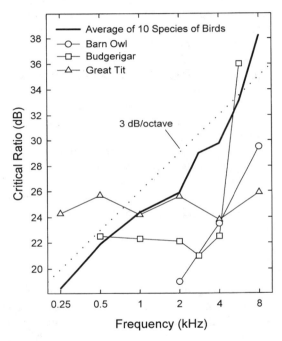

FIGURE 7.4. Average critical ratios from ten species of birds (canaries, *Serinus canarius*; cockatiels, *Nymphicus hollandicus*; zebra finches, *Taeniopygia guttata*; pigeons, *Columba livia*; song sparrows, *Melospiza melodia*; swamp sparrows, *Melospiza georgiana*; starlings, *Sturnus vulgaris*; red-winged blackbirds, *Agelaius phoeniceus*; brown-headed cowbirds, *Molothrus ater*; and firefinches, *Lagonosticta senegala*) are from Okanoya and Dooling 1987, 1988; Hienz and Sachs 1987; Dooling et al. 1986; Langemann, Klump, and Dooling 1995; Lohr and Dooling 1999. Barn owls (*Tyto alba*) are from Dyson, Klump, and Gauger 1998. Budgerigars (*Melopsittacus undulatus*) are from Dooling and Saunders 1975a; Dooling and Searcy 1979; Hashino and Sokabe 1989; Hashino, Sokabe, and Miyamoto 1988; Okanoya and Dooling 1987; Saunders, Rintelmann, and Bock 1979. Great tits (*Parus major*) are from Langemann, Gauger, and Klump 1998.

budgerigar (*M. undulatus*), the great tit (*Parus major*), and the barn owl (*T. alba*). These three species, in different ways, show a significant departure from the pattern observed in other birds and other vertebrates. One can only speculate as to the adaptive value of these divergent critical ratio functions. Contact calls in budgerigars are used for long distance communication and recognition. The energy in these calls falls exclusively in the frequency region with the smallest critical ratios. Perhaps the auditory system may have evolved design features for communication over long distances or under noisy conditions (Klump 1996).

In the case of the great tit (*P. major*), critical masking ratios are relatively flat over the entire range of hearing (Langemann et al. 1998). This means that at high frequencies, the great tit has unusually low critical masking ratios compared to other birds, suggesting an ability to communicate very efficiently at high frequencies. Interestingly, the great tit produces an aerial predator alarm call that is pure tone-like and falls in the range of 8 kHz. Presumably this call has been selected to be inaudible to the great tit's chief predator, the European sparrowhawk (*Accipiter nisus*) (Klump et al. 1986). Finally, in the case of the barn owl (*T. alba*), critical ratios are extremely small at 2 kHz and increase at the rate of 5 dB/octave rather than 2–3 dB/octave as in other birds. What is different about barn owls is that this frequency region is used by the owls in localizing and capturing prey (Konishi 1973a). One supposes that the unusually small critical ratios at these frequencies are the result of selection pressures related to this predatory lifestyle.

3.2. Directly Measured Critical Bands

The notion that the auditory system behaves as a bank of bandpass filters under certain masking conditions has been widely accepted in hearing since Fletcher's (1940) experiments measuring the threshold of pure tones in noises of different bandwidths. Critical bands have been measured directly by band-widening techniques in only two species—the budgerigar (*M. undulatus*) (Saunders et al. 1978a) and the starling (*Sturnus vulgaris*) (Langemann et al. 1995). For starlings, directly measured critical bandwidths and critical ratio bandwidths are almost identical (Buus et al. 1995; Langemann et al. 1995). Moreover, these estimates of the starling's auditory filter are in close agreement with estimates from the CMF (critical modulation frequency) (Klump, pers. comm.). Figure 7.5 shows the results of these measures for the starling and the budgerigar. Note that all of these measures are highly consistent in showing the bandwidths of the starling auditory system double with each doubling of frequency over the most important frequencies of hearing for this bird. In the budgerigar, on the other hand, the critical band function and the critical ratio function have roughly similar shapes but neither increases at the rate of 3 dB/octave (Dooling and Searcy 1979; Saunders et al. 1979; Okanoya and Dooling

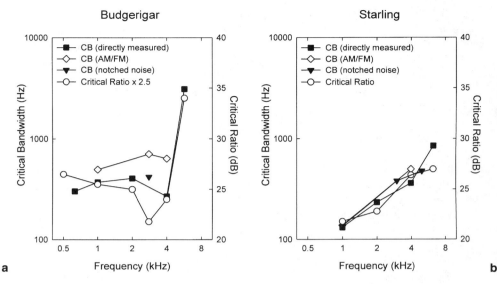

FIGURE 7.5. (**a**) Critical bands and critical ratios for budgerigars (*Melopsittacus undulatus*) are from Saunders, Denny, and Bock 1978; Saunders, Rinlelmann, and Bock 1979. Critical bands from detection of AM and FM are from Nespor, Dooling, and Triblehorn 1996. Critical bands from notched noise maskers are from Lin, Dooling, and Dent 1997. (**b**) Critical bands and critical ratios for starlings (*Sturnus vulgaris*) are from Langemann, Klump, and Dooling 1995. Critical bands from detection of AM and FM are from Klump (pers. comm.). Critical bands from notched noise maskers are from Marean et al. 1998.

1987). Moreover, in contrast to the starling, directly measured critical bands appear to be about 2.5 times larger than bandwidths estimated from critical ratios, as is the case with humans (Dooling and Searcy 1979).

For different reasons, both the critical ratio and the directly measured critical bands have been criticized as estimates of the size of the auditory filter (see, for example, Patterson and Moore 1986). In both cases, for instance, such measures provide an indication of the size but not the shape of the auditory filter. Moreover, there are problems with off-frequency listening, or detection of signals through filters that are not centered at the signal frequency. For these reasons and more, other measures of the shape and width of the auditory spectral filters have gained favor over the years.

3.3. Filter Bandwidths Measured by Notched-Noise Maskers

The notched noise method of estimating auditory filter size and shape circumvents many of the problems of the above methods. Here, the threshold of a pure tone signal is measured as a function of the width of a spectral notch in a broadband noise masker. Assuming the listener's auditory filter is centered on the pure tone, increasingly more noise will pass through the auditory filter as the width of the notch is narrowed. Notched noise methods have now also been used to measure auditory filter widths in two species: starlings, *S. vulgaris*, and budgerigars, *M. undulatus*. In the starling, Marean et al. (1998) measured auditory filter widths at 1, 3, and 5 kHz and obtained estimates of the filter width using the Polyfit procedure of Rosen and Baker (1994). The ERBs (equivalent rectangular bandwidths) (Glasberg and Moore 1990) for the starling are about 169, 404, and 550 Hz at 1, 3, and 5 kHz respectively. These bandwidths compare well with the filter band-widths of the starling obtained from the critical ratio, the critical band, and also the CMF (see below) at 1 kHz, and are only slightly larger at 3 and 5 kHz (Buus et al. 1995). A similar procedure has also recently been used to measure auditory filter widths in the budgerigar (Lin et al. 1997) at its frequency of best hearing—2.86 kHz. Budgerigars show only a slightly larger ERB than they do when critical bandwidths are measured by band narrowing procedures.

3.4. Detection of Amplitude Modulation and Frequency Modulation

The vocalizations of many bird species are tonal and characterized by a species-typical pattern of amplitude and frequency modulation (Kroodsma and Miller 1996). It is of some interest, therefore, to know how well birds can hear such modulation. There are also reasons directly related to the mechanisms of hearing. Previous work with human listeners (Zwicker 1952) suggests that an estimate of the size of the auditory filter may be reflected in the differential sensitivity to detection of amplitude modulation (AM) and frequency modulation (FM) in tonal stimuli. In humans, the detectability of AM is relatively constant across a wide range of modulation frequencies and only decreases at very high modulation frequencies. The detectability of FM is relatively poor at very low modulation frequencies and improves with increasing modulation frequency. At low modulation frequencies, the detectability of AM is better than the detectability of FM and the modulation frequency at which the detectability of AM and FM become similar is called the critical modulation frequency, or CMF (Schorer 1986).

These differences in detectability between AM and FM have led to speculations about the mechanisms of frequency perception in the vertebrate auditory system. Under some stimulus conditions, such as near threshold where the modulation index is small (i.e., less than 0.3), an AM tone and a FM tone will have components (i.e., carrier and two sidebands) that are identical in frequency and amplitude but differ only in the relative phase of the components. When the sidebands of the carrier that are generated by this modulation fall outside the critical bandwidth, then FM and AM become equally detectable. This suggests that humans are sensitive to phase effects only when the spectral components lie within a single critical band (Zwicker 1952; Schorer 1986). Though the mechanisms underlying this phenomenon are unresolved (Moore and Sek 1994a, 1994b), it is thought that an estimate of the auditory filter width can be obtained by doubling the CMF (where the two functions intersect).

The results of AM and FM detection for budgerigars (*M. undulatus*) at 2.86 kHz (the most sensitive point in the audibility curve) give a CMF estimate of 354 Hz (Nespor et al. 1996). This would correspond to a critical bandwidth of about 700 Hz for the budgerigar at 2.86 kHz, which is considerably larger than bandwidth estimates from either critical ratios, directly measured critical bands, or notched-noise procedures (Saunders et al. 1978, 1979). Despite the frequency (Lin et al. 1997) and amplitude-modulated characteristics of their tonal vocalizations, thresholds for detecting FM or AM in budgerigars are not much different than those reported for humans. As with humans, FM thresholds are worse than AM thresholds at low modulation frequencies and become more similar at high modulation frequencies. Interestingly, AM thresholds for budgerigars do not drop off at very high modulation rates as they do for humans. Perhaps birds are more sensitive to temporal fine structure generated when the modulation rate becomes a significant fraction of the value of the carrier frequency.

The CMF also has been measured at several frequencies in the starling (Klump, pers. comm.). Thus, it is possible in two species of birds to compare measures of the auditory filter from tone thresholds in broadband noise (critical ratios), directly measured critical bandwidths (band narrowing procedures), masking by spectrally notched noises, and the CMF. These comparative results are shown in Figure 7.5. In the starling, *S. vulgaris*, there is a close correspondence between bandwidth estimates obtained from critical ratios, critical bands, notched noise procedures, and the CMF. Moreover, these measures all show a monotonic increase of about 3 dB/octave in the starling's range of best hearing. By contrast, estimates for budgerigars, *M. undulatus*, show a pattern that is different in several ways. First, bandwidth estimates from critical ratios underestimate the size of the critical band by a factor of 2.5 as they do in humans. Second, there is no 3 dB/octave increase in bandwidth at lower frequencies (500–3,000 Hz). Third, bandwidth estimates taken from the CMF are considerably larger than other estimates from other methods. The significance of these differences between species and between estimates of filter bandwidths (in the case of budgerigars) is not yet clear. These results do underscore how much there is yet to learn about frequency analysis by the avian auditory system.

3.5. Psychophysical Tuning Curves and Nonsimultaneous Masking

Another measure of frequency analysis comes from psychophysical tuning curves. As with critical ratios, psychophysical tuning curves and other tone-on-tone masking data are fairly easy to obtain in animals. They also provide some indication of the shape of the auditory filter and, where direct comparisons have been made, are strongly correlated with these other measures of filter bandwidths. There have been a number of such studies in birds using several different methods. These include traditional studies using a fixed, low-level tone in a simultaneous masking paradigm (Saunders and Rintelmann 1978; Saunders et al. 1979), using fixed, high-level maskers, or using variable tones in a forward or backward masking paradigm (Dooling and Searcy 1985a). The data for birds suggest that their auditory filters are more symmetrical than in mammals, more narrowly tuned in the region of high sensitivity, and vary less as a function of level than in mammals (Saunders and Rintelmann 1978; Saunders et al. 1979; Dooling and Searcy 1985a). In general, the behavioral measures of tuning curves and their consistency across level parallel the findings from single unit studies showing narrower, more symmetrical tuning, and less change over level than in mammals (summarized in Manley 1990).

3.6. Perception of Spectral Contrast

A final measure of frequency selectivity comes from measures of spectral contrast using a class of sounds having rippled spectral profiles. These sounds, called log-rippled noises, are becoming popular because of their noted parallel with sine wave gratings used in studies of vision. The sounds are generated either by imposing a sinusoidal spectral envelope onto broadband noise in the frequency domain, or by algebraically adding frequency components with amplitudes determined by a sinusoidal envelope. The spectral envelope is sinusoidal when frequency is represented on a logarithmic scale and sinusoidal envelopes are expressed in units of cycles/octave. In vision, this class of stimuli purports to allow for a linear systems analysis of visual function as long as the underlying principle of linearity of summation is not violated (De Valois and De Valois 1988). Whether a linear systems analysis approach will prove useful for understanding complex signal processing in the auditory system remains to be seen. This approach has recently been used for comparative explorations of auditory system function at different levels and in different species (Shamma and Versnel 1995; Shreiner and Calhoun 1995) and there are strong arguments for studying rippled spectra in humans based on their parallels with the vowel sounds of human speech (Hillier 1991; Summers and Leek 1994).

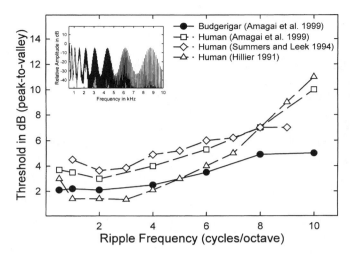

FIGURE 7.6. Inset shows a log-rippled stimulus at two cycles per octave and 15 dB of modulation depth. Figure shows thresholds for budgerigars (*Melopsittacus undulatus*) (*closed circles*) and humans (*Homo sapiens*) from Amagai et al. 1999.

A recent experiment compared thresholds for modulation detection in budgerigars and humans (Fig. 7.6). Both budgerigars (*M. undulatus*) and humans show best thresholds at low ripple frequencies where the spacing of peaks in the spectrum is large relative to the critical bandwidth. Budgerigars show thresholds that are better than those of humans at high ripple frequencies (Amagai et al. 1999). If modulation detection is affected by ripple frequency relative to the critical bandwidth of the auditory system, these species differences may be a reflection of the smaller critical ratio bandwidths of budgerigars at 3 kHz compared with humans (Dooling and Saunders 1975a).

4. Auditory Frequency and Intensity Discrimination

4.1. Frequency Discrimination

Field studies have long shown that frequency is an important cue for song recognition in many species (Falls 1963; Bremond 1968, 1975; Fletcher and Smith 1978; Nelson and Marler 1990). Psychophysical studies of frequency discrimination in a number of avian species support the notion that birds are quite sensitive to changes in the frequency of acoustic signals. Figure 7.7 shows frequency difference limens across a broad frequency range for seven species of birds. Frequency difference limens in the range of 1–4 kHz show that, on average, birds can probably discriminate better than about a

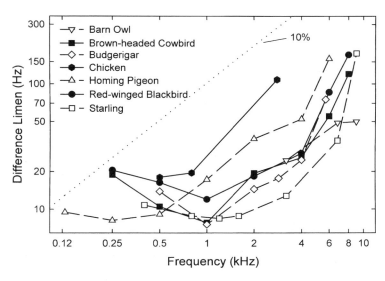

FIGURE 7.7. Average frequency difference limens from seven species of birds. Brown-headed cowbirds (*Molothrus ater*), red-winged blackbirds (*Agelaius phoeniceus*), and homing pigeons (*Columba livia*) are from Sinnott, Sachs, and Hienz 1980. Chickens (*Gallus gallus*) are from Gray and Rubel 1985. Budgerigars (*Melopsittacus undulatus*) are from Dooling and Saunders 1975a. Barn owls (*Tyto alba*) are from Quine and Konishi 1974. Starlings (*Sturnus vulgaris*) are from Kuhn, Leppelsack, and Schwartzkopff 1980. *Dashed line* represents an increase of 10%.

1% change in frequency. Human thresholds for detecting changes in frequency are usually slightly better than those of birds at all test frequencies (Wier et al. 1977). However, there is some evidence that frequency difference limens in birds may change less as duration decreases than is the case for humans (Moore 1973; Dent and Dooling 1998). It remains surprising that birds do so well given that their basilar papilla is an order of magnitude shorter than the human basilar membrane (Schwartzkopff 1973; Manley and Gleich 1991).

4.2. Intensity Discrimination

The greater number of hair cells across the width of the bird basilar papilla compared to the mammalian basilar membrane has led to speculation that the avian ear may be capable of a greater power of intensity discrimination compared to mammals (Pumphrey 1961). Psychophysical data from five species of birds suggest that this is not the case, however. Intensity difference limens indicate quite clearly that birds are not particularly sensitive to changes in intensity—certainly not more sensitive than are humans (see, for example, Dooling and Saunders 1975b; Hienz et al. 1980; Okanoya and

Dooling 1985). Birds show intensity discrimination thresholds that are similar to those measured in other nonhuman vertebrates (Fay 1988), falling in the range of 1–4 dB.

It is perhaps not surprising that birds do not excel in intensity discrimination when one considers a bird's lifestyle and habitat. For most birds, nonhomogeneous environments along with constant variation in distance, elevation, and azimuth between source and receiver must surely render sound intensity an unreliable acoustic dimension with which to encode species or individual identity, motivational state, or even distance. However, the kind of repetitive amplitude modulation so characteristic of bird song (as, for example, the trilled portion of a song) may be useful for enhancing information encoding by signal redundancy (Richards and Wiley 1980).

4.3. Complex Sound Discrimination

Perhaps more than in any other vertebrate, birds present the hearing scientist with some predicaments familiar from human work. What is the relation between the perception of simple sounds such as tones and noises, the perception of complex natural sounds such as speech, and the perception of complex synthetic sounds often modeled after speech, music, or other complex, natural sounds? What is the relation between hearing and listening? What is the relationship between synthetic and analytic listening? While these problems have yet to be solved even for humans, recent experiments with birds are now providing some insights into these issues from a comparative perspective.

Perhaps it is worth reflecting that birds occupy a special place in this framework because there has been a long-standing fascination with the fact that they produce, learn, and use complex species-specific acoustic signals for communication as do humans. There is probably more known about the perception of complex sounds in birds under both natural and laboratory conditions, including species-specific vocalizations, than in any other nonhuman organism. As with human speech, however, perceptual approaches can be aimed at several levels. Stimuli can range from individual phonemes and words to complete sentences and running speech. The present review focuses primarily on the limits of auditory detection and resolving power in birds. These thresholds have relevance for both the mechanisms of hearing in birds and the perception of complex sounds used in communication and species and individual recognition.

The topic of complex sound perception in birds using natural stimuli alone would require an entire volume, in part because, as is the case with humans, it would engage processes such as memory, learning, stimulus familiarity, and attention, to name a few. It is also true that complex synthetic sounds such as rippled noises and harmonic complexes are proving increasingly useful as probes in assessing the spectral or temporal resolution of the auditory system. What follows below is not intended to be exhaustive, but rather an attempt to highlight and provide an overview of some of the more recent areas of focus in the study of the perception of complex sounds by birds.

4.3.1. Comodulation Masking Release

Complex sounds often cover a broad frequency range and thus involve more than one filter bandwidth in the auditory system. The auditory system has the ability to compare the outputs of widely separated auditory filters. The simplest of these cases is related to the situation described above where a pure tone is masked by a noise. These tests include comodulation masking release (CMR) (Hall et al. 1984) and modulation detection interference (Yost 1992). While critical bands and critical ratios have shown that the energy surrounding a signal's frequency is most effective in masking that signal, recent work has shown that under certain conditions a considerable amount of unmasking can occur when modulated noise is added outside the critical bandwidth (Hall et al. 1984). In other words, the human auditory system is capable of using spectrally distant information to enhance detection. The usual paradigm for studying CMR is one in which a pure tone signal is masked by a narrow band noise. Narrow band noises are characterized by fluctuations in overall amplitude over time. A second band of noise, outside the spectral region of the target band of noise, can improve the signal-to-noise ratio if it is modulated in parallel with the target band. The data from these experiments show that information in one frequency region can be used to aid detection of a signal in noise in a totally separate frequency region.

Working with starlings (*S. vulgaris*), Klump and his colleagues (Klump and Langemann 1995; Hamann et al. 1999) have now shown a similar effect in birds. Figure 7.8 shows the release from masking of a pure tone when the masking noise is coherently amplitude modulated (comodulated) over the total range of the spectrum compared to masking by an unmodulated noise of the same bandwidth and overall energy. There is over 10 dB of masking release in this situation with the broadest masker bandwidths, which is on the order of that found in humans. Moreover, masking release increases with increasing masker bandwidths, as is the case with humans. The ability to make across-filter comparisons of temporal envelopes may be a general vertebrate feature of auditory pattern analysis, which enhances the ability to extract signals from noise and for separating different sound sources.

4.3.2. Pattern Perception and the Perception of Calls and Songs

Bird vocalizations include some of the most complex communication signals known, ranging from short notes and calls to longer, more elaborate songs. These signals convey important information used in individual and species recognition, mate selection, territorial defense, and predator avoidance (for a recent comprehensive treatment of these issues see Kroodsma

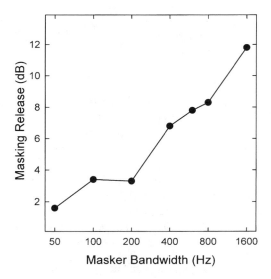

FIGURE 7.8. From Klump and Langemann 1995. Amount of masking release is shown as a function of the masker bandwidth at a test frequency of 2 kHz. Modulator was a 50 Hz lowpass noise. Graph shows median values for five starlings determined after the release from masking in each subject was calculated separately.

and Miller 1996). There is no doubt that birds can and do perceive the complexity in these vocalizations—especially their own species-specific vocalizations. Not surprisingly, numerous studies have shown that the acoustic categories of songs and calls that emerge from even a cursory examination of a species' repertoire can also form natural perceptual categories (Dooling et al. 1987). In some cases, it is clear that these categories are influenced by learning and can be sustained by extremely fine grain, multidimensional acoustic differences between categories (Brown et al. 1988). This is demonstrable not only at the level of individual calls and individual elements and notes that make up longer, more complex songs, but also at the level of organization of songs (Farabaugh and Dooling 1996).

A large body of evidence from field studies and song learning studies shows that birds are sensitive to suprasegmental (e.g., number and ordering of notes and syllables, tempo, and number of syllable repetitions) features of their species' songs (Kroodsma and Miller 1982, 1996). However, it has only been recently shown that this form of auditory pattern perception has been addressed in laboratory studies under highly controlled conditions. Working with starlings, Hulse and his colleagues have shown that starlings are quite adept at the kind of serial pattern perception that appears to underlie such phenomena as music perception in humans. In categorizing rising and falling pitch patterns, starlings use both relative and absolute pitch features of the patterns but, in contrast to humans, absolute frequency cues generally predominate (Hulse and Cynx 1985, 1986; Page et al. 1989; MacDougall-Shackleton and Hulse 1996).

Along similar lines, budgerigars (*M. undulatus*) were tested on rising and falling pitch patterns composed of five pure tones, each 25 ms in duration, played in sequence with a 5 ms intertone interval. Results showed that budgerigars are quite sensitive to frequency range but relatively insensitive to pitch contour (Dooling et al. 1987). That is, they were relatively insensitive to whether successive tones in the complex rose in frequency, stayed the same, or fell in frequency. This is the exact opposite of what humans report when listening to these patterns (Dooling et al. 1987). These results address the differential salience of particular features of complex tonal patterns, show that some easily discriminable changes in frequency are more salient than others, and indicate that this effect is context dependent.

A more recent study examined the ability of budgerigars to discriminate frequency changes in single elements of a seven-element tonal sequence. This work was modeled after work on humans exploring the discriminability of frequency changes in single elements of word-length, 10-tone patterns (Watson et al. 1975, 1976). In humans and budgerigars, knowing when the frequency change would occur (low uncertainty trials) in the sequence resulted in discrimination thresholds that were as good as thresholds for single tones presented in isolation (Dooling and Saunders 1975a). If the location of the frequency change was changed from trial to trial (high uncertainty), however, discrimination thresholds were much worse for the humans but not for budgerigars (Dent and Dooling 1998). The birds were unaffected by changing the level of uncertainty in the task, suggesting that they could be hearing the tonal patterns as a whole, perhaps as humans hear words. If so, then the roles of learning strategies or differences in the focusing of auditory attention also become important.

4.3.3. Auditory Scene Analysis

The pioneering work of Bregman and his colleagues (Bregman and Campbell 1971; Bregman 1990) as well as others, show that humans routinely segregate concurrent sounds into separate auditory objects. These auditory objects can be identified by a unique set of acoustic features such as spatial location (the well-known "cocktail party effect"), spectral quality, pattern, and pitch. It is a very interesting question as to whether such a phenomenon occurs in a nonhuman animal. This question has recently been directly tested in starlings using both natural and synthetic stimuli. Hulse and co-workers (1997) showed that starlings (*S. vulgaris*) could be trained to identify a sample of one species' bird song presented concurrently with a sample of another species' bird song. Moreover, these birds could learn to discriminate among many samples of the songs of two individual starlings and could maintain that discrimination when songs of a third starling were digitally added to both song sets and songs from additional starlings were added as further background distracters (Wisniewski and Hulse 1997). When testing the birds on tonal patterns, starlings were found to maintain discrimination (i.e., segregate) galloping from isochronous tonal patterns in the face of large frequency differences in components of the stimulus pattern (MacDougall-Shackleton et al. 1998). Taken together, these results with starlings using both natural and synthetic stimuli suggest that auditory scene analysis may play a general role in auditory perception in birds and other nonhuman vertebrates that must parse the world into auditory objects.

4.3.4. Speech Perception

Perhaps the most studied complex sounds in the field of auditory perception are those of human speech. Here again, birds provide an interesting contrast on several levels. It is well known that several nonhuman mammals are quite sensitive to the acoustic features of speech sounds that define phonetic categories, suggesting that speech sounds may have evolved to exploit the psychoacoustic sensitivities of the mammalian auditory system (Kuhl 1989). Interestingly, there are a number of birds that are well known for their ability to mimic the sounds of speech with great fidelity, strongly suggesting these birds may hear complex speech sounds as humans do (Banta and Pepperburg 1995).

A series of studies have shown that budgerigars (*M. undulatus*), popular cage birds and speech mimics, perceive natural vowel tokens /i/, /a/, /e/, and /u/ in phonetically appropriate categories in spite of variation in talker, pitch contour, and gender (Dooling and Brown 1990; Dooling 1992b). Working with some of the familiar synthetic speech continua, budgerigars have been shown to exhibit perceptual boundaries near the human boundaries for /ba/-/pa/, /da/-/ta/, /ga/-/ka/, /ra/-/la/, and /ba/-/wa/ (Dooling et al. 1989, 1995; Dent et al. 1997b). The perception of speech sound categories is not unique to budgerigars, however. Zebra finches (*Taeniopygia guttata*), starlings (*S. vulgaris*), japanese quail (*Coturnix coturnix japonica*), red-winged blackbirds (*Agelaius phoeniceus*), and pigeons (*C. livia*) also discriminate and categorize speech sounds in a way similar to humans (Hienz et al. 1981; Kleunder et al. 1987; Dooling 1992b; Dooling et al. 1995).

5. Auditory Time Analysis

It is well accepted that the auditory system, perhaps more than any other sensory system, is time oriented (Viemeister and Plack 1993). Traditionally, the important questions for the auditory system concern what is the shortest time interval that can be perceived on the one hand, and how the auditory system might cope with environmentally induced temporal smearing of important biological sounds on the other. The stereotyped, punctate nature of bird song magnifies these issues. Long-standing evidence from several domains including song learning (Greenewalt 1968), cochlear anatomy (Pumphrey 1961; Schwartzkopff 1968, 1973), and single-unit recordings (Konishi 1969) has been used to examine whether birds are capable of an unusual degree of temporal resolving power. Single-unit recordings in both the periphery and the forebrain of starlings (*S. vulgaris*) show similar sensitivity to that found psychoacoustically when measuring temporal resolution (Buchfellner et al. 1989; Klump and Gleich 1991; Gleich and Klump 1995).

As is clear from other reviews of this problem, a fundamental issue here concerns the definition of temporal resolving power. Over the years, several psychophysical approaches have been developed to assess the sensitivity to temporal changes in an acoustic signal while minimizing or eliminating concomitant spectral or amplitude changes. Many of the traditional measures of auditory time resolution involve changes in relatively slow temporal cues such as envelope cues. On these measures, there is no evidence that birds are significantly better than other vertebrates at detecting changes in the temporal aspects of an acoustic signal. In fact, the similarity between avian and mammalian sensitivity to temporal change is quite remarkable. On the other hand, several new measures involving the discrimination of temporal fine structure show that birds do possess some sort of enhanced temporal resolving power.

5.1. Maximum Temporal Integration

Temporal integration usually refers to the relation between the threshold for detection of a sound and the duration of that sound. It is a measure of the auditory system's ability to sum acoustical energy over time. For a number of vertebrates including humans (Fay 1988), the ability to detect the occurrence of a sound improves as the duration of the sound is increased from a few milliseconds up to about 200 ms. This number is known as the time constant (Plomp and Bouman 1959). Increasing the duration of sound beyond 200 ms does not render the sound easier to hear.

Maximum temporal integration has now been measured in six species of birds. These results are shown in Figure 7.9, along with data from humans. Thus, the time constant of the avian auditory system is similar to that found for higher vertebrates (including humans).

5.2. Minimum Temporal Integration—Gap Detection and the Temporal Modulation Transfer Function (TMTF)

The gap detection threshold is a measure of the minimum detectable interval between two sounds, analogous in some ways to the familiar measure of sensory acuity in the visual system involving the minimum separation

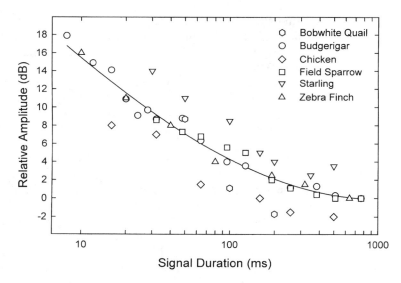

FIGURE 7.9. Maximum temporal integration functions for six species of birds. Data are plotted relative to threshold in decibels of the longest stimulus tested at that frequency. Budgerigars (*Melopsittacus undulatus*) (2.86 kHz pure tones) and field sparrows (*Spizella pusilla*) (2.86 kHz pure tones) are from Dooling 1979 and Dooling and Searcy 1985b. Zebra finches (*Taeniopygia guttata*) (3 kHz pure tones) are from Okanoya and Dooling 1990. Starlings (*Sturnus vulgaris*) (2 kHz pure tones) are from Klump and Maier 1990. Chickens (*Gallus gallus*) (1.6–2.5 kHz pure tones) are from Saunders and Salvi 1993. Bobwhite quail (*Colinus virgianus*) (3.5 kHz pure tones) are from Barton et al. 1984. *Solid line* represents third order regression function.

between two lines. The minimally detectable gap is measured in noise to minimize spectral cues arising from energy splatter when a sound onset and offset is abrupt. The ability to detect a gap in noise has been measured in a number of vertebrates including humans and they all fall, quite consistently, in the range of about 2–4 ms (Fay 1988). Gap detection thresholds in noise have now also been measured in four species of birds as a function of level (Klump and Maier 1989; Okanoya and Dooling 1990; Klump et al. 1998). These thresholds are shown in Figure 7.10. Consistent with the data from other vertebrates, the best thresholds for birds approach the range of 2–3 ms. Interestingly, though, there appears to be much less of an effect of level on gap detection thresholds for birds compared with humans (Plomp 1964). At low sound pressure levels, birds are clearly more sensitive than humans (and other mammals as well) are at detecting gaps in noise.

Discrimination of time intervals marked by tonal stimuli rather than by broadband noises affords a somewhat different view of auditory system function in humans (Williams and Perrott 1972; Divenyi and Danner 1977;

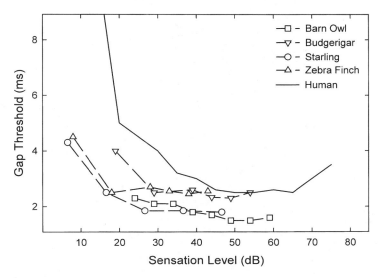

FIGURE 7.10. Minimum temporal integration functions for four species of birds are normalized by sensation levels for broadband noise. Budgerigar (*Melopsittacus undulatus*) and zebra finch (*Taeniopygia guttata*) gap detection thresholds are from Okanoya and Dooling 1990. Starling (*Sturnus vulgaris*) gap detection thresholds are from Klump and Maier 1989. Barn owl (*Tyto alba*) gap detection thresholds are from Klump et al. 1998. Human gap detection thresholds are from Plomp 1964.

Moore and Glasberg 1988; Formby and Forrest 1991). Recent work has shown that changing the frequency of one of the tonal markers, usually the post-gap marker, has a profound effect on the detection or discrimination of the gap (Formby et al. 1996; Hienz et al. 1996). In general, the assumption is that gap detection thresholds worsen as a function of the frequency separation of the first (pre-gap) sinusoid and the second (post-gap) sinusoid because of differential attenuation of the markers by the auditory filter. Otherwise said, as the post-gap sinusoid moves outside of the critical band (channel) centered on the pre-gap sinusoid, information about the end of the gap is diminished. Interestingly, when budgerigars (*M. undulatus*) were tested on this task, they behaved differently than human listeners. Figure 7.11 shows the results of humans and birds tested on this task. Both humans and budgerigars showed gap discrimination thresholds of about 2 ms when the frequency of the first tone was the same as the frequency of the second tone. When the frequency of the post-gap sinusoid was varied either above or below that of the first tone, human performance dropped dramatically, while the performance of budgerigars was much less affected (Amagai et al. 1997). These results suggest differences in temporal processing between birds and humans when the spectral characteristics of the stimuli are processed in different channels.

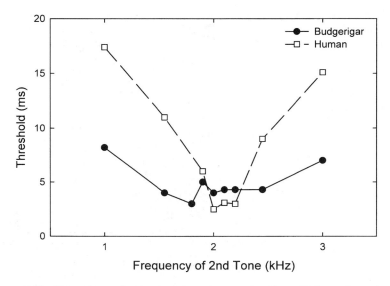

FIGURE 7.11. Detection of silent temporal gaps in sinusoidal markers by the budgerigar (*Melopsittacus undulatus*) are shown. Threshold values for temporal gap detection for budgerigars and humans are plotted against the frequency of the second tone when the first tone was 2 kHz. From Amagai et al. 1997.

The temporal modulation transfer function, or TMTF, is another measure of temporal resolving power that assesses minimum integration time and has some parallels with the flicker fusion frequency threshold in visual psychophysics. This approach has advantages over measures of gap detection, especially for modeling purposes (Viemeister 1979). Using broadband noise to minimize spectral cues, thresholds for the detection of modulation are measured at different modulation frequencies from a few Hertz to several thousand Hertz. As modulation frequency increases, the auditory system has difficulty following the intensity changes and requires a greater depth of modulation for detection (Viemeister 1979).

There are now comparative data for several species of birds on the TMTF. Figure 7.12 shows the results from budgerigars (*M. undulatus*) (Dooling and Searcy 1981), starlings (*S. vulgaris*) (Klump and Okanoya 1991), barn owls (*T. alba*) (Dent et al. 1999c) and humans (*Homo sapiens*) (Bacon and Viemeister 1985). At low modulation frequencies, humans and other mammals (Salvi et al. 1982; Moody 1994) are more sensitive to changes in intensity than are birds. However, as modulation frequency is increased, the ability to detect modulation decreases faster for mammals compared to birds.

There are several ways to interpret these results. One interpretation involves computing a time constant from modulation functions, thereby treating the auditory system as a lowpass filter. With this approach, the

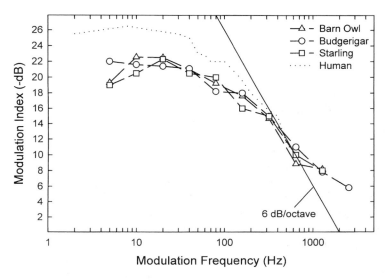

FIGURE 7.12. Temporal modulation transfer functions are shown for three species of birds. Average data from budgerigars (*Melopsittacus undulatus*) for broadband noise are from Dooling and Searcy 1981. Data from starlings (*Sturnus vulgaris*) for continuous broadband noise are from Klump and Okanoya 1991. Average data for barn owls (*Tyto alba*) for gated (800 ms bursts) broadband noise are from Dent et al. 1999c. Average data from humans for a continuous broadband noise are from Bacon and Viemeister 1985. *Solid line* represents a decrease of 6 dB/octave.

lowpass bandwidth (3-dB down point) of the human auditory system would be about 30 Hz, while that of the birds would be about 130 Hz. This results in a time constant for the bird auditory system that is about one-quarter that of the human auditory system. Said another way, the bird's auditory system begins to falter when trying to resolve events happening faster than once every 1–2 ms, while the human auditory system begins to lose sensitivity to changes occurring faster than every 4–5 ms. On the other hand, humans are generally more sensitive to intensive changes than birds (which is evident at low modulation frequencies). Taken together, these species differences conspire in a way that birds and mammals are roughly equal in the ability to detect modulation at high modulation frequencies.

5.3. Discrimination of Auditory Duration

Duration discrimination is another fundamental measure of the perception of auditory temporal abilities. Budgerigars (*M. undulatus*) (Dooling and Haskell 1978), starlings (*S. vulgaris*) (Maier and Klump 1990), and pigeons (*C. livia*) (Kinchla 1970) are able to discriminate a change in the duration of pure tones in the range of about a 10–20% change. These results are slightly worse, but not dramatically so, than typically found in other verte-brates (Fay 1988). One must conclude that birds are not unusually sensitive to changes in duration of acoustic signals presented in isolation.

5.4. Forward and Backward Masking and the Temporal Window

Nonsimultaneous masking refers to masking effects that either precede or follow a masker in time. The mechanisms responsible for nonsimultaneous masking effects are complex and not fully understood. Presumably, when a tone follows a masker, improvement in thresholds is due to a decrease in auditory fatigue following the masker (Plomp 1964). The masking of a tone that precedes the masker suggests more central processes, since the masker obscures some events initiated by the tone but not completed by the time of onset of the masker (Lynn and Small 1977). Nonsimultaneous masking effects have been examined in budgerigars (*M. undulatus*) using both tones and noises. In terms of the time course of masking, for tones preceding a 70-dB masker (backward masking) budgerigars show no masking at inter-vals of 10 ms or greater. For tones following the same masker (forward masking), however, thresholds do not return to unmasked levels until the interval between masker and tone is 100 ms or longer (Dooling and Searcy 1980b). Combining both forward and backward masking has led to a method of measuring the "temporal window," which is loosely analogous to the notched noise method of measuring the spectral filter. Following on work in humans by Moore and his colleagues (Moore et al. 1988) that mea-sured the size and shape of the temporal window, similar measures in budgerigars reveal that the temporal window is roughly the same size as that of humans but more symmetrical (Lin et al. 1998).

5.5. Harmonic Stimuli

All of the above-cited measures of temporal resolution involve detecting changes in relatively slow temporal characteristics of a sound referred to as the envelope characteristics. However, there are other temporal features of complex sounds that can be very subtle and can occur at higher frequen-cies. These are referred to as temporal fine structure. The basic property of most vibrating objects is that they have a fundamental frequency and many harmonics. For this reason, harmonic complexes are proving to be a useful class of sounds for probing the temporal bases of a number of perceptual phenomena in humans including pitch, timbre, speech perception, music perception, and auditory scene analysis, among others (Yost 1992). Sounds with similar characteristics are now being used to study hearing in birds. While it is the melodic, tonal quality of many bird vocalizations that usually catches the attention, there are a number of species, such as the zebra finch

(*T. guttata*), that produce primarily harmonic songs and calls. For this reason, zebra finches, along with other birds, have now been tested on various kinds of harmonic complexes.

5.5.1. Cosine or Linear-Rippled Noise

Perhaps the simplest measure involving these complex stimuli uses broadband noise, where the threshold for detecting modulation at harmonic multiples of a fundamental frequency is measured. These sounds are created by delaying a portion of wideband noise and adding it back to the undelayed original noise. Such sounds are called linear-rippled noise, repetition noise, or cosine noise (Bilsen et al. 1975; Yost and Hill 1978; Fay et al. 1983). This stimulus has spectral peaks at integer multiples of $1/T$, where T is the time of delay. Varying the attenuation of the delayed sound before adding it back to the undelayed original controls the amount of spectral modulation in linear-rippled stimuli. For humans and probably goldfish, the salience of repetition pitch increases as the depth of modulation increases (Fay et al. 1983; Yost et al. 1996). The details of linear-rippled noise processing are important for theories of vertebrate hearing because models and theories of pitch must be able to account for the pitches and pitch-strength of rippled noises. Linear-rippled noises have been used in studies of a number of vertebrates. The prevailing notion is that the auditory system is performing a time-domain waveform analysis similar to an autocorrelation (Fay et al. 1983; Shofner and Yost 1994, 1995; Yost 1996; Yost et al. 1996). Recently, these stimuli have been used in the investigation of hearing in the budgerigar (*M. undulatus*) (Amagai et al. 1999). Figure 7.13 shows modulation detection thresholds for budgerigars tested on broadband, linear-rippled noises. For comparison, results are also shown for humans and chinchillas (*Chinchilla lanigera*) tested on similar noises. Budgerigars are much more sensitive than chinchillas and as good or better than humans over a wide range of ripple frequencies.

5.5.2. The Mistuned Harmonic

Mistuned harmonics produce a variety of sensations in human listeners and recent literature (Moore et al. 1985a, 1985b, 1986; Hartmann et al. 1990) has focused on the subjective pitch of the mistuned components, the perception of roughness, or other specific cues to inharmonicity. Using harmonic complexes with a 570 Hz fundamental (modeled after the naturally occurring contact call of zebra finches, *T. guttata*), Lohr and Dooling (1998) measured the threshold for detecting inharmonicity in zebra finches, budgerigars (*M. undulatus*), and humans (*H. sapiens*). These results are shown in Figure 7.14. Thresholds for detecting inharmonicity in birds were almost an order of magnitude lower than for humans. Moreover, zebra finches (which produce predominantly harmonic vocalizations) were more sensitive than budgerigars (which produce predominantly tonal vocalizations). Zebra finches and

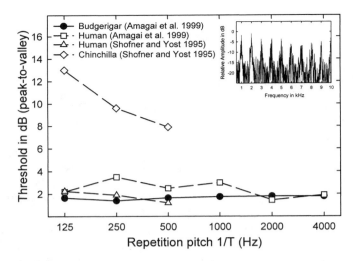

FIGURE 7.13. Inset shows a linear-rippled stimulus with 999 iterations, a 1 ms delay and 0 dB attenuation. Figure shows linear-rippled stimulus thresholds from Amagai et al. 1999 for budgerigars (*Melopsittacus undulatus*) (*closed circles*). Open symbols show results from several studies on humans. In addition, results from chinchillas (*Chinchilla lanigera*) are presented.

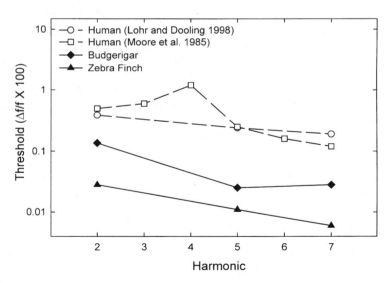

FIGURE 7.14. Thresholds for detecting inharmonicity in humans and two species of birds. Bird data are from Lohr and Dooling 1998 for a fundamental of 570 Hz. Human data are from Lohr and Dooling 1998 for 570 Hz and Moore et al. 1985b for 400 Hz. Mistuning threshold is plotted as a percentage of harmonic frequency (log scale).

humans also were tested on harmonic complexes where all components began at zero phase or at random phases. While this phase manipulation had little effect on human thresholds, zebra finches were significantly worse at detecting harmonic mistuning with random stimuli, suggesting these birds may be especially sensitive to relative phase information in a complex stimulus. A recent study of single-unit responses in the zebra finch auditory forebrain to complex harmonic stimuli shows that extremely precise preservation of temporal cues in the auditory forebrain is necessary for a full response to complex, learned, species-specific vocalizations (Theunissen and Doupe 1998).

Interestingly, mistuning of a harmonic causes a number of changes in a harmonic complex, including changes in the temporal fine structure of the sound. The above findings are further supported by a more recent series of investigations in which budgerigars were tested on harmonic complexes known as Schroeder maskers, which were designed to hold spectral and intensive information constant while changing component phases (Schroeder 1970). The avian auditory system appears to respond somewhat differently than the human auditory system to these sounds when they are used as maskers (Dent et al. 1999b; Leek et al. 2000). Moreover, budgerigars, as well as other birds, were able to discriminate different envelope shapes at remarkably high fundamental frequencies up to 800–900 Hz (requiring fine structure temporal analysis over fundamental periods of 1–2 ms), while humans fail to discriminate envelope shapes of the same stimuli with fundamental frequencies higher than about 250 Hz (Dent et al. 1999a). These species differences parallel those found in the measurement of the temporal modulation transfer function in birds and humans. At the present time, it is unclear what auditory system mechanism could support such fine temporal resolution.

5.5.3. Timbre Discrimination

Differences in the relative amplitude of components in a harmonic stimulus bring about changes in the perceptual quality, or "timbre," of the stimulus. Such differences also lead to changes in the temporal fine structure of the sound. The ability to detect changes in timbre has been tested in birds, both with natural vocalizations and standard harmonic complexes. Cynx et al. (1990) demonstrated that zebra finches (*T. guttata*) were able to detect an amplitude change of 5–10 dB in the second harmonic of a song syllable (fundamental = 615 Hz), with some individuals showing a response to changes as small as 2 dB. Using stimuli and procedures modeled after a class of human perceptual studies known collectively as profile analysis, Lohr and Dooling (1998) tested both birds and humans on their ability to detect changes in amplitude of single components of a standard harmonic stimulus. Both zebra finches and budgerigars (*M. undulatus*) had thresholds on the order of 1.5–2 dB for detecting changes in amplitude of the fifth harmonic in a stimulus with a 570 Hz fundamental. Human thresholds, while quite variable, averaged higher, at 3.8 dB. Zebra finches are known to produce highly consistent amplitude relationships in their harmonic vocalizations, and the specific relative amplitude relationships differ among individuals (Williams et al. 1989). An enhanced ability to detect such differences might serve to facilitate individual recognition in bird vocalizations, given that these differences may serve as reliable cues to individual identity.

6. Hearing and Vocal Communication

In nature, the auditory system of any animal is faced with the task of detecting and discriminating important biological signals against a background of environmental noise. There has been a long-term interest in the possibility of environmental influences on the design of animal vocalizations, especially birds (see, for example, Morton 1970, 1975; Waser and Waser 1977; Wiley and Richards 1978; Marten and Marler 1977; Marten et al. 1977). There are many factors influencing the maximum distance over which a biologically meaningful sound may be heard and whether certain characteristics of the sound are more prone to environmental influences than others. Consideration must be given to the location and intensity with which the signaler vocalizes, the sound-attenuating and sound-modifying characteristics of the environment, and the location and sensitivity of the receiver. Assessing the relative importance of these three factors is a difficult problem, but the issue of environmental influence on vocal signals clearly cannot be adequately examined unless data are available for all three factors.

The vocal output levels of a number of songbirds and non-songbirds has been measured and maximum peak sound pressure levels of 90–100 dB a meter away from a bird are not uncommon (Brackenbury 1979). Moreover, it is likely that, at least in some birds, the control of vocal intensity is both under voluntary control and reflexively responsive to environmental noise conditions. Manabe et al. (1998) trained budgerigars (*M. undulatus*) using operant conditioning to produce contact calls for food. Through differential reinforcement of vocal intensity, birds modified their vocal intensity over a 5–10 dB range. Moreover, when background noise was introduced, birds exhibited a reflexive increase in vocal intensity first studied in humans and known as the Lombard effect (Lane and Tranel 1971). These results, as well as earlier results from *C. coturnix japonica* (Potash 1972) and zebra finches (*T. guttata*) (Cynx et al. 1998), show that birds can control their vocal intensity over at least a limited range.

With knowledge of a bird's masking patterns, it is possible to generate a family of functions describing the maximum communication distance between two birds for a given habitat (excess attenuation) and vocal intensity (see, for example, Dooling 1982). Critical ratio data for a number of species over the last ten years (see Fig. 7.4) strengthen the value of such an approach. Until recently, there were almost no masking data for birds other than budgerigars. New data also now show that birds exhibit spatial release from masking. That is, there is an improvement in audibility when the signal and masker are delivered from different locations in space, as in humans and other mammals (Hirsh 1971; Durlach and Colburn 1978). Budgerigars, in spite of their small interaural distance, show a spatial release from masking of about the same magnitude as seen in humans (Dent et al. 1997a). Thus, when the noise and signal are from different locations, birds may gain a significant advantage in signal to noise ratio simply by turning their heads.

With the above new data and the recent plethora of laboratory masking data from songbirds, it is worth revisiting the issue of sound transmission in the environment and the maximum distance over which one bird can be expected to detect the song of another bird against a background of noise. Figure 7.15 shows a simple situation for the song sparrow (*Melospiza melodia*) based on the detection of pure tones in broadband noise. The spectrum level of masking noise is plotted as a function of maximum distance at which song detection could occur. For a reference, a typical song sparrow territory diameter is also shown. This is an extremely simple set of relations,

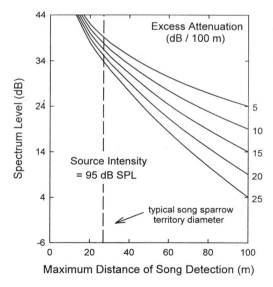

FIGURE 7.15. With the signal intensity assumed constant at 95 dB SPL, maximum transmission distance is calculated given background noise level at different values of excess attenuation (dB/100 m). Background noise level is given as spectrum level (per cycle energy distribution over the entire band of noise), and the amount of masking assuming a critical ratio of 20 dB.

and there are a huge number of factors that bear on this relation including comodulation masking and spatial release from masking as described by Klump (1996), specific characteristics of the habitat, and thresholds for the detection of more complex sounds such as vocalizations in noise, to name a few. Any worthwhile model of how environmental or anthropogenic noise might effect acoustic communication in nature must accommodate these issues.

7. Recovery of Hearing Following Hair Cell Regeneration

Finally, the widespread occurrence of hair cell regeneration in nonmammalian vertebrates represents one of the most important recent findings in auditory science (Ryals and Rubel 1988). Behavioral studies of hearing in birds over the past decade have now provided the first critical tests of the extent to which hearing is restored following the regeneration of hair cells in the peripheral auditory system. In general, the results to date from several species show that noise- or drug-induced damage to hair cells in the basilar papilla of birds can result in both temporary and permanent threshold shifts (Rubel and Ryals 1982; Saunders et al. 1991; Corwin 1992; Cotanche et al. 1994; Tsue et al. 1994). While there can be rather large species differences, hair cell regeneration appears to result in almost complete recovery of absolute thresholds, measures of auditory filter width, frequency difference limens, and intensity difference limens (Hashino et al. 1988; Hashino and Sokabe 1989; Marean et al. 1993; Niemiec et al. 1994; Saunders and Salvi 1995; Dooling et al. 1997; Ryals et al. 1999).

The current significance for hair cell regeneration and subsequent recovery of hearing in birds is the implication that these findings have for human hearing impairment. The question is whether hair cell regeneration results in functional recovery of normal auditory and vocal behavior. Many species of birds (songbirds and parrots) must learn their vocalizations and, like humans, rely on auditory feedback to produce these vocalizations. A recent study has shown that, in budgerigars (*M. undulatus*), the ability to discriminate and classify complex, species-specific vocalizations returns following hair cell regeneration (Dooling et al. 1997). Moreover, the ability to produce learned, species-specific vocalizations with precision also returns following hair cell regeneration. Interestingly, while the precision in vocal production is initially affected by hearing loss from hair-cell damage, this precision recovers long before the papilla is fully repopulated with new, functional hair cells (Dooling et al. 1997). This suggests that only a small amount of hearing (or, more accurately, hearing recovery) is necessary to guide nearly normal vocal precision in vocal production.

8. Conclusions

This review describes what is currently known about the hearing capabilities of reptiles and birds. In the case of reptiles, the data are limited. Nevertheless, one is struck by the extraordinary diversity of anatomical structures and hearing capabilities. Because of a lack of direct behavioral measures in this group of vertebrates, understanding of hearing in reptiles is still in its infancy. There are a number of both proximate and ultimate questions about hearing in reptiles that await more direct measures of detection, discrimination, and auditory perception in the reptiles.

In the case of birds, however, the wealth of data on detection, discrimination, and perception of both simple and complex sounds rivals that available in humans. In fact, in some cases, the only appropriate comparison is with humans since equivalent tests have not been conducted on other animals. As an example, because birds are the acoustic virtuosos of the animal world, more is known about the relation between hearing and vocalizations in this group compared to other vertebrate groups.

There are a number of themes that emerge from the review in this chapter. First, there are clearly some general aspects of avian hearing that birds share with reptiles, such as a relatively narrow hearing range and poor hearing at high frequencies. Second, given the known differences between the avian and mammalian auditory systems, the similarity in hearing performance between birds and humans in the frequency range of 1–5 kHz is striking. On the one hand, this argues for the physics of sound constraining hearing in humans and birds in similar ways. On the other hand, results from recent tests now point to a number of emerging areas of investigation where there may be rather significant differences between birds and mammals such as spatial hearing, temporal processing, the tradeoff between time and frequency perception, the salience of complex sounds, and the recovery of hearing following hair cell regeneration.

Future studies of hearing in birds should focus on these emerging areas of differences between birds and mammals to encourage investigations of the auditory mechanisms that account for these differences and to provide insight into their evolution. The understanding of hearing and its adaptation in the reptiles is still today severely compromised by the lack of behavioral data. An enormous amount of anatomical and physiological data collected over the past few decades on basic hearing sensitivity in reptiles lies in waiting for sophisticated behavioral tests of frequency, intensity, and time discrimination and more complicated tests of sound localization, the masking of complex sounds, comodulation masking release, and pattern perception to truly understand the evolution and adaptive significance of the reptilian hearing organ and auditory system.

Acknowledgments. We thank Marjorie Leek, Otto Gleich, Georg Klump, Arthur Popper, Richard Fay, and Catherine Carr for helpful comments on earlier versions of this chapter.

References

Aleksandrov LI, Dmitrieva LP (1992) Development of auditory sensitivity of altricial birds: absolute thresholds of the generation of evoked potentials. Neurosci Behav Physiol 22:132–137.

Amagai S, Dooling RJ, Formby C, Forrest TG (1997) Discrimination of silent temporal gaps in sinusoidal markers by the budgerigar (*Melopsittacus undulatus*). J Acoust Soc Am 101:3124.

Amagai S, Dooling RJ, Shamma S, Kidd TL, Lohr B (1999) Detection of modulation in spectral envelopes by budgerigars (*Melopsittacus undulatus*). J Acoust Soc Am 105:2029–2035.

Art JJ, Fettiplace R (1984) Efferent desensitization of auditory nerve fibre responses in the cochlea of the turtle *Pseudemys scripta elegans*. J Physiol 356:507–523.

Bacon SP, Viemeister NF (1985) Temporal modulation transfer functions in normal-hearing and hearing-impaired listeners. Audiology 24:117–134.

Banta P, Pepperberg I (1995) Learned English vocalizations as a model for studying budgerigar (*Melopsittacus undulatus*) warble song. In: Burrows M, Matheson T, Newland PL, Schuppe H (eds) Nervous Systems and Behavior Proceedings of the 4th International Congress of Neuroethology. New York: Thieme, p. 335.

Barton LA, Bailey ED, Gatehouse RW (1984) Audibility curve of bobwhite quail (*Colinus virginianus*). J Auditory Res 24:87–97.

Békésy GV (1960) Experiments in Hearing. New York: McGraw-Hill.

Bilsen FA, ten Kate JH, Buunen TJF, Raatgever J (1975) Responses of single units in the cochlear nucleus of the cat to cosine noise. J Acoust Soc Am 58:858–866.

Brackenbury JH (1979) Power capabilities of the avian sound-producing system. J Exp Biol 78:163–166.

Bregman AS (1990) Auditory Scene Analysis. Cambridge, MA: MIT Press.

Bregman AS, Campbell J (1971) Primary auditory stream segregation and perception of order in rapid sequences of tones. J Exp Psychol 89:244–249.

Bremond JC (1968) Recherches sur la semantique et les elements vecteurs d'information dans les signaux acoustiques du rouge-gorge (*Erithacus rubecula* L.). Terre Vie 2:109–220.

Bremond JC (1975) Specific recognition in the song of Bonelli's warbler (*Phylloscopus bonelli*). Behaviour 58:99–116.

Brodkorb P (1971) Origin and evolution of birds. In: Farner DS, King JR, Parkes KC (eds) Avian Biology, Vol. 1. New York: Academic Press, pp. 19–55.

Brown SD, Dooling RJ, O'Grady K (1988) Perceptual organization of acoustic stimuli by budgerigars (*Melopsittacus undulatus*): III Contact calls. J Comp Psychol 102:236–247.

Buchfellner E, Leppelsack HJ, Klump GM, Hausler U (1989) Gap-detection in the starling (*Sturnus vulgaris*): II: Coding of gaps by forebrain neurons. J Comp Physiol A 164:539–549.

Buus S, Klump GM, Gleich O, Langemann U (1995) An excitation-pattern model for the starling (*Sturnus vulgaris*). J Acoust Soc Am 98:112–124.

Campbell HW (1969) The effects of temperature on the auditory sensitivity of lizards. Physiol Zool 42:183–210.

Carroll RL (1987) Vertebrate Paleontology and Evolution. New York: WH Freeman.

Cohen SM, Stebbins WC, Moody DB (1978) Audibility thresholds of the blue jay. Auk 95:563–568.

Coles RB, Konishi M, Pettigrew JD (1987) Hearing and echolocation in the Australian grey swiftlet, *Collocalia spodiopygia*. J Exp Biol 129:365–371.

Corwin JT (1992) Regeneration in the auditory system. Exp Neurol 115:7–12.

Cotanche DA, Lee KH, Stone JS, Picard DA (1994) Hair cell regeneration in the bird cochlea following noise damage or ototoxic drug damage. Anat Embryol 189:1–18.

Crawford AC, Fettiplace R (1980) The frequency selectivity of auditory nerve fibres and hair cells in the cochlea of the turtle. J Physiol 306:79–125.

Cynx J, Williams H, Nottebohm F (1990) Timbre discrimination in zebra finch (*Taeniopygia guttata*) song syllables. J Comp Psychol 104:303–308.

Cynx J, Lewis R, Tavel B, Tse H (1998) Amplitude regulation of vocalizations in noise by a songbird, *Taeniopygia guttata*. Anim Behav 56:107–113.

Dent ML, Dooling RJ (1998) Frequency difference limens in budgerigars (*Melopsittacus undulatus*) as a function of tone duration. Paper presented at Midwinter Meeting of the Association for Research in Otolaryngology, St Petersburg, FL.

Dent ML, Larsen ON, Dooling RJ (1997) Free-field binaural unmasking in budgerigars (*Melopsittacus undulatus*). Behav Neurosci 111:590–598.

Dent ML, Brittan-Powell EF, Dooling RJ, Pierce A (1997) Perception of synthetic /ba/-/wa/ speech continuum by budgerigars (*Melopsittacus undulatus*). J Acoust Soc Am 102:1891–1897.

Dent ML, Dooling RJ, Leek MR (1999) Perception of harmonic complexes in budgerigars (*Melopsittacus undulatus*). J Acoust Soc Am 105:1319.

Dent ML, Dooling RJ, Leek MR, Summers, V (1999) Masking by harmonic complexes with different phase spectra in budgerigars (*Melopsittacus undulatus*). Paper presented at Midwinter Meeting of the Association for Research in Otolaryngology, St Petersburg, FL.

Dent ML, Klump GM, Schwenzfeier C (1999) Temporal modulation transfer functions in the barn owl (*Tyto alba*). Paper presented at Midwinter Meeting of the Association for Research in Otolaryngology, St Petersburg, FL.

De Valois RL, De Valois KK (1988) Spatial Vision. New York: Oxford University Press.

Divenyi PL, Danner WF (1977) Discrimination of time intervals marked by brief acoustic pulses of various intensities and spectra. Percept Psychophys 21:125–142.

Dooling RJ (1979) Temporal summation of pure tones in birds. J Acoust Soc Am 65:1058–1060.

Dooling RJ (1980) Behavior and psychophysics of hearing in birds. In: Popper AN, Fay RR (eds) Comparative Studies of Hearing in Vertebrates. New York: Springer-Verlag, pp. 261–288.

Dooling RJ (1982) Auditory perception in birds. In: Kroodsma DE, Miller EH (eds) Acoustic Communication in Birds, Vol 1. New York: Academic Press, pp. 95–130.

Dooling RJ (1992a) Hearing in Birds. In: Webster DB, Fay RR, Popper AN (eds) The Evolutionary Biology of Hearing. New York: Springer-Verlag, pp. 545–559.

Dooling RJ (1992b) Perception of speech sounds by birds. In: Cazals Y, Demany L, Horner K (eds) Advances in Biosciences: Auditory Physiology and Perception. London: Pergamon, pp. 407–413.

Dooling RJ, Brown SD (1990) Speech perception by budgerigars (*Melopsittacus undulatus*): spoken vowels. Percept Psychophys 47:568–574.

Dooling RJ, Haskell RJ (1978) Auditory duration discrimination in the parakeet (*Melopsittacus undulatus*). J Acoust Soc Am 63:1640–1642.

Dooling RJ, Saunders JC (1975a) Hearing in the parakeet (*Melopsittacus undulatus*): absolute thresholds, critical ratios, frequency difference limens, and vocalizations. J Comp Physiol 88:1–20.

Dooling RJ, Saunders JC (1975b) Auditory intensity discrimination in the parakeet (*Melopsittacus undulatus*). J Acoust Soc Am 58:1308–1310.

Dooling RJ, Searcy MH (1979) The relation among critical ratios, critical bands, and intensity difference limens in the parakeet (*Melopsittacus undulatus*). Bull Psychon Soc 13:300–302.

Dooling RJ, Searcy MH (1980b) Forward and backward auditory masking in the parakeet (*Melopsittacus undulatus*). Hearing Res 3:279–284.

Dooling RJ, Searcy MH (1981) Amplitude modulation thresholds for the parakeet (*Melopsittacus undulatus*). J Comp Physiol 143:383–388.

Dooling RJ, Searcy MH (1985a) Nonsimultaneous auditory masking in the budgerigar (*Melopsittacus undulatus*). J Comp Psychol 99:226–230.

Dooling RJ, Searcy MH (1985b) Temporal integration of acoustic signals by the budgerigar (*Melopsittacus undulatus*). J Acoust Soc Am 77:1917–1920.

Dooling RJ, Zoloth SR, Baylis JR (1978) Auditory sensitivity, equal loudness, temporal resolving power and vocalizations in the house finch (*Carpodacus mexicanus*). J Comp Physiol Psych 92:867–876.

Dooling RJ, Peters SS, Searcy MH (1979) Auditory sensitivity and vocalizations of the field sparrow (*Spizella pusilla*). Bull Psychon Soc 14:106–108.

Dooling RJ, Okanoya K, Downing J, Hulse S (1986) Hearing in the starling (*Sturnus vulgaris*): absolute thresholds and critical ratios. Bull Psychon Soc 24: 462–464.

Dooling RJ, Park TJ, Brown SD, Okanoya K, Soli SD (1987) Perceptual organization of acoustic stimuli by budgerigars (*Melopsittacus undulatus*) II: Vocal signals. J Comp Psychol 101:367–381.

Dooling RJ, Okanoya K, Brown SD (1989) Speech perception by budgerigars (*Melopsittacus undulatus*): the voiced-voiceless distinction. Percept Psychophys 46:65–71.

Dooling RJ, Best CT, Brown SD (1995) Discrimination of synthetic full-formant and sinewave /ra-la/ continua by budgerigars (*Melopsittacus undulatus*) and zebra finches (*Taeniopygia guttata*). J Acoust Soc Am 97:1839–1846.

Dooling RJ, Ryals BM, Manabe K (1997) Recovery of hearing and vocal behavior after hair-cell regeneration. Proc Natl Acad Sci U S A 94:14206–14210.

Durlach N, Colburn HS (1978) Binaural phenomena. In: Carterette C, Friedman MP (eds) Handbook of Perception, Vol. IV: Hearing. New York: Academic Press, pp. 365–466.

Dyson ML, Klump GM, Gauger B (1998) Absolute hearing thresholds and critical masking ratios in the European barn owl: a comparison with other owls. J Comp Physiol A 182:695–702.

Eatock RA, Manley GA (1981) Auditory nerve fibre activity in the tokay gecko: II, temperature effect on tuning. J Comp Physiol A 142:219–226.

Eatock RA, Manley GA, Pawson L (1981) Auditory nerve fibre activity in the tokay gecko: I, implications for cochlear processing. J Comp Physiol A 142:203–218.

Falls JB (1963) Properties of bird song eliciting responses from territorial males. Proc Int Ornith Congr 13:359–371.

Farabaugh SM, Dooling RJ (1996) Acoustic communication in parrots: laboratory and field studies of budgerigars, *Melopsittacus undulatus*. In: Kroodsma DE, Miller EH (eds) Ecology and Evolution of Acoustic Communication in Birds. Ithaca: Cornell University Press, pp. 97–117.

Fay RR (1988) Hearing in Vertebrates: A Psychophysics Databook. Winnetka, IL: Hill-Fay Associates.

Fay RR, Yost WA, Coombs S (1983) Psychophysics and neurophysiology of repetition noise processing in a vertebrate auditory system. Hear Res 12:31–55.

Fedducia A (1980) The Age of Birds. Cambridge, MA: Harvard University Press.

Fletcher H (1940) Auditory patterns. Rev Mod Phys 12:47–65.

Fletcher LE, Smith DG (1978) Some parameters of song important in conspecific recognition by gray catbirds. Auk 95:338–347.

Formby C, Forrest TG (1991) Detection of silent temporal gaps in sinusoidal markers. J Acoust Soc Am 89:830–837.

Formby C, Sherlock LP, Forrest TG (1996) An asymmetric Roex filter model for describing detection of silent temporal gaps in sinusoidal markers. Aud Neurosci 3:1–20.

Gans C, Wever EG (1972) The ear and hearing in Amphisbaenia (Reptilia). J Exp Zool 179:17–34.

Gans C, Wever EG (1976) The ear and hearing in *Sphenodon punctatus*. Proc Natl Acad Sci U S A 73:4244–4246.

Glasberg BR, Moore BCJ (1990) Derivation of auditory filter shapes from notched-noise data. Hear Res 47:103–138.

Gleich O, Klump GM (1995) Temporal modulation transfer functions in the European starling (*Sturnus vulgaris*): II. Responses of auditory-nerve fibers. Hear Res 82:81–92.

Goerdel-Leich A, Schwartzkopff J (1984) The auditory threshold of the pigeon (*Columba livia*) by heart-rate conditioning. Naturwiss 71:S98.

Gray L, Rubel EW (1985) Development of auditory thresholds and frequency difference limens in chickens. In: Gottlieb G, Krasnegor NA (eds) Measurement of Audition and Vision in the First Year of Postnatal Life: A Methodological Overview. Norwood NJ: Ablex, pp. 145–165.

Greenewalt CH (1968) Bird Song: Acoustics and Physiology. Washington DC: Smithsonian Institute Press.

Greenwood DD (1961a) Auditory masking and the critical band. J Acoust Soc Am 33:484–502.

Greenwood DD (1961b) Critical bandwidth and the frequency coordinates of the basilar membrane. J Acoust Soc Am 33:1344–1356.

Griffin DR (1954) Acoustic orientation in the oil bird, *Steatornis*. Proc Natl Acad Sci U S A 39:885–893.

Gulick WL, Zwick H (1966) Auditory sensitivity of the turtle. Psychol Rec 16:47–53.

Hall JW, Haggard MP, Fernandes MA (1984) Detection in noise by spectro-temporal pattern analysis. J Acoust Soc Am 76:50–56.

Hamann I, Klump GM, Fichtel C, Langemann U (1999) CMR in a songbird studied with narrow-band maskers. Paper presented at Midwinter Meeting of the Association for Research in Otolaryngology, St Petersburg, FL.

Harrison JB, Furumoto L (1971) Pigeon audiograms: comparison of evoked potential and behavioral thresholds in individual birds. J Aud Res 11:3342.

Hartline PH (1971a) Physiological basis for detection of sound and vibration in snakes. J Exp Biol 54:349–371.

Hartline PH (1971b) Mid-brain responses of the auditory and somatic vibration systems in snakes. J Exp Biol 54:373–390.

Hartline PH, Campbell HW (1969) Auditory and vibratory responses in the mid-brains of snakes. Science 163:1221–1223.

Hartmann WM, McAdams S, Smith BK (1990) Hearing a mistuned harmonic in an otherwise periodic complex tone. J Acoust Soc Am 88:1712–1724.

Hashino E, Okanoya K (1989) Auditory sensitivity in the zebra finch (*Poephila guttata castanotis*). J Acoust Soc Jpn 10:1–2.

Hashino E, Sokabe M (1989) Hearing loss in the budgerigar (*Melopsittacus undulatus*). J Acoust Soc Am 85:289–294.

Hashino E, Sokabe M, Miyamoto K (1988) Frequency specific susceptibility to acoustic trauma in the budgerigar. J Acoust Soc Am 83:2450–2452.

Hedges SB, Poling LL (1999) A molecular phylogeny of reptiles. Science 283: 998–1001.

Heise GA (1953) Auditory thresholds in the pigeon. Amer J Psychol 66:1–19.

Hienz MG, Goldstein MH Jr, Formby C (1996) Temporal gap detection thresholds in sinusoidal markers simulated with a multi-channel, multi-resolution model of the auditory periphery. Aud Neurosci 3:35–56.

Hienz RD, Sachs MB (1987) Effects of noise on pure-tone thresholds in blackbirds (*Agelaius phoeniceus* and *Molothrus ater*) and pigeon (*Columba livia*). J Comp Psychol 101:16–24.

Hienz RD, Sinnott JM, Sachs MB (1977) Auditory sensitivity of the redwing blackbird and the brown-headed cowbird. J Comp Physiol Psych 91:1365–1376.

Hienz RD, Sinnott JM, Sachs MB (1980) Auditory intensity discrimination in blackbirds and pigeons. J Comp Physiol Psych 94:993–1002.

Hienz RD, Sachs MB, Sinnott JM (1981) Discrimination of steady-state vowels by blackbirds and pigeons. J Acoust Soc Am 70:699–706.

Hillier DA (1991) Auditory processing of sinusoidal spectral envelopes. Unpublished doctoral dissertation, Washington University, St. Louis, MO.

Hirsh IJ (1971) Masking of speech and auditory localization. Audiology 10:110–114.

Hulse SH, Cynx J (1985) Relative pitch perception is constrained by absolute pitch in songbirds (*Mimus, Molothrus, Sturnus*). J Comp Psychol 99:176–196.

Hulse SH, Cynx J (1986) Interval and contour in serial pitch perception by a Passerine bird, the European starling (*Sturnus vulgaris*). J Comp Psychol 100:215–228.

Hulse SH, MacDougall-Shackleton SA, Wisniewski AB (1997) Auditory scene analysis by songbirds: stream segregation of birdsong by European starlings (*Sturnus vulgaris*). J Comp Psychol 111:3–13.

Kinchla J (1970) Discrimination of two auditory durations by pigeons. Percept Psychophys 8:299–307.

Kleunder KR, Diehl RL, Killeen PR (1987) Japanese quail can learn phonetic categories. Science 237:1195–1197.

Klinke R, Pause M (1980) Discharge properties of primary auditory fibres in *Caiman crocodilus*: comparisons and contrasts to the mammalian auditory nerve. Exp Brain Res 38:137–150.

Klump GM (1996) Bird communication in the noisy world. In: Kroodsma DE, Miller EH (eds) Ecology and Evolution of Acoustic Communication in Birds. Ithaca, NY: Cornell University Press, pp. 321–338.

Klump GM, Gleich O (1991) Gap detection in the European starling (*Sturnus vulgaris*). III. Processing in the peripheral auditory system. J Comp Physiol A 169:469–476.

Klump GM, Langemann U (1995) Comodulation masking release in a songbird. Hear Res 87:157–164.

Klump GM, Maier EH (1989) Gap detection in the starling (*Sturnus vulgaris*), I: Psychophysical thresholds. J Comp Physiol 164:531–538.

Klump GM, Maier EH (1990) Temporal summation in the starling (*Sturnus vulgaris*). J Comp Psychol 104:94–100.

Klump GM, Okanoya K (1991) Temporal modulation transfer functions in the European starling (*Sturnus vulgaris*). Hear Res 52:1–12.

Klump GM, Kretzschmar E, Curio E (1986) The hearing of an avian predator and its avian prey. Behav Ecol Sociobiol 18:317–323.

Klump GM, Dooling RJ, Fay RR, Stebbins WC (1995) Methods in Comparative Psychoacoustics. Basel: Birkhauser Verlag.

Klump GM, Schwenzfeier C, Dent ML (1998) Gap detection in the barn owl. Paper presented at Midwinter Meeting of the Association for Research in Otolaryngology, St Petersburg, FL.

Konishi M (1969) Time resolution by single auditory neurons in birds. Nature 222:566–567.

Konishi M (1970) Comparative neurophysiological studies of hearing and vocalizations on song birds. Z Vergl Physiol 66:257–272.

Konishi M (1973a) How the barn owl tracks its prey. Am Sci 61:414–424.

Konishi M (1973b) Locatable and nonlocatable acoustic signals for barn owls. Am Nat 107:775–785.

Konishi M, Knudsen EI (1979) The oilbird: hearing and echolocation. Science 204:425–427.

Köppl C, Manley GA (1992) Functional consequences of morphological trends in the evolution of lizard hearing organs. In: Fay RR, Popper AN, Webster DB (eds) The Evolutionary Biology of Hearing. New York: Springer-Verlag, pp. 489–509.

Kreithen ML, Quine DM (1979) Infrasound detection by the homing pigeon: a behavioral audiogram. J Comp Physiol 129:1–4.

Kroodsma DE, Miller EH (1982) Acoustic Communication in Birds, Vol 2: Song Learning and Its Consequences. New York: Academic Press.

Kroodsma DE, Miller EH (1996) Ecology and Evolution of Acoustic Communication in Birds. Ithaca, NY: Cornell University Press.

Kuhl PK (1989) On babies, birds, modules, and mechanisms: a comparative approach to the acquisition of vocal communication. In: Dooling RJ, Hulse SH (eds) The Comparative Psychology of Audition. Hillsdale, NJ: Lawrence Erlbaum Associates, pp. 379–419.

Kuhn A, Leppelsack H-J, Schwartzkopff J (1980) Measurement of frequency discrimination in the starling (*Sturnus vulgaris*) by conditioning of heart rate. Naturwiss 67:102.

Kuhn A, Muller CM, Leppelsack H-J, Schwartzkopff J (1982) Heart rate conditioning used for determination of auditory thresholds in the starling. Naturwiss 69:245–256.

Lane HL, Tranel B (1971) The Lombard sign and the role of hearing in speech. J Speech Hear Res 14:677–709.

Langemann U, Klump GM, Dooling RJ (1995) Critical bands and critical-ratio bandwidth in the European starling. Hear Res 84:167–176.

Langemann U, Gauger B, Klump GM (1998) Auditory sensitivity in the great tit: perception of signals in the presence and absence of noise. Anim Behav 56:763–769.

Lin JY, Dooling RJ, Dent ML (1997) Auditory filter shapes in the budgerigar (*Melopsittacus undulatus*) derived from notched-noise maskers. J Acoust Soc Am 101:3125.

Lin JY, Dooling RJ, Lohr B, Leek MR (1998) The temporal resolution of the avian auditory system. Paper presented at Midwinter Meeting of the Association for Research in Otolaryngology, St Petersburg, FL.

Linzenbold A, Dooling RJ, Ryals BM (1993) A behavioral audibility curve for the Japanese quail (*Coturnix coturnix japonica*). Paper presented at Midwinter Meeting of the Association for Research in Otolaryngology, St Petersburg, FL.

Lohr B, Dooling RJ (1998) Detection of changes in timbre and harmonicity in complex sounds by zebra finches (*Taeniopygia guttata*) and budgerigars (*Melopsittacus undulatus*). J Comp Psychol 112:36–47.

Lohr B, Dooling RJ (1999) Hearing in the red-billed firefinch (*Lagonosticta senegala*): an estrildid finch with narrowband vocalizations. Paper Presented at Midwinter meeting of the Association for Research in Otolaryngology, St Petersburg, FL.

Lynn GK, Small AM (1977) Interactions of backward and forward matching. J Acoust Soc Am 61:185–189.

MacDougall-Shackleton SA, Hulse SH (1996) Concurrent absolute and relative pitch processing by European starlings (*Sturnus vulgaris*). J Comp Psychol 110:139–146.

MacDougall-Shackleton SA, Hulse SH, Gentner TQ, White W (1998) Auditory scene analysis by European starlings (*Sturnus vulgaris*): perceptual segregation of tone sequences. J Acoust Soc Am 103:3581–3587.

Maier EH, Klump GM (1990) Auditory duration discrimination in the European starling (*Sturnus vulgaris*). J Acoust Soc Am 88:616–621.

Maiorana VA, Schleidt WM (1972) The auditory sensitivity of the turkey. J Aud Res 12:203–207.

Manabe K, Sadr EI, Dooling RJ (1998) Control of vocal intensity in budgerigars (*Melopsittacus undulatus*): differential reinforcement of vocal intensity and the Lombard effect. J Acoust Soc Am 103:1190–1198.

Manley GA (1970) Frequency sensitivity of auditory neurons in the caiman cochlear nucleus. Z Vergl Physiol 66:251–256.

Manley GA (1972) Frequency response of the ear of the Tokay gecko. J Exp Zool 181:159–168.

Manley GA (1990) Peripheral Hearing Mechanisms in Reptiles and Birds. Heidelberg, New York: Springer-Verlag.

Manley GA, Gleich O (1991) Evolution and specialization of function in the avian auditory periphery. In: Fay RR, Popper AN, Webster DB (eds) The Evolutionary Biology of Hearing. Heidelberg: Springer-Verlag.

Manley GA, Köppl C, Yates GK (1997) Activity of primary auditory neurons in the cochlear ganglion of the emu *Dromaius novaehollandiae*: spontaneous discharge, frequency tuning, and phase locking. J Acoust Soc Am 101:1560–1573.

Marean GC, Burt JM, Beecher MD, Rubel EW (1993) Hair cell regeneration in European starling (*Sturnus vulgarus*): recovery of pure-tone detection thresholds. Hear Res 71:125–136.

Marean GC, Burt JM, Beecher MD, Rubel EW (1998) Auditory perception following hair cell regeneration in European starling (*Sturnus vulgarus*): frequency and temporal resolution. J Acoust Soc Am 103:3567–3580.

Marten K, Marler P (1977) Sound transmission and its significance for animal vocalization. I. Temperate habitats. Behav Ecol Sociobiol 2:271–290.

Marten K, Quine D, Marler P (1977) Sound transmission and its significance for animal vocalization. II. Tropical forest habitats. Behav Ecol Sociobiol 2:291–302.

Moody DB (1994) Detection and discrimination of amplitude-modulated signals by macaque monkeys. J Acoust Soc Am 95:3499–3510.

Moore BCJ (1973) Frequency difference limens for short-duration tones. J Acoust Soc Am 54:610–619.

Moore BCJ, Glasberg BR (1988) Gap detection with sinusoids and noise in normal, impaired, and electrically stimulated ears. J Acoust Soc Am 83:1093–1101.

Moore BCJ, Sek A (1994a) Discrimination of modulation type (amplitude modulation or frequency modulation) with and without background noise. J Acoust Soc Am 96:726–732.

Moore BCJ, Sek A (1994b) Effects of carrier frequency and background noise on the detection of mixed modulation. J Acoust Soc Am 96:741–751.

Moore BCJ, Glasberg BR, Peters RW (1985) Relative dominance of individual partials in determining the pitch of complex tones. J Acoust Soc Am 77:1853–1860.

Moore BCJ, Peters RW, Glasberg BR (1985) Thresholds for the detection of inharmonicity in complex tones. J Acoust Soc Am 77:1861–1867.

Moore BCJ, Glasberg BR, Peters RW (1986) Thresholds for hearing mistuned partials as separate tones in harmonic complexes. J Acoust Soc Am 80:479–483.

Moore BCJ, Glasberg BR, Plack CJ, Biswas AK (1988) The shape of the ear's temporal window. J Acoust Soc Am 83:1102–1116.

Morton ES (1970) Ecological sources of selection on avian sounds. Ph.D. Thesis, Yale University, New Haven, CT.

Morton ES (1975) Ecological sources of selection on avian sounds. Am Nat 109:17–34.

Nelson DA, Marler P (1990) The perception of birdsong and an ecological concept of signal space. In: Stebbins WC, Berkley MA (eds) Comparative Perception. Vol 2: Complex Signals. New York: Wiley, pp. 443–478.

Nespor AN, Dooling RJ, Triblehorn JD (1996) Discrimination of frequency modulation (FM) and amplitude modulation (AM) by budgerigars (*Melopsittacus undulatus*). J Acoust Soc Am 100:2753.

Nieboer E, Van der Paardt M (1977) Hearing of the African wood owl, *Strix woodfordii*. Neth J Zool 27:227–229.

Niemiec AJ, Raphael Y, Moody DB (1994) Return of auditory function following structural regeneration after acoustic trauma: behavioral measures from quail. Hear Res 79:1–16.

Okanoya K, Dooling RJ (1985) Colony differences in auditory thresholds in the canary. J Acoust Soc Am 78:1170–1176.

Okanoya K, Dooling RJ (1987) Hearing in Passerine and Psittacine birds: a comparative study of masked and absolute auditory thresholds. J Comp Psychol 101:7–15.

Okanoya K, Dooling RJ (1988) Hearing in the swamp sparrow (*Melospiza georgiana*) and the song sparrow (*Melospiza melodia*). Anim Behav 36:726–732.

Okanoya K, Dooling RJ (1990) Detection of gaps in noise by budgerigars (*Melopsittacus undulatus*) and zebra finches (*Poephila guttata*). Hear Res 50:185–192.

Page SC, Hulse SH, Cynx J (1989) Relative pitch perception in the European starling (*Sturnus vulgaris*): further evidence for an elusive phenomenon. J Exp Psychol Anim Behav Process 15:137–146.

Patterson RD, Moore BCJ (1986) Auditory filters and excitation patterns as representations of frequency resolution. In: Moore BCJ (ed) Frequency Selectivity in Hearing. London: Academic Press, pp. 123–177.

Patterson WC (1966) Hearing in the turtle. J Aud Res 6:453–464.

Pettigrew JD, Larsen ON (1990) Directional hearing in the plains-wanderer, *Pedionomus torquatus*. In: Rowe M, Aitkin L (eds) Information Processing in Mammalian Auditory and Tactile Systems. New York: Wiley-Liss, pp. 179–190.

Plomp R (1964) Rate decay of auditory sensation. J Acoust Soc Am 36:277–282.

Plomp R, Bouman MA (1959) Relation between hearing threshold and duration for tone pulses. J Acoust Soc Am 31:749–758.

Potash LM (1972) Noise induced changes in calls of the Japanese quail. Psychon Sci 26:252–254.

Pumphrey RJ (1961) Sensory organs: hearing. In: Marshall AJ (ed) Biology and Comparative Anatomy of Birds. New York: Academic Press, pp. 69–86.

Quine DB (1978) Infrasound detection and ultra low frequency discrimination in the homing pigeon (*Columba livia*). J Acoust Soc Am 63:S75.

Quine DB, Konishi M (1974) Absolute frequency discrimination in the barn owl. J Comp Physiol 93:347–360.

Richards DG, Wiley RH (1980) Reverberations and amplitude fluctuations in the propagation of sound in a forest: implications for animal communication. Am Nat 115:381–399.

Romer AS (1966) Vertebrate Paleontology, 3rd ed. Chicago: University of Chicago Press.

Rosen S, Baker RJ (1994) Characterizing auditory filter nonlinearity. Hear Res 73:231–243.

Rosowski JJ, Graybeal A (1991) What did *Morganucodon* hear? Biol J Linn Soc 101:131–168.

Rubel EW, Ryals BM (1982) Patterns of hair cell loss in chick basilar papilla after intense auditory stimulation: exposure, duration and survival time. Acta Otolaryngol 93:31–41.

Ryals BM, Rubel EW (1988) Hair cell regeneration after acoustic trauma in adult *Coturnix* quail. Science 240:1774–1776.

Ryals BM, Dooling RJ, Westbrook E, Dent ML, MacKenzie A, Larsen ON (1999) Avian species differences in susceptibility to noise exposure. Hear Res 131:71–88.

Salvi RJ, Giraudi DM, Henderson D, Hamernik RP (1982) Detection of sinusoidally amplitude modulated noise by the chinchilla. J Acoust Soc Am 71:424–429.

Sams-Dodd F, Capranica RR (1994) Representation of acoustic signals in the eighth nerve of the Tokay gecko: I. pure tones. Hear Res 76:16–30.

Saunders JC, Dooling RJ (1974) Noise-induced threshold shift in the parakeet (*Melopsittacus undulatus*). Proc Natl Acad Sci U S A 71:1962–1965.

Saunders JC, Pallone RL (1980) Frequency selectivity in the parakeet studied by isointensity masking contours. J Exp Biol 87:331–342.

Saunders JC, Rintelmann WF (1978) Frequency selectivity in man: the relation between critical band, critical ratio, and psychophysical tuning curves. Paper presented at Midwinter Meeting of the Association for Research in Otolaryngology, St Petersburg, FL.

Saunders JC, Salvi RJ (1993) Psychoacoustics of normal adult chickens: thresholds and temporal integration. J Acoust Soc Am 94:83–90.

Saunders JC, Salvi RJ (1995) Pure tone masking patterns in adult chicken before and after recovery from acoustic trauma. J Acoust Soc Am 98:1365–1371.

Saunders JC, Denny RM, Bock GR (1978) Critical bands in the parakeet (*Melopsittacus undulatus*). J Comp Physiol 125:359–365.

Saunders JC, Rintelmann WF, Bock G (1979) Frequency selectivity in bird and man: a comparison among critical ratios, critical bands, and psychophysical tuning curves. Hear Res 1:303–323.

Saunders JC, Cohen YE, Szymko YM (1991) The structural and functional consequences of acoustic injury in the cochlea and peripheral auditory system: a five year update. J Acoust Soc Am 90:136–146.

Schermuly L, Klinke R (1985) Change of characteristic frequencies of pigeon primary auditory afferents with temperature. J Comp Physiol A 156:209–211.

Schorer E (1986) Critical modulation frequency based on detection of AM versus FM tones. J Acoust Soc Am 79:1054–1057.

Schroeder MR (1970) Synthesis of low peak-factor signals and binary sequences with low autocorrelation. IEEE Trans Info Theory 16:85–89.

Schwartzkopff J (1949) Über Sitz und Leistung von Gehör und Vibrationssinn bei Vögeln. Z Vergl Physiol 31:527–603.

Schwartzkopff J (1968) Structure and function of the ear and the auditory brain areas in birds. In: DeReuck AVS, Knight J (eds) Hearing Mechanisms in Vertebrates. Boston, MA: Little, Brown, pp. 41–59.

Schwartzkopff J (1973) Mechanoreception. In: Farner DS, King JR, Parks KC (eds) Avian Biology, Vol 3. New York: Academic Press, pp. 417–477.

Shamma SA, Versnel H (1995) Ripple analysis in ferret primary auditory cortex I. Response characteristics of single units to sinusoidally rippled spectra. Aud Neurosci 1:233–254.

Shofner WP, Yost WA (1994) Repetition pitch: auditory processing of rippled noise in the chinchilla. Paper presented at the 10th International Symposium on Hearing, Irsec, Germany.

Shofner WP, Yost WA (1995) Discrimination of rippled spectrum noise from flat-spectrum wideband noise by chinchillas. Aud Neurosci 1:127–138.

Shreiner CE, Calhoun BM (1995) Spectral envelope coding in cat primary auditory cortex. Aud Neurosci 1:39–61.

Sinnott JM, Sachs MB, Hienz RD (1980) Aspects of frequency discrimination in passerine birds and pigeons. J Comp Physiol Psychol 94:401–415.

Smolders JWT, Klinke R (1984) Effects of temperature on the properties of primary auditory fibres of the spectacled caiman, *Caiman crocodilus* (L.). J Comp Physiol 155:19–30.

Starck D (1978) Vergleichende Anatomie der Wirbeltiere, Band 1. Berlin: Springer-Verlag.

Stebbins WC (1970) Studies of hearing and hearing loss in the monkey. In: Stebbins WC (ed) Animal Psychophysics: The Design and Conduct of Sensory Experiments. New York: Appleton, pp. 41–66.

Suga N, Campbell HW (1967) Frequency sensitivity of single auditory neurons in the gecko *Coleonyx variegatus*. Science 157:88–90.

Summers V, Leek MR (1994) The internal representation of spectral contrast in hearing-impaired listeners. J Acoust Soc Am 95:3518–3528.

Theunissen FE, Doupe AJ (1998) Temporal and spectral sensitivity of complex neurons in the nucleus Hvc of male zebra finches. J Neurosci 18:3786–3802.

Tsue TT, Oesterle EC, Rubel EW (1994) Hair cell regeneration in the inner ear. Otolaryngol Head Neck Surg 111:281–301.

Trainer JE (1946) The auditory acuity of certain birds. Ph.D. Thesis, Cornell University, Ithaca, NY.

Van Dijk T (1973) A comparative study of hearing in owls of the family Strigidae. Neth J Zool 23:131–167.

Viemeister NF (1979) Temporal modulation transfer functions based upon modulation thresholds. J Acoust Soc Am 66:1364–1380.

Viemeister NF, Plack CJ (1993) Time analysis. In: Yost WA, Popper AN, Fay RR (eds) Human Psychophysics. New York: Springer-Verlag, pp. 116–154.

Waser PM, Waser MS (1977) Experimental studies of primate vocalization: specializations for long-distance propagation. Z Tierpsychol 43:239–263.

Watson CS, Wroton HW, Kelly WJ, Benbassat CA (1975) Factors in the discrimination of tonal patterns. I. Component frequency, temporal position, and silent intervals. J Acoust Soc Am 57:1175–1185.

Watson CS, Kelly WJ, Wroton HW (1976) Factors in the discrimination of tonal patterns. II. Selective attention and learning under various levels of stimulus uncertainty. J Acoust Soc Am 60:1176–1186.

Wever EG (1971) Hearing in the Crocodilia. Proc Natl Acad Sci U S A 68:1498–1500.

Wever EG (1978) The Reptile Ear. Princeton, NJ: Princeton University Press.

Wever EG, Vernon JA (1956) The sensitivity of the turtle's ear as shown by its electrical potentials. Proc Natl Acad Sci U S A 42:213–220.

Wever EG, Vernon JA (1957) Auditory responses in the spectacled caiman. J Cell Comp Physiol 50:333–339.

Wever EG, Vernon JA (1960) The problem of hearing in snakes. J Aud Res 1:77–83.

Wier CC, Jesteadt W, Green DM (1977) Frequency discrimination as a function of frequency and sensation level. J Acoust Soc Am 61:177–184.

Wiley RH, Richards DG (1978) Physical constraints on acoustic communication in the atmosphere: implications for the evolution of animal vocalizations. Behav Ecol Sociobiol 3:69–94.

Williams H, Cynx J, Nottebohm F (1989) Timbre control in zebra finch (*Taeniopygia guttata*) song syllables. J Comp Psychol 103:366–380.

Williams KN, Perrott DR (1972) Temporal resolution of tonal pulses. J Acoust Soc Am 51:644–647.

Wisniewski AB, Hulse SH (1997) Auditory scene analysis in European starlings (*Sturnus vulgaris*): discrimination of song segments, their segregation from multiple and reversed conspecific songs, and evidence for conspecific categorization. J Comp Psychol 111:337–350.

Yodlowski ML (1980) Infrasonic sensitivity in pigeons (*Columba livia*). Ph.D. Thesis, Rockefeller University, New York.

Yost WA (1992) Auditory perception and sound source determination. Curr Dir Psychol Sci 1:179–184.

Yost WA (1996) Pitch of iterated rippled noise. J Acoust Soc Am 100:511–518.

Yost WA, Hill R (1978) Strength of pitches associated with ripple noise. J Acoust Soc Am 64:485–492.

Yost WA, Patterson R, Sheft S (1996) Pitch and pitch discrimination of broadband signals with rippled power spectra. J Acoust Soc Am 63:1166–1173.

Zwicker E (1952) Die Grenzen der Horbarkeit der Amplitudenmodulation und der Frequenzmodulation eines Tones. Acustica 2:125–133.

Index

Most species in the book are listed in this index under their common name. Scientific names are given in the index and referred to the common name.